Selected Titles in This Series

611 **D. Ginzburg, I. Piatetski-Shapiro, and S. Rallis,** L functions for the orthogonal group, 1997

610 **Mark Hovey, John H. Palmieri, and Neil P. Strickland,** Axiomatic stable homotopy theory, 1997

609 **Liviu I. Nicolaescu,** Generalized symplectic geometries and the index of families of elliptic problems, 1997

608 **Christina Q. He and Michel L. Lapidus,** Generalized Minkowski content, spectrum of fractal drums, fractal strings, and the Riemann zeta-functions, 1997

607 **Adele Zucchi,** Operators of class C_0 with spectra in multiply connected regions, 1997

606 **Moshé Flato, Jacques C. H. Simon, and Erik Taflin,** Asymptotic completeness, global existence and the infrared problem for the Maxwell-Dirac equations, 1997

605 **Liangqing Li,** Classification of simple C^*-algebras: Inductive limits of matrix algebras over trees, 1997

604 **Hajnal Andréka, Steven Givant, and István Németi,** Decision problems for equational theories of relation algebras, 1997

603 **Bruce N. Allison, Saeid Azam, Stephen Berman, Yun Gao, and Arturo Pianzola,** Extended affine Lie algebras and their root systems, 1997

602 **Igor Fulman,** Crossed products of von Neumann algebras by equivalence relations and their subalgebras, 1997

601 **Jack E. Graver and Mark E. Watkins,** Locally finite, planar, edge-transitive graphs, 1997

600 **Ambar Sengupta,** Gauge theory on compact surfaces, 1997

599 **Tai-Ping Liu and Yanni Zeng,** Large time behavior of solutions for general quasilinear hyperbolic-parabolic systems of conservation laws, 1997

598 **Valentina Barucci, David E. Dobbs, and Marco Fontana,** Maximality properties in numerical semigroups and applications to one-dimensional analytically irreducible local domains, 1997

597 **Ragnar-Olaf Buchweitz and John J. Millson,** CR-geometry and deformations of isolated singularities, 1997

596 **Paul S. Bourdon and Joel H. Shapiro,** Cyclic phenomena for composition operators, 1997

595 **Eldar Straume,** Compact connected Lie transformation groups on spheres with low cohomogeneity, II, 1997

594 **Solomon Friedberg and Hervé Jacquet,** The fundamental lemma for the Shalika subgroup of $GL(4)$, 1996

593 **Ajit Iqbal Singh,** Completely positive hypergroup actions, 1996

592 **P. Kirk and E. Klassen,** Analytic deformations of the spectrum of a family of Dirac operators on an odd-dimensional manifold with boundary, 1996

591 **Edward Cline, Brian Parshall, and Leonard Scott,** Stratifying endomorphism algebras, 1996

590 **Chris Jantzen,** Degenerate principal series for symplectic and odd-orthogonal groups, 1996

589 **James Damon,** Higher multiplicities and almost free divisors and complete intersections, 1996

588 **Dihua Jiang,** Degree 16 Standard L-function of $GSp(2) \times GSp(2)$, 1996

587 **Stéphane Jaffard and Yves Meyer,** Wavelet methods for pointwise regularity and local oscillations of functions, 1996

586 **Siegfried Echterhoff,** Crossed products with continuous trace, 1996

585 **Gilles Pisier,** The operator Hilbert space OH, complex interpolation and tensor norms, 1996

584 **Wayne W. Barrett, Charles R. Johnson, and Raphael Loewy,** The real positive definite completion problem: Cycle completability, 1996

583 **Jin Nakagawa,** Orders of a quartic field, 1996

(Continued in the back of this publication)

L Functions for the Orthogonal Group

MEMOIRS
of the
American Mathematical Society

Number 611

L Functions for the Orthogonal Group

D. Ginzburg
I. Piatetski-Shapiro
S. Rallis

July 1997 • Volume 128 • Number 611 (third of 4 numbers) • ISSN 0065-9266

American Mathematical Society
Providence, Rhode Island

1991 *Mathematics Subject Classification.*
Primary 11Exx.

Library of Congress Cataloging-in-Publication Data

Ginzburg, D. (David)
L functions for the orthogonal group / D. Ginzburg, I. Piatetski-Shapiro, S. Rallis.
p. cm. — (Memoirs of the American Mathematical Society, ISSN 0065-9266 ; no. 611)
"July 1997, volume 128, number 611 (third of 4 numbers)."
Includes bibliographical references.
ISBN 0-8218-0543-6 (alk. paper)
1. L-functions. 2. Automorphic functions. 3. Linear algebraic groups. I. Pi͡atet͡skiĭ-Shapiro, I. I. (Il′i͡a Iosifovich), 1929– . II. Rallis, Stephen, 1942– . III. Title. IV. Series.
QA3.A57 no. 611
[QA247]
510 s—dc21
[512′.74] 97-12294
CIP

Memoirs of the American Mathematical Society

This journal is devoted entirely to research in pure and applied mathematics.

Subscription information. The 1997 subscription begins with number 595 and consists of six mailings, each containing one or more numbers. Subscription prices for 1997 are \$414 list, \$331 institutional member. A late charge of 10% of the subscription price will be imposed on orders received from nonmembers after January 1 of the subscription year. Subscribers outside the United States and India must pay a postage surcharge of \$30; subscribers in India must pay a postage surcharge of \$43. Expedited delivery to destinations in North America \$35; elsewhere \$110. Each number may be ordered separately; *please specify number* when ordering an individual number. For prices and titles of recently released numbers, see the New Publications sections of the *Notices of the American Mathematical Society.*

Back number information. For back issues see the *AMS Catalog of Publications.*

Subscriptions and orders should be addressed to the American Mathematical Society, P. O. Box 5904, Boston, MA 02206-5904. *All orders must be accompanied by payment.* Other correspondence should be addressed to Box 6248, Providence, RI 02940-6248.

Memoirs of the American Mathematical Society is published bimonthly (each volume consisting usually of more than one number) by the American Mathematical Society at 201 Charles Street, Providence, RI 02904-2294. Periodicals postage paid at Providence, RI. Postmaster: Send address changes to Memoirs, American Mathematical Society, P. O. Box 6248, Providence, RI 02940-6248.

This publication is indexed in *Science Citation Index*®, *SciSearch*®, *Research Alert*®, *CompuMath Citation Index*®, *Current Contents*®*/Physical, Chemical & Earth Sciences.*
Printed in the United States of America.

♾ The paper used in this book is acid-free and falls within the guidelines established to ensure permanence and durability.
Visit the AMS homepage at URL: `http://www.ams.org/`

10 9 8 7 6 5 4 3 2 1 02 01 00 99 98 97

Table of Contents

Abstract

In this Memoir we establish global Rankin Selberg integrals which determine the standard L function for the group $GL_r \times G'$ where G' is an isometry group of a nondegenerate symmetric form. The class of automorphic representations considered here is for any pair $\Pi_1 \otimes \Pi_2$ where Π_1 is generic cuspidal for $GL_r(\mathbb{A})$ and Π_2 is cuspidal for $G'(\mathbb{A})$. The constructon of these L functions involves the use of certain new "models" of local representations; these models generalize the usual generic models. Also we compute local unramified factors in a new way using geometric ideas.

1991 *Mathematics Subject Classification.* 11E.
Key words and phrases. Automorphic L functions.
Partially supported by NSF

§0. Introduction

The construction of L functions for classical groups has its origin in the work of Godemont-Jacquet[1]. In [G-J] there is an integral presentation of the standard L function of a cuspidal irreducible automorphic module for the group $GL_n(\mathbb{A})$. In modern language such an integral presentation is a special case of the Rankin Selberg integral technique. By the standard L-function of GL_n we mean the L function associated to irreducible cuspidal representation Π of $GL_n(\mathbb{A})$ and the representation $\omega_1 =$ the first fundamental representation of the L group of $GL_n(\cong GL_n(\mathbb{C}))$. In [G-PS-R] the idea of "doubling" the group is used to construct Rankin Selberg integrals which represent the standard L function of the classical groups (in such an instance we mean those groups corresponding to the isometry group of some Hermitian, symmetric or skew symmetric form. Again the standard L function is associated to the first fundamental representation of the L-group of the given classical object. In any case there is no assumption about the cuspidal automorphic representation used in the construction of the doubled integral mentioned above.

The next level of generalization in the study of Rankin Selberg integrals is to construct the standard L function for the product group $G_1 \times G_2$ where each of G_1 or G_2 is a classical group. Specifically for reasons stated below we require one of the groups to be GL_n. Thus $G_1 \times G_2 = GL_n \times G_2$ where G_2 is classical.

The case where $G_2 = GL_r$ (r not necessarily n here) has been studied in [J-PS-S]. In particular the Rankin Selberg construction given in [J-PS-S] requires that both automorphic irreducible cuspidal modules be automorphic "generic" modules. This is simply the condition that requires the automorphic module to admit nonzero "nondegenerate" Fourier coefficients on the group $N_+(\mathbb{A})/N_+(k)[N_+, N_+](\mathbb{A})$ where $N_+ =$ upper triangular subgroup with 1 on the diagonal. We note that each irreducible cuspidal representation of GL admits such a property. Moreover the basic Rankin Selberg integral determining this family of L functions is based on a certain model adapted to these Whittaker models in a simple fashion. This model is also tied to the construction of a derivative of a local representation as given in [B-Z]. In any case we consider the subgroup $N_+^{(r)}$ of N_+ given explicitly

$$g = \begin{bmatrix} \begin{matrix} 1 & & 0 \\ & \ddots & \\ 0 & & 1 \end{matrix} & y_{ij} \\ 0 & \begin{matrix} 1 & x_{12} & & * \\ & & \ddots & x_{\ell-1,\ell} \\ & & & 1 \end{matrix} \end{bmatrix}, r + \ell = n$$

We consider the global unitary character ψ_r on N_+^r defined by $g \rightsquigarrow \psi(y_{r1} + x_{12} + \cdots + x_{\ell-1\ell})$. Then the stabilizer $GL_r(\psi_r) = \{ \begin{bmatrix} A & \begin{matrix} * \\ \vdots \\ * \end{matrix} \\ 0 \cdots 0 & 1 \end{bmatrix} | A \in GL_{r-1}, *$ arbitrary $\}$ = the mirabolic subgroup of GL_r. Then starting with a cuspidal automorphic representation Π of GL_n and $f_\pi \in \Pi$ we consider the ψ_r Fourier coefficient of f_π

[1]Received by the editor July 20, 1995.

given by the formula

$$(f_\pi^{\psi_r})(\left[\begin{smallmatrix} g & 0 \\ 0 & I_\ell \end{smallmatrix}\right]) = \int\limits_{N_+^{(r)}(k)\backslash N_+^{(r)}(\mathbb{A})} f_\pi(u\left[\begin{smallmatrix} g & 0 \\ 0 & I_\ell \end{smallmatrix}\right])\psi_r(u)du$$

We also allow here the specific case where $\ell = 0 (r = n)$ so that $f_\pi^{\psi_r} = f_\pi$ in this case. Then given any cuspidal automorphic representation σ of $GL_{r-1}(\mathbb{A})$ viewed as a subgroup of $\mathrm{Stab}_{GL_r}(\psi_r)$ we form the basic pairing between the space π and the space σ via the following type of integral.

$$\int\limits_{GL_{r-1}(k)\backslash GL_{r-1}(\mathbb{A})} f_\pi^{\psi_r}(\left[\begin{smallmatrix} g & 0 \\ 0 & I \end{smallmatrix}\right])f_\sigma(g)||detg||^s dg$$

This family of integrals defines an element in the space

$$Hom_{GL_n(\mathbb{A})}(\Pi, \mathrm{Ind}_{GL_{r-1}(\mathbb{A})\times N_+^{(r)}}^{GL_n(\mathbb{A})}(\check{\sigma}||^s \otimes \overline{\psi}_r)$$

Here Ind represents noncompact induction and $\check{\sigma}$ = the contragredient representation of σ and $\overline{\psi}_r = \overline{\psi_r(u)}$ (complex conjugate). This specific family of integrals represents the standard L function for the product group $GL_n \times GL_{r-1}$ given by

$$L(\Pi \otimes \sigma, s)$$

The integral above "unwinds" to an integral involving the "Whittaker or generic" models of both Π and σ. The point here is that the basic integral involving Π and σ involves a specific embedding of Π into $\mathrm{Ind}_{GL_{r-1}(\mathbb{A})\ltimes N_+^{(r)}}^{GL_n(\mathbb{A})}(\check{\sigma}|\,|^s \otimes \overline{\psi}_r)$. This in fact determines a product formula of the global integral into local Euler factors. In this paper we place an emphasis on this specific type of family of models to construct a generalization of this case $(GL_n \times GL_r)$ to the case $(GL_n \times O(V))$ where $O(V)$ is any isometry group of a symmetric form.

Before we begin the discussion of the general integral we note that the basic point in studying these "tensor product" L functions is to infer certain specific information about the poles. Specifically the tensor L function above $L(\Pi \otimes \sigma, s)$ has a continuation to a holomorphic function in $\mathbb{C}$. This point plays an important role in the determination of the discrete (non cuspidal) spectrum of $GL_n(\mathbb{A})$ (See [M-W]).

We return now to the general case of the construction of the standard L function for the group $GL_r \times O(V)$ (where $O(V)$ = isometry group of an symmetric form). The importance of such a construction is associated to the problem of functoriality. Namely given a group $O(V)$ it is possible to define a functorial lifting from $O(V)$ to $GL_{\dim V}$ (if $\dim V$ even) or $GL_{\dim V-1}$ (if $\dim V$ odd). In specific terms it is possible to associate an L group $O(V)^L$ and a homomorphism of $O(V)^L$ to the semi direct product $GL_{\dim V}(\mathbb{C}) \ltimes \sigma$ (where σ is an involution on $GL_{\dim V}(\mathbb{C})$ whose fixed point set equals $O(n,\mathbb{C})$ (in the case $\dim V$ even). The twisting here by σ is needed to define appropriately the lift from $O(V)^L$ into the *twisted* $GL_{\dim V}(\mathbb{C})$. The issue of functoriality here is to show that there is a lifting of an automorphic cuspidal representation of $O(V)(\mathbb{A})$ to an automorphic representation of $GL_{\dim V}(\mathbb{A})$ which is predicted by the functorial map at all local primes where the data is unramified principal series. In any case there seems to be at this point (for this specific example) two diverse approaches to the problem. First there is the expected use of trace formula techniques to establish an identity between traces of cuspidal kernels on the groups $O(V)(\mathbb{A})$ and the twisted $(GL_{\dim(V)} \ltimes \sigma)(\mathbb{A})$. From these results it is possible to deduce the evidence of the functorial lifting (at least at the level of L-packets). See [W] for recent results. The second approach to the functorial problem is the application of the Converse Theorem. Here we use the fact that for a given cuspidal irreducible representation Π of $O(V)(\mathbb{A})$ we have almost perfect knowledge of the family of twisted L functions $L(\Pi\otimes\sigma, s)$ where σ ranges over all those cuspidal representations of $GL_r(\mathbb{A})$ $(r \leq \dim V - 1)$ which are unramified at all finite primes. The basic difficulty is the determination of poles of the tensor L function $L(\Pi\otimes\sigma, s)$. The use of the Converse Theorem ([C-PS]) implies the existence of a weak global functorial lifting. That is, we know that there is an automorphic representation Π' of $GL_{\dim V}(\mathbb{A})$ which is equivalent at all finite unramified primes to the functorial lift of the given Π of $O(V)$. After existence of this weak functorial lifting in order to establish the local lifting of representations we do not require exact knowledge of the poles. We need only to show the finiteness of poles of the tensor L functions and the construction of local γ factors. The point of the present work is to establish the beginning of this theory for an arbitrary automorphic module of $O(V)(\mathbb{A})$.

The starting point for this work is the existence of certain families of models for a representation of $O(V)(\mathbb{A})$ which is a proper generalization of the model constructed in the GL_n case above. The data we start with is the orthogonal group $O(V)$ associated to a form q_V on a vector space V defined over a number field K. We consider a splitting of the space

$$V = R \oplus S \oplus R^*$$

where R, R^* isotropic subspaces nonsingularly paired by q_V and S the orthogonal complement of $R \oplus R^*$ and $S \neq \{0\}$. Then we consider the maximal parabolic subgroup P_R of $O(V)$ stabilizing the space R. This parabolic has a Levi decomposition of the form

$$GL(R) \times O(S, q_V|_S) \ltimes U_R.$$

Here $GL(R) =$ the general linear group of the space $R, O(S, q_V|_S) =$ the isometry group of the q_V restricted to S and $U_R =$ the unipotent radical of P_R. Then U_R can be fit into the exact sequence of groups.

$$1 \to \Lambda^2(R) \to U_R \underset{p_{R,S}}{\to} Hom(R,S) \to 1.$$

Here $\Lambda^2(R) =$ the second exterior power of R and $Hom(R,S)$ are taken with the under lying vector space structure as groups. In any case where $S \neq 0$ we know that the set of all unitary characters on U_R is given by the group (through the usual duality construction) $Hom(S, R)$

$$[\Lambda^2(R)\backslash U_R]^\wedge \cong [Hom(R,S)]^\wedge \underset{\langle\ ,\ \rangle}{\cong} Hom(S,R)$$

Then we consider a very specific type of character on U_R. Namely by the identification $\langle\ ,\ \rangle$ above we consider a rank one operator in $Hom(S, R) \cong S \otimes R^*$ having the form $\xi \otimes v^*$ where ξ has nonzero q_V length. Then we consider the character $\psi_{\xi\otimes v^*}$ on U_R defined via $\xi \otimes v^* : u \rightsquigarrow \psi(\langle P_{R,S}(u)|\xi\otimes v^*\rangle)$ (ψ a nontrivial additive character on the local field k). Then the stabilizer of $\psi_{\xi\otimes v^*}$ in the group $GL(R) \times O(S, q_V|_S)$ consists of the product $GL(R)^{v^*} \times O(S, q_V|_S)^\xi$, which is isomorphic to the product of a mirabolic subgroup of $GL(R)$ and the orthogonal group $O((\xi)^\perp, q_V|_{(\xi)^\perp}) =$ isometry group of the form q_V restricted to $(\xi)^\perp$, the q_V perpendicular complement to ξ in S. Then we know that there is an irreducible module of $GL(R)^{v^*}$ has the form

$$Z(\psi_{v^*}) = \mathrm{ind}_{N_R}^{GL(R)^{v^*}}(\psi_{v^*})$$

where N_R is a maximal unipotent subgroup (of a Borel subgroup) of $GL(R)^{v^*}$ and ψ_{v^*} is a generic character on N_R (i.e. a Whittaker type character). Also we let σ be any admissible irreducible module for $O((\xi)^\perp, q_V|_{(\xi)^\perp})$. Then we form the local compactly induced module

$$X_\sigma^{\xi\otimes v^*} = \mathrm{ind}_{(GL(R)^{v^*}\times O(\xi)^\perp, q_V|_{(\xi)^\perp})\ltimes U_R}^{O(V)}((Z(\psi_{v^*})\otimes\sigma)\otimes\psi_{\xi\otimes v^*})$$

The basic starting point in this paper is that the above module is a *multiplicity free representation* of the group $O(V)$. This means that for Π an admissible irreducible module for $O(V)$

$$Hom_{O(V)}(X_\sigma^{\xi\otimes v^*}, \Pi)$$

has dimension equal to at most 1. We prove this statement in [R] for the case when the field k is non Archimedean. We note that in the case when $R = \{0\}$ then $P_R = O(V)$ and the module (ξ, v^* do not exist in this case)

$$X_\sigma = \mathrm{ind}_{O((\xi)^\perp, q_V|_{(\xi)^\perp})}^{O(V)}(\sigma)$$

is the substitute for $X_\sigma^{\xi\otimes v^*}$. The *multiplicity one statement is also valid for* X_σ. This is the generalization of the well known branching rule of $O(n)$ to $O(n-1)$ in the category of finite dimensional representations of the groups $(O(n), O(n-1))$.

The basic problem is to use the above models for $O(V)$ in order to construct a Rankin Selberg integral which represents the standard L function $L(\Pi\otimes\sigma, s)$.

We now fix σ as a irreducible cuspidal representation of $O(V)(\mathbb{A})$ and τ as an irreducible cuspidal representation of $GL_r(\mathbb{A})$ and construct a family of Rankin Selberg integrals built from this data.

Then we take any orthogonal space (W, q_W) which admits a decomposition of the form

$$W = R_1 \oplus S_1 \oplus R_1^*$$

where R_1 and R_1^* isotropic subspaces nonsingularly paired by q_W and S_1 the q_W orthogonal complement to $R_1 \oplus R_1^*$. Here we have $GL_r = GL(R_1)$. Then we *require* that W admits another decomposition

$$W = R_2 \oplus S_2 \oplus R_2^*$$

of a similar nature where $O(V)$ is isomorphic to the orthogonal group $O((\xi)^\perp, q_{S_2}|_{\xi^\perp})$ where ξ is a nonzero vector in S_2 with $q_{S_2}(\xi) \neq 0$.

Then we consider any automorphic irreducible representation ρ of $O(W)(\mathbb{A})$. As in the construction of the Rankin Selberg integrals for $GL_r \times GL_{r'}$ above we choose $f_\rho \in \rho$. Then we form the Fourier coefficient of f_ρ relative to the group $N_{R_2} \ltimes U_{R_2}$ and the additive character $(n, u) \to \Psi_{\xi \otimes v^*}(n, u) = \psi_{v^*}(n)\psi_{\xi\otimes v^*}(\langle P_{R,S}(u)|\xi \otimes v^*\rangle)$. given by

$$f_\rho^{\Psi_{\xi\otimes v^*}}(g) = \int_{N_{R_2}U_{R_2}(k)\backslash N_{R_2}U_{R_2}(\mathbb{A})} f_\rho(nug)\Psi_{\xi\otimes v^*}(nu)dndu$$

Then we form a basic pairing between the representation ρ of $O(W)(\mathbb{A})$ and σ of $O(V)(\mathbb{A})$ via a period integral of the form:

$$\mathbf{P}_{\xi\otimes v^*}(f_\rho, f_\sigma) = \int_{O(V)(k)\backslash O(V)(\mathbb{A})} f_\rho^{\Psi_{\xi\otimes v^*}}(h) f_\sigma(h)dh$$

Here $f_\sigma \in \sigma$, the space of σ. The integral defines a $O(W)(\mathbb{A})$ equivariant pairing in the space

$$Hom_{O(W)(\mathbb{A})}(\rho, \tilde{X}_\sigma^{\xi\otimes v^*}(\mathbb{A}))$$

Here

$$\tilde{X}_\sigma^{\xi\otimes v^*}(\mathbb{A}) = \bigotimes_v \left[(X_\sigma^{\Xi\otimes v^*})_v^\sim\right]$$

Note $(X_\sigma^{\xi\otimes v^*})_v^\sim$ is just the local non compact induced module associated to $(X_\sigma^{\xi\otimes v^*})_v$. The uniqueness principle above asserts that the above space has dimension at most equal to a finite number, namely

$$\dim(Hom_{O(W)(\infty)}(\rho_\infty, \tilde{X}_\sigma^{\xi\otimes v^*}(\infty)))$$

where $\rho_\infty = \bigotimes_{v \text{ Archimedean}} \rho_v$, etc...

We note that the period integral $\mathbf{P}_{\xi\otimes v^*}(\ ,\ )$ does not depend on any $GL_r(\mathbb{A})$ module yet.

We note that the use of the model above was first presented by Novodvorsky in [N] to construct the L function for $GL_1 \times O(V)$ (with $\dim V$ odd). Later Ginzburg in [G] gave the general construction for the tensor L function for $GL_r \times O(V)$ using this model. In both cases all the data is generic. Then Furosawa in [F] constructed the L function of $O(3,2) \times GL_2$ using again such a model; in [F] there is no assumption in genericity. In [B-F-G] there is again a use of the above model in order to represent L functions of the functorial lifting from Sp_n to $SO(2n+2)$. We note Murase and Sugano in [M-S] have used a specific case of the model above to construct the standard L function for $O(V) \times GL_1$ (with no generic condition on $O(V)$).

At this point noting that $GL(R_1) = GL_r$ we choose for ρ a very specific type of Eisenstein series representation to substitute in the above integral.

Namely we consider the induced module (normalized induction)

$$\operatorname{ind}_{GL(R_1)(\mathbb{A}) \times O(S_1)(\mathbb{A}) \ltimes U_{R_1}(\mathbb{A})}^{O(W)(\mathbb{A})} (\tau \otimes |\ |^s \otimes \tilde{\rho} \otimes \mathbb{1})$$

Here τ is a cuspidal representation of $GL_r(\mathbb{A})$, $|\ |^s = |detx|^s$ and $\tilde{\rho}$ some automorphic irreducible module for $O(S_1)(\mathbb{A})$ (not necessarily cuspidal). Then we form the associated family of Eisenstein series:

$$\mathcal{E}(f_s(\tau, \tilde{\rho},), x)$$

associated to an analytic section $f_s(\tau, \tilde{\rho},)$ in the induced module above.

In the above period integral we let $f_\rho = \mathcal{E}(f_s(\tau, \tilde{\rho},),\ )$. Thus the basic point of this paper is the determination of

$$\mathbf{P}_{\xi \otimes v^*}(\mathcal{E}(f_s(\tau, \tilde{\rho},\ ), f_\sigma)$$

The above considerations imply that this period is essentially given by an Euler product. In **Theorem** 4.3 we evaluate explicitly the Eulerian nature of the integral. But even more explicitly we evaluate $\mathbf{P}_{\xi \otimes v^*}$ in the case where $\dim R_2 = \dim R_1 - 1$. In fact we assume that $S_2 \supseteq (\xi)^\perp = V \supseteq S_1$. Then we prove

Theorem 4.5. *For a suitable finite set S of primes there exists the identity:*

$$\mathbf{P}_{\xi \otimes v^*}(\mathcal{E}(f_s(\tau, \tilde{\rho},\ ), f_\sigma) = \left(\frac{L^S(\tau \otimes \sigma, s + \frac{1}{2})}{L^S(\tau \otimes \tilde{\rho}, s+1) L^S(\tau, {}_{sym^2}^{\Lambda^2}, 2s+1)} \right) \int_{O(S_1)(\mathbb{A}_S) \backslash O(V)(\mathbb{A}_S)} \mathbf{P}_{\tilde{\rho}}((f_s)^{\Psi_{\xi \otimes v^*}} * h, f_\sigma * h) dh$$

Here $\mathbf{P}_{\tilde{\rho}}$ denotes the $O(S_1)(\mathbb{A})$ invariant pairing between σ and $\tilde{\rho}$ (given by an integral similar to $\mathbf{P}_{\xi \otimes v^*}$ case). The term $(f_s)^{\Psi_{\xi \otimes v^*}}$ represents an integral determined from the formation data of the form (for an appropriate Weyl element w_S)

$$\int_{(U_{R_2} \cap N_{R_1})(\mathbb{A}) l \backslash U_{R_2}(\mathbb{A})} W_\tau^{\psi_r}(f_S)[w_S u g] \psi_\xi(u) du$$

In any case $(f_s)^{\Psi_{\xi\otimes v^*}}$ when restricted to $O(S_1)(\mathbb{A})$ transforms according to the irreducible module $\tilde{\rho}$; thus we can form the period integral $\mathbf{P}_\rho$.

Also we note here that $L(\tau\otimes\sigma,\ )$ and $L(\tau\otimes\tilde{\rho},\ )$ represent the standard L functions associated to the group $GL_r \times O(V)(\mathbb{A})$ and $GL_r \times O(S_1)(\mathbb{A})$ respectively.

Moreover $L(\tau, \overset{\Lambda^2}{_{sym^2}},\ )$ is either the exterior square or symmetric square L function of GL_r (associated to τ). Specifically it becomes Λ^2 if W is even dimensional and sym^2 if W is odd dimensional. The L^S means here the restricted L function built on the primes away from the set S.

We note that we in fact expect a stronger version of **Theorem** 4.5. Namely we expect that we can choose the data f_s, f_τ and the set S so that the integral in **Theorem** 4.5 can be replaced by a term of the type:

$$\rho_\infty(s)\mathbf{P}_{\tilde{\rho}}((f_s), f_\sigma)$$

Here $\rho_\infty(s)$ is nonvanishing at $s \in \mathbb{C}$ and the data in $\mathbf{P}_{\tilde{\rho}}$ is independent of s and in fact equals a nonzero multiple of $\mathbb{P}_{\tilde{\rho}}(f_{\tilde{\rho}}, f_\sigma)$ for an appropriate $f_{\tilde{\rho}} \in \tilde{\rho}$. Here $f_{\tilde{\rho}}$ can be chosen so that $\mathbb{P}_{\tilde{\rho}}(f_{\tilde{\rho}}, f_\sigma) \not\equiv 0$.

We note here that the denominator L factor $L^S(\tau \otimes \tilde{\rho}, s+1)L^S(\tau, \overset{\Lambda^2}{_{sym^2}}, 2s+1)$ should possess the expected property of normalizing the Eisenstein series $\mathcal{E}(f_s(\tau, \tilde{\rho},\ ),\ )$.

Thus the main issue of the paper is the evaluation of $\mathbf{P}_{\xi\otimes v^*}(\mathcal{E}(f_s, (\tau, \tilde{\rho},\ ), f_\sigma)$ in terms of the Euler decomposition predicted by the general uniqueness principle stated above.

Thus there is a basic local integral involved in the calculation of

$$\mathbf{P}_{\xi\otimes v^*}(\mathcal{E}(f_s(\tau, \tilde{\rho},\ ), f_\sigma)$$

where all the data is unramified, i.e. the local representations are all spherical (class one) and the local components or vectors in these representations are the K fixed vector suitably normalized. *Then the second main Theme of this paper involves the determination of the local embedding*

$$Hom_{O(V))}(X_{\sigma_v}^{\xi\otimes v^*}, \Pi_v)$$

where σ_v and Π_v are unramified modules. Specifically we must dualize here and study the embedding

$$Hom_{O(V)}(\Pi_v^\sim, (X_{\sigma_v}^{\xi\otimes v^*})^\sim)$$

We want to establish enough information about the specific image of the $K_{O(V)}$ fixed vector of Π_v^ν in the space $(X_{\sigma_v}^{\xi\otimes v^*})^\sim$ to calculate the local integrals.

Specifically the primary goal here should be to establish a *Casselman-Shalika* formula for this specific vector. We however use a geometric method (that should be equivalent to determination of a Casselman-Shalika formula) which yields a very simple conceptual way to evaluate the above integrals. The starting point is the following type of local integral. We first let $P_{\tilde{\rho}_v}$ be a local component of $\mathbb{P}_{\tilde{\rho}}$ determined by the uniqueness principle above. Then we consider

$$\int_{O(S_1)\backslash O(V)} \mathbb{P}_{\tilde{\rho}_v}((f_s)_v^{\Psi_{\xi\otimes v^*}} * h_v, w * h_v) dh_v$$

Here $w \in \sigma_v$. Then the primary issue is to evaluate the above integral when all the data is unramified.

Our method of proof here is given in three steps. (I) to use the geometry of the situation to determine exactly the poles (in s) of the above integral (II) to determine the local γ factors associated to certain functional equations defined by the above integral and (III) to use (I) and (II) to find the exact zeroes (in s) of the above integral.

In order to avoid a lot of notation here we describe the method of proof in the 3 steps above in a simple example (where $R_2 = \{0\}$). Thus the above integral equals

$$\int_{O(S_1)\backslash O(V)} \mathbb{P}_{\tilde{\rho}_v}((f_s)_v * h_v, w * h_v) dh_v$$

Specifically the geometry here is the determination of the $O(V)$ orbits in the projective space $P_{R_1}\backslash O(W)$. In such a case there exists $2 O(V)$ orbits in this projective space. This determine an exact sequence of $O(V)$ modules

$$O \to \mathrm{ind}_{O(S_1)}^{O(V)}(\tilde{\rho}) \to \mathrm{ind}_{GL_1 \times O(S_1) \ltimes U_1}^{O(W)}(|\ |^s \otimes \tilde{\rho} \otimes \mathbb{1})$$
$$\to \mathrm{ind}_{GL_1 \times O(S_1') \ltimes U_1'}^{O(V)}(|\ |^{s+\frac{1}{2}} \otimes \tilde{\rho}|_{O(S_1')} \otimes \mathbb{1}) \to O$$

The last term is induction from a parabolic subgroup of $O(V)$ which stabilizes an isotropic line in the space V. It suffices to say here that S_1' is the *codimension one* subspace of S_1 and $\tilde{\rho}|_{O(S_1')}$ represents the restriction to $O(S_1')$.

Then the basic integral above involves f_s which is the unique $K_{O(W)}$ fixed vector in the middle step of the exact sequence and w the unique $K_{O(V)}$ fixed vector in σ.

The idea here is to use a variation of the Bernstein polynomial technique. Namely we need to find an operator T_s on the middle piece of the exact sequence above so that $T_s(f_s)$ (for all s) has support in $\mathrm{ind}_{O(S_1)}^{O(V)}(\tilde{\rho})$ (not dependent on s). We also require that the operator T_s can be adjointed across in the integral above in such a way that w is an eigenvector for ${}^tT_s$, i.e. ${}^tT_s(w) = \lambda_\sigma(s)w$ where $\lambda_\sigma(s)$ is a specific polynomial in s, dependent on π. Thus we deduce an identity of the form:

$$\int_{O(S_1)\backslash O(V)} \mathbb{P}_{\tilde{\rho}_v}(T_s(f_s) * h_v, w * h_v) dh_v = \lambda_\sigma(s)$$
$$\int_{O(S_1)\backslash O(V)} \mathbb{P}_{\tilde{\rho}_v}(f_s * h_v, w * h_v) dh_v$$

This identity determines the poles of our given integral (i.e. contained in the set of zeroes of the polynomial $\lambda_\sigma(s)$); the key point here is that $T_s(f_s)$ is supported in the first component of the exact sequence and hence the left hand side of the above identity in an analytic function in s. Thus we have completed (I) above. The issue becomes to determine the zeroes of the left hand side.

In this specific example we construct T_s in an elementary way and determine the zeroes at the same time. We start with the Hecke algebra $\mathcal{H}_V = \mathcal{H}(O(V)//K_{O(V)}) =$ the space of compactly supported $K_{O(V)}$ biinvariant functions on the group $O(V)$. Then we tensor this algebra with the Laurent series $\mathbb{C}[X, X^{-1}]$ to form

$$A_V = \mathbb{C}[X, X^{-1}] \otimes \mathcal{H}_V$$

Then A_V acts on the middle piece of the exact sequence via the formula: $X \otimes \phi \rightsquigarrow (h_s * \varphi)q^{-s}$. Then to produce T_s above we look at the ideal = the *annihilator of the boundary component* =

$$I_V^{supp} = \{f \in A_V | f \text{ annihilates the last step in the exact sequence (for all } s)\}.$$

In particular $I_V^{supp} \neq A_V$ and in fact contains a distinguished element T_s. Then we prove T_s acts on w via a very specific polynomial $\lambda_\sigma(s)$. We note $\mathcal{H}_V$ can be explicitly described via the Satake isomorphisms, i.e. $f \rightsquigarrow tr\sigma'(f)$ where σ' is an arbitrary spherical representation of $O(V)$. Then $\lambda_\sigma(s)$ equals the inverse $(L(\sigma, st, s))^{-1}$, where $L(\sigma, st, s)$ is the standard L function of σ defined in the usual way (this is generally true with a small modification in certain examples, see §1 and §2 for precise statement). However the *remarkable property that T_s admits is the following identity*:

$$T_s(f_s) = (L(\tilde{\rho}, st, s+1))^{-1} f_s^0$$

where f_s^0 is supported as a function in $\operatorname{ind}_{O(S_1)}^{O(V)}(\tilde{\rho})$ on the set $O(S_1) \cdot K_{O(V)}$ and equals $\tilde{\rho}(m)(v_0)$ where $m \in O(S_1)$ and v_0 is the unique fixed vector of $K_{O(S_1)}$ in $\tilde{\rho}$. With this specific condition we deduce easily that

$$\int_{O(S_1)\backslash O(V)} \mathbf{P}_{\tilde{\rho}_v}(f_s^0 * h_v | w * h_v) dh_v \equiv \mathbf{P}_{\tilde{\rho}_v}(v_0, w)$$

Here the basic integral (with unramified data) equals

$$\left[\frac{L(\sigma, st, s)}{L(\tilde{\rho}, st, s+1)}\right] \mathbf{P}_{\tilde{\rho}_v}(v_0, w)$$

The proof of the explicit formula above for $T_s(f_s)$ is based on the idea of "local" doubling alluded to at the beginning of this section. We have not followed steps (II) and (III) given above. We in fact use these steps in the more *general case.* The use of the support idea as given above is also used in the general case. In any event we prove by the support idea (§5) that

$$\int_{O(S_1)\backslash O(V)} \mathbf{P}_{\tilde{\rho}_v}((T_s(f_s)^{\Psi_{\xi\otimes V^*}} * h_v, w * h_v) dh_V =$$

$$[L(\sigma \otimes \tilde{\rho}, st, s)]^{-1} \Big(\int_{O(S_1)\backslash O(V)} \mathbf{P}_{\tilde{\rho}}((f_s)^{\Psi_{\xi\otimes v^*}} * h_v, w * h_v) dh_v \Big)$$

Thus we have that the poles of our given general integral is completely controlled by the poles of the local L function

$$L(\sigma \otimes \tilde{\rho}, st, s)$$

The point now is to determine the zeroes of the basic integral above. For this we look at certain other data attached to these integrals. We start with a $O(W)$ spherical representation $\Pi_v = \Pi_v(\chi_v)$ where χ_v is an appropriate quasi character on a torus associated to a Borel subgroup; there exists $\tau_v, \tilde{\rho}_v$ both spherical so that Π_v generically is equivalent to

$$\mathrm{ind}_{GL(R_1)\times O(S_1)\ltimes U_{R_1}}^{O(W)}(\tau_v \otimes \tilde{\rho}_v \otimes \mathbb{1})$$

Similarly we assume that σ_v is an unramified and $\sigma_v = \sigma_v(\gamma_v)$, γ_v another quasi character. The *multiplicity one statement about the pair* $(\Pi_v(\chi_v), \sigma_v(\gamma_v))$ is simply that

$$\dim Hom_{O(W)}(\Pi_v(\chi_v), \tilde{X}_{\sigma_v(\gamma_v)}^{\xi\otimes v^*}) \leq 1.$$

One minor technical point here is that we really require this uniqueness in the special orthogonal case for the construction of this manuscript; however the ensuing discussion still remains valid and illustrates the type of arguments used in the paper. This uniqueness implies a family of functional equations of the following nature. Namely, consider any pair of intertwining operators

$$I_{w_1} \otimes I_{w_2} : \Pi_v(\chi_v) \otimes \sigma_v(\gamma_v) \rightsquigarrow \Pi_v(w_1(\chi_v)) \otimes \sigma_v(w_2(\gamma_v))$$

constructed in the usual way. ((w_1, w_2) is a Weyl element in the group $O(W) \times O(V)$). Then there is a rational function $\epsilon(\chi_v, \gamma_v, w_1, w_2)$ so that

$$\int\limits_{O(S_1)\backslash O(V)} \mathbf{P}_{\tilde{\rho}_v}(I_{w_1}(f_{\chi_v})^{\Psi_{\xi\otimes v^*}} * h_1, I_{w_2}(w) * h_1)dh_1 =$$

$$\gamma(\chi_v, \gamma_v, w_1, w_2)(\int\limits_{O(S_1)\backslash O(V)} \mathbf{P}_{\tilde{\rho}_v}((f_{\chi_v})^{\Psi_{\xi\otimes v^*}} * h_1, w * h_1)dh_1)$$

Then we assert that the knowledge of the data $\gamma(\chi_v, \gamma_v, w_1, e)$ for all w_1 as well as the complete set of poles of the given integral (with unramified data) completely determines the given integrals (§6, 7, and 8). We let the basic integral

$$\int\limits_{O(S_1)\backslash O(V)} \mathbf{P}_{\tilde{\rho}_v}((f_{\chi_v})^{\Psi_{\xi\otimes v^*}} * h_1, w * h_1)dh_1$$

equal the ratio

$$\ell(\chi, \gamma) = \frac{P(\chi, \gamma)}{Q(\chi, \gamma)}$$

where $Q(\chi, \gamma) = L(\tau \otimes \tilde{\rho}, st, 1/2)^{-1}$ and $P(\chi, \gamma)$ is some rational function in χ and γ. In any case we deduce that

$$\gamma(\chi_v, \gamma_v, w_1, e) = (\frac{P(w_1(\chi_v), \gamma_v)Q(\chi_v, \gamma_v)}{P(\chi_v, \gamma_v)Q(w_1(\chi_v), \gamma_v)})$$

The explicit evaluation of γ and Q shows that P satisfies a certain type of invariance property under the Weyl group of $O(W)$. i.e. in the χ_v variable. However the invariance property for P still does not suffice for the exact evaluation of P. What we need (**Lemmas** 6.1 - 6.3) is the expansion of $\ell(\chi, \gamma)$ as a sum of Weyl elements w in $O(V)$ of an expansion similar to the one given for either a spherical function or a Whittaker vector. Basically this sum is a calculation based on the Casselman-Shalika formula. In any case the key idea that allows us to evaluate P is the fact that the term $[L(\tau' \otimes \tilde{\rho}, st, \frac{1}{2})]^{-1}$ when projected onto the skew invariants of the Weyl group $W_{O(W)}$ of $O(W)$ in the ring

$$\mathbb{C}[q^{\langle \chi, \rangle}, q^{-\langle \chi, \rangle}]$$

we obtain $\Delta_{O(W)}(\chi)$ = the Weyl denominator $= \sum_{W \in W_{O(W)}} sgn(w) q^{\langle w\rho^{\vee}, \chi \rangle}$. (here τ' is an arbitrary unramified representation of $GL(R_{i-1})$).

Then the remaining issue for the paper is the determination of the $\gamma(\chi_v, \gamma_v, w_1, e)$ factors. This problem is perhaps of separate interest by itself (§8). What is quite remarkable here is how simple the actual evaluation of γ is in this case. We note first by the usual Gindinkin Karpelevic formula it is possible to evaluate γ only for those w_1 which are simple reflections relative to a fixed root system of $O(W)$. In particular by suitable choice of R_1 and R_2 (differing by one dimension) it suffices to compute the γ for those simple reflections that belong to $GL(R_1)$ and $O(S_2)$. We first note it is possible to determine the γ factor for the pair $(O(V), O(S_2))$ (a codimension one pair which is the simple first case coming from the doubling construction). This follows simply by the direct formula for the integral in this case . Then we prove that the $\gamma(\chi_v, \gamma_v, w_1, e)$ (for a Weyl element w_1 in $O(S_2)$) in fact equals the γ factor associated to the same w_1 relative to the pair $(O(V), O(W))$ (**Theorem** 8.1). This determines a simple inductive way to compute $\gamma(\chi_v, \gamma_v, w_1, e)$ for such w_1. On the other hand we prove (**Theorem** 8.1) that for a w_α (for a simple root in $GL(R_1)$) the $\gamma(\chi_v, \gamma_v, w_\alpha, e)$ equals the ϵ factor for w_α associated to the Whittaker-Jacquet integral for the group $GL(R_1)$. Thus we deduce (through multiplicativity) a very elegant formula for the general $\gamma(\chi_v, \gamma_v, w_1, e)$.

One of the consequences in the determination of the above unramified integral is an explicit determination of the term $\ell(\chi, \gamma)$ or even more correctly the term $\ell(\chi, \gamma)\ell(\chi^{-1}, \gamma^{-1})$. The import of such a quantity comes from the following construction. In the case where $\dim R_2 = 0$ we know that there is at most one element in the space

$$Hom_{O(V)}(\pi(\chi), \sigma(\gamma))$$

This means that we also can write the integral (for arbitrary data) in the form

(**Corollary** 3 to **Lemma** 7.2)

$$T_{\chi\otimes\gamma}(\varphi_1\otimes\varphi_2) = \int_{B_{O(S_2)}\times B_{O(V)}} \varphi_1 * \varphi_2[b_{O(S_2)}w_0\eta b_{O(V)}^{-1}]$$
$$(\chi^{-1}\delta^{1/2}_{B_{O(S_2)}})(b_{O(S_2)})(\gamma^{-1}\delta^{1/2}_{B_{O(V)}})(b_{O(V)})d_\ell(b_{O(S_2)})d_\ell(b_{O(V)})$$

Here $\varphi_1\otimes\varphi_2 \in$ the space $S(O(S_2)\times O(V)) = S(O(S_2))\otimes S(O(V)) =$ locally constant functions on $O(S_2)\times O(V)$ with compact support. The groups $B_{O(V)}$ and $B_{O(S_2)}$ are Borel subgroups of $O(S_2)$ and $O(V)$ suitably placed. The element $w_0\eta$ determines the *unique open orbit* of $B_{O(S_2)}\times B_{O(V)}$ acting on $O(S_2)$ (through the left and right action). Now $*$ is the convolution of the function φ_2 into φ_1 and finally $\delta_{B_{O(S_2)}}$ and $\delta_{B_{O(V)}}$ represent the modular functions on the Borel subgroups $B_{O(S_2)}$ and $B_{O(V)}$. Moreover d_ℓ represents usual left invariant measure on a group. Then we call $T_{\chi\otimes\gamma}$ the spherical functional associated to the unique (up to scalar) element in $Hom_{O(V)}(\pi(\chi),\sigma(\gamma))$.

In fact from such data it is possible to construct a spherical distribution on the group $O(S_2)\times O(V)$ which is invariant under the left and right action of the group $O(V)^\wedge$ (that is, $O(V)^\wedge$ is the diagonally embedded $\{(h,h)\in O(S_2)\times O(V)|h\in O(V)\}$. The formula for the distribution is given by the following:

$$\langle(\varphi_1\otimes\varphi_2)*T_{\chi\otimes\gamma}|T_{\chi^{-1}\otimes\gamma^{-1}}\rangle.$$

Here $\varphi_1\otimes\varphi_2$ represents the convolution of $\varphi_1\otimes\varphi_2$ into the function $T_{\chi\otimes\gamma}$; thus $(\varphi_1\otimes\varphi_2)*T_{\chi\otimes\gamma}$ represents a $K_{O(S_2)}\times K_{O(V)}$ finite function in $\pi(\chi)\otimes\sigma(\gamma)$ and hence taking the scalar product with $T_{\chi^{-1}\otimes\gamma^{-1}}$ makes sense. Then we prove

Theorem 7.1. *If $\varphi_1\otimes\varphi_2$ belongs to the Hecke algebra $\mathcal{H}(O(S_2)\otimes O(V)//K_{O(S_2)}\times K_{O(V)})$ then $\langle(\varphi_1\otimes\varphi_2)*T_{\chi\otimes\gamma}|T_{\chi^{-1}\otimes\gamma^{-1}}\rangle = \ell(\chi,\gamma)\ell(\chi^{-1},\gamma^{-1})Sat(\varphi_1\otimes\varphi_2)(\chi\otimes\gamma) = r(\chi,\gamma)$*

$$L(\pi(\chi)\otimes\sigma(\gamma),1/2)[L(\pi(\chi),Ad,1)L(\sigma(\gamma),Ad,1)]^{-1}$$
$$Sat(\varphi_1\otimes\varphi_2)(\chi\otimes\gamma)$$

with $r(\chi,\gamma)$ having no zeroes or poles in (χ,γ)!

Here $L(\pi(\chi)\otimes\sigma(\gamma),s)$ is the local L function associated to the group $O(S_2)\times O(V)$ and the pair $(\pi(\chi),\sigma(\gamma))$. We note the dual group associated to the product $O(S_2)\times O(V)$ is of the form $SO(n,n)(\mathbb{C})\times Sp_n(\mathbb{C})$ (or vice versa) in the case where both $O(S_2)$ and $O(V)$ are split groups. Moreover $L(\pi(\chi),Ad,s)$ and $L(\sigma(\gamma),Ad,s)$ represent the "adjoint" L functions for $O(S_2)$ and $O(V)$. Also $Sat(\varphi_1\otimes\varphi_2)(\ )$ represents the Satake transform of the pair $\varphi_1\otimes\varphi_2$.

The significance of **Theorem 7.1** is that we expect the nonvanishing of the ratio

$$\frac{L(\pi(x)\otimes\sigma(\gamma),\frac{1}{2})}{L(\pi(x),\mathrm{Ad},1)L(\sigma(\gamma),\mathrm{Ad},1)}$$

to be an "approximation" to the condition of the nonvanishing of $\mathrm{Hom}_{O(V)}(\pi(x), \sigma(\gamma))$.

We note that the above numerical criteria is essentially a local analogue of the *Gross-Prasad conjecture* [GP] which give the precise conditions of when to expect

$$\mathrm{Hom}_{O(V)}(\pi, \sigma) \neq 0$$

for a pair Π and σ of irreducible admissible representations.

§1. Basic Data

We are given a vector space V with a nondegenerate quadratic form q. We let W be a vector space of the form $V \oplus V_1$ provided with the quadratic form $\tilde{q} = q \oplus q_1$, which is again assumed to be nondegenerate.

Then we let $O(W, \tilde{q}), O(V_1, q_1)$ and $O(V, q)$ be the isometry groups of the forms $\tilde{q}, q_1$ and q. In particular there exists an embedding

$$O(V, q) \times O(V_1, q_1) \hookrightarrow O(W, \tilde{q})$$

In particular we consider the case where there exists an embedding $i : V \hookrightarrow V_1$ in such a way that the form q_1 restricted to $V(i(V)$ in $V_1)$ equals $-q$. With this condition $q_1 = -q \oplus q_2$ where q_2 is the form q_1 restricted to the orthogonal q_1 *complement* of $i(V)$ in V_1 (we let V be the space of q_2). This implies that there exists an embedding

$$O(V, q) \hookrightarrow O(V_1, q_1)$$

Then we know that the Witt index $(\tilde{q})$ = Witt index $(q \oplus -q \oplus q_2) = \dim q +$ Witt index (q_2).

We let X be the variety of isotropic subspaces of W of dimension equal to Witt index $(\tilde{q})$.

Then X is a homogeneous space under the standard action of the group $O(W, \tilde{q})$.

Under the action of $O(V, q) \times O(V_1, q_1)$ the variety X admits a finite number of orbits. In fact the set

$$X_\gamma = \{S \in X | \dim(S \cap V_1) = \gamma\}$$

is an orbit under $O(V, q) \times O(V_1, q_1)$ and all the $O(V, q) \times O(V_1, q_1)$ orbits in X have this form.

We note that in the case $\gamma > 0$ and X_γ is non empty then the stabilizer of a subspace $S \in X_\gamma$ in the group $O(V, q) \times O(V_1, q_1)$ is negligible in the sense that it contains as a normal subgroup the unipotent radical of a parabolic subgroup of $O(V, q) \times O(V_1, q_1)$.

In the case $\gamma = 0$ and X_0 nonempty, then Witt index $(q_2) = 0$. This implies that q_2 is anisotropic. Moreover the subspace $S = \{(v, i(v)) | v \in V\} \in X_0$ and *Stab*(S) in $O(V, q) \times O(V_1, q_1)$ equals $\{(g, g) | g \in O(V, q)\} \cdot \{(1, g_2) | g_2 \in O(V_2, q_2)\}$. Thus *Stab*$(S) \cong O(V, q)^\Delta \times O(V_2, q_2)$. Here $O(V, q)^\Delta$ is the diagonally embedded subgroup in $O(V, q) \times O(V, q)$ given above.

Then we choose the subspace $S \in X$ having the form

$$\{(v, i(v)) | v \in V\} \oplus S \cap V_2$$

where $S \cap V_2$ is a *maximal isotropic* subspace of V_2 relative to q_2. In particular the stabilizer of S in $O(W, \tilde{q})$ is the parabolic subgroup

$$P_S = GL(S) \times O((S \cap V_2)^\perp / S \cap V_2, q_2) \ltimes U_S.$$

Here $(S \cap V_2)^{\perp}$ is the q_2 orthogonal complement to $S \cap V_2$ in

V_2 Moreover U_S is the unipotent radical of P_S.

On the other hand if $S \cap V_2 \neq 0$ the stabilizer of S in $O(V, q) \times O(V_1, q_1)$ is the group

$$O(V, q)^{\Delta} \times \{(1, g) | g \in P_S \cap O(V_1, q_1)\}.$$

Note here that $P_S \cap O(V_1, q_1)$ is the subgroup of $O(V_1, q_1)$ having the form

$$GL(S \cap V_2) \times O((S \cap V_2)^{\perp}/S \cap V_2, q_2) \ltimes U_{S \cap V_2}$$

Here $U_{S \cap V_2}$ is the unipotent radical of the parabolic subgroup of $O(V_1, q_1)$ stabilizing the isotropic subspace $S \cap V_2$ (in V_1).

At this point we let k be a local field. Then we consider the analytic family of characters

$$\chi_s : P_S \to \mathbb{C}^x$$

given by

$$\chi_s((g_1, g_2) \cdot u) = |detg_1|^s$$

where $g_1 \in GL(S), g_2 \in O((S \cap V_2)^{\perp}/S \cap V_2), q_2)$ and $u \in U_S$. Then we form the induced module (normalized induction)

$$I(s) = ind_{P_S}^{O(W, \tilde{q})}(\chi_s)$$

Then it follows that when we *restrict* $I(s)$ to the open $O(V, q) \times O(V_1, q_1)$ orbit in $P_S \backslash O(W, \tilde{q})$ we obtain functions in the space (non-compactly induced)

$$Ind_{O(V,q)^{\Delta} \times P_S \cap O(V_1, q_1)}^{O(V,q) \times O(V_1, q_1)} \qquad (1 \otimes |\ |^{s+\lambda})$$

Here $|\ |^{s+\lambda}$ is character on $GL(S \cap V_2)$ given by $g \rightsquigarrow |\det g|^{s+\lambda}$ with $\lambda = \frac{1}{2}(\dim S - \dim S \cap V_2)$.

We note here that (compactly induced)

$$ind_{O(V,q)^{\Delta} \times P_S \cap O(V_1, q_1)}^{O(V,q) \times O(V_1, q_1)} \qquad (1 \otimes |\ |^{s+\lambda})$$

is a subspace of $I(s)$.

We let $f_s \in I(s)$. Then given irreducible admissible representation $\mathbf{\Pi}$ of $O(V, q)$ we form the family of integrals

$$\Lambda(f_s, v)(g_1) = \int_{O(V,q)} f_s(g, g_1) \mathbf{\Pi}(g) v dg$$

where $v \in V_{\pi}$, the space of $\mathbf{\Pi}$.

As a generalization of the results in [G-PS-R] we deduce that $\Lambda(f_s, v) \epsilon$

$$I(\mathbf{\Pi}, s) = Ind_{GL(S \cap V_2) \times O(V,q) \times O((S \cap V_2)^{\perp}/S \cap V_2, q_2) \ltimes U_{S \cap V_2}}^{O(V_1, q_1)} (|\ |^s \otimes \mathbf{\Pi} \otimes 1)$$

(normalized induction). We note that in the case where $(S \cap V_2)^{\perp} = S \cap V_2$ then $Ind = ind$ here.

From the comments above, we deduce that every element of $ind(\mathbf{\Pi}, s)$ can be obtained as a finite linear combination of $\Lambda(f_s, v)$ for a family of v in V and a finite family of

$$f_s \in ind_{O(V,q)^{\Delta} \times P_S \cap O(V_1,q_1)}^{O(V,q) \times O(V_1,q_1)} \qquad (1 \otimes |\ |^{s+\lambda})$$

It is possible to compute $\Lambda(f_s, v)$ more concretely under certain circumstances.

Namely, let (V_2, q_2) be a split form (hence $\dim V_2$ is even). Then $(W, \tilde{q})$ is a split form.

In such instance let $K_{O(W,\tilde{q})}$ be the standard maximal compact subgroup (i.e. stabilizer in $O(W, \tilde{q})$ of a maximal lattice adapted to $\tilde{q}$).

Let Φ_s be the element in $I(s)$ which is $K_{O(W,\tilde{q})}$ invariant and normalized so that $\Phi_s(e) = 1$.

Then as a generalization of [PS-R] (see also [G-PS-R]) we have the following fact.

Theorem 1.1. *Let (V_2, q_2) be a nonzero split form. Let (V, q) be a quasi-split form. Let $\mathbf{\Pi}$ be a representation of $O(V, q)$ which*

admits a $K_{O(V,q)}$ fixed vector v_0. Then $\Lambda(\Phi_s, v_0)$ is a $K_{O(V_1,q_1)}$ fixed vector in $I(\mathbf{\Pi}, s)$. Moreover

$$\Lambda(\Phi_s, v_0)(k) = \left[\frac{L(\mathbf{\Pi}, st, s + \frac{1}{2} + \frac{1}{4}\dim V_2)}{d_V(s + \frac{1}{4}\dim V_2)}\right] v_0$$

where $L(\mathbf{\Pi}, st, \ )$ represents the standard L function associated to

the L group of $SO(V, q)$ (= the identity component of $O(V, q)$) and $d_V(\ )$ the normalizing factor of the group $O(V, q)$ given in [PS-R].

Remark 1. In the above construction it is possible to replace $I(s)$ by

$$\tilde{I}(s) = ind_{P_S}^{SO(W,\tilde{q})}(\chi_s)$$

Moreover starting with an admissible irreducible representation σ of $SO(V, q)$ we can form the integral

$$\tilde{\Lambda}(f_s, v)(g_1) = \int_{SO(V,q)} f_s(g, g_1)\sigma(g) v dg$$

with $v \in V_\sigma$, the space of σ. Then $\tilde{\Lambda}(f_s, v) \in \tilde{I}(\sigma, s)$

$$= ind_{GL(S \cap V_2) \times SO(V,q) \ltimes U_{S \cap V_2}}^{SO(V_1,q_1)} (|\ |^s \otimes \sigma \otimes 1)$$

and with the same hypotheses as in **Theorem 1** above $\tilde{\Lambda}(\Phi_s, v)$ is a $K_{SO(V_1,q_1)}$ fixed vector in $\tilde{I}(\sigma, s)$ and admits exactly the same formula as in **Theorem 1** when restricted to $K_{SO(V_1,q_1)}$ (with (V_2, q_2) a split form here).

We are going to apply **Remark 1** to compute a certain zeta integral.

Namely we consider a *sequence of codimension one vector spaces*

$$V \hookrightarrow T \hookrightarrow V_1$$

and nondegenerate quadratic forms q, q_T and q_1 associated to V, T and V_1 respectively so that

$$(V_1, q_1) = (T \oplus T^{\perp}, q_T \oplus q_1|_{T^{\perp}}) = (V \oplus V^{\perp}, q \oplus q_2|V^{\perp}).$$

Here $T^{\perp}$ and $V^{\perp}$ represent perpendicular complements in V_1 (relative to q_1).

Moreover we assume that q_1/q is a split hyperbolic plane.

Remark 2. If $W = V \oplus V_1$ as above, then $W = S_1 \oplus \tilde{S}_1 \oplus \{(0, \xi)|\xi \in i(V)^{\perp} \text{ in } V_1\}$ where

$$S_1 = \{(v, i(v))|v \in V\} \text{ and } \tilde{S}_1 = \{(v, -i(v))|v \in V\}.$$

Here S_1 and $\tilde{S}_1$ are isotropic subspaces of W which are nonsingularly paired by $\tilde{q}$. Moreover the space $i(V)^{\perp}$ in V_1 splits an orthogonal direct sum $\langle \xi_1 \rangle \oplus \langle \xi_1^{\#} \rangle$ where $T = V \oplus \langle \xi_1 \rangle$ (orthogonal). Then we choose the vectors ξ_1 and $\xi_1^{\#}$ so that $q_1(\xi_1) = -q_1(\xi_1^{\#})$. Thus the space W is the *orthogonal direct sum* (relative to $\tilde{q}$).

$$T_1 \oplus T_1^{\#}$$

where $T_1 = \{(v, 0)|v \in V\} \oplus \{(0, \langle \xi_1 \rangle\}$ and $T_{\#} = \{(0, i(v))|v \in V\} \oplus \{(0, \langle \xi_1^{\#} \rangle)\}$. Then $(T_1, \tilde{q}|_{T_1}) \cong (T, q_T)$ and $(T_1^{\#}, \tilde{q}|_{T_1^{\#}}) \cong (T, -q_T)$. In particular the maximal isotropic subspace $S = S_1 \oplus \{(0, \xi_1 + \xi_1^{\#})) = \{z + i(z)|z \in T_1\}$. Here i is the embedding $T_1 \to T_1^{\#}$ given by the map defined by the following data:

$$(2) \qquad \begin{aligned} &(v, 0) \to (0, i(v)), v \in V \\ &(0, \xi_1) \to (0, \xi_1^{\#}), \{\xi_1, \xi_1^{\#}\} \in i(V)^{\perp} \text{ in } V_1. \end{aligned}$$

In any case there is the embedding of the group $O(T_1, \tilde{q}|_{T_1}) \times O(T_1^{\#}, \tilde{q}|_{T_1^{\#}})(\cong O(T, q_T) \times O(T, q_T))$ in the group $O(W, \tilde{q})$. Then the stabilizer of S in $O(T, q_T) \times O(T, q_T)$ equals $O(T, q_T)^{\Delta}$ (= the diagonal embedding of $O(T, q_T)$.

We denote the corresponding special orthogonal groups

$$G = SO(V_1, q_1) \supset H = SO(T, q_T) \supset M = SO(V, q).$$

Following the notation above we let σ and $\mathbf{\Pi}$ be irreducible admissible representations of M and H respectively.

Form $\tilde{I}(\sigma, s)$ the G module as above.

We let $\langle | \rangle_\sigma$ denote a M-invariant pairing on the space $\sigma \otimes \mathbf{\Pi}$. That is, $< 1 >_\sigma$ is a bilinear pairing on $\sigma \otimes \mathbf{\Pi}$ satisfying

$$\langle \sigma(m)v | \mathbf{\Pi}(m)w \rangle_\sigma = \langle v | w \rangle_\sigma$$

for $m \in M$ and $v \in V_\sigma$ and $w \in V_{\mathbf{\Pi}}$.

Let P_1 be the parabolic subgroup of G which fixes the line $S \cap V_2 = (\xi_1 + \xi_1^{\#})$ in V_1. Then $P_1 \cong GL_1 \times M \times U_{S \cap V_2}$. Here $GL_1 \underset{\lambda}{\cong} k^\times$ acts as t on $\xi_1 + \xi_1^{\#}$ and t^{-1} on $\xi_1 - \xi_1^{\#}$.

Let w_0 be a Weyl group element of G so that w_0 and the parabolic subgroup $GL_1 \times M \times U_{S \cap V_2}$ generates G. In particular $w_0 M w_0^{-1} = M$ and $w_0 \lambda(t) w_0^{-1} = \lambda(t^{-1})$ (where $t \rightsquigarrow \lambda(t)$ is the embedding of GL_1 into G). Moreover we can choose w_0 with the property that $w_0(\xi_1 + \xi_1^{\#}) = \xi_1 - \xi_1^{\#}$ and $w_0(\xi_1 - \xi_1^{\#}) = \xi_1 + \xi_1^{\#}$. Thus w_0 lies in H.

Then we know from above that

$$\tilde{\Lambda}(f_s, v)(mg) = \sigma(m)\tilde{\Lambda}(f_s, v)(g)$$

for all $m \in M, g \in G$. Here σ is the M module given above. Thus the function $h \rightsquigarrow \tilde{\Lambda}(f_s, v)(h)$ lie in the H induced module

$$Ind_M^H(\sigma)$$

Then if we consider $V_s \in \tilde{I}(\sigma, s)$ the function

$$h \rightsquigarrow \langle V_s(h) | \mathbf{\Pi}(h)w \rangle_\sigma$$

defines a M invariant function on the left. Thus we can define an H invariant pairing between $\tilde{I}(\sigma, s)$ and $\mathbf{\Pi}$ via the integral

$$\int_{(M \backslash H)} \langle V_s(h) | \mathbf{\Pi}(h)w \rangle_\sigma dh$$

We show below that this integral is convergent for $Re(s)$ sufficiently large and in fact is rational in q^{-s} (see **Remark 3**).

The problem here is to compute the above integral more explicitly.

That is, if $V_s = \tilde{\Lambda}(f_s, v)$ we deduce that the above integral equals

$$\int_{(M \backslash H)} \langle \tilde{\Lambda}(f_s, v)(h) | \mathbf{\Pi}(h)w \rangle_\sigma dh =$$

$$\int_{M \backslash H} (\langle \int_M f_s(g, g_1)\sigma(g)v | \mathbf{\Pi}(g_1)w \rangle_\sigma dg) dg_1$$

$$= \int_H f_s(1, g) \langle v | \mathbf{\Pi}(g)w \rangle_\sigma dg$$

This identity follows from the fact that we can expand

$$\int_M f_s(g, g_1)\sigma(g)(v)dg =$$
$$\sum_{\xi \in V_\sigma} (\int_M f_s(g, g_1)\langle \sigma(g)v|\xi\rangle dg)\hat{\xi}$$

Here $\{\xi\}$ forms a basis of V_σ and $\{\hat{\xi}\}$ is the dual basis of $V_{\sigma^\nu} = (V_\sigma)^\nu$ (here $\langle , \rangle$ denotes the usual pairing between σ and $\check{\sigma}$) . The we substitute the right hand side of the integral over M into $\langle \mid \rangle_\sigma$ above and deduce that

$$\langle \int_M f_s(1, g^{-1}g_1)\sigma(g)vdg|\mathbf{\Pi}(g_1)w\rangle_\sigma =$$
$$\int_M f_s(1, g^{-1}g_1)(\sum_{\xi \in V_\sigma} \langle \sigma(g)v|\hat{\xi}\rangle\langle \xi|\mathbf{\Pi}(g_1)w\rangle_\sigma)dg$$
$$= \int_M f_s(1, g^{-1}g_1)\langle \sigma(g)v|\mathbf{\Pi}(g_1)w\rangle_\sigma dg =$$
$$= \int_M f_s(1, g^{-1}g_1)\langle \sigma(g)v|\mathbf{\Pi}(g_1)w\rangle_\sigma dg$$
$$= \int_M f_s(1, g^{-1}g_1)\langle v|\mathbf{\Pi}(g^{-1}g_1)w\rangle_\sigma dg$$

Then

$$\int_{M\backslash H} (\int_M f_s(1, g^{-1}g_1)\langle v|\mathbf{\Pi}(g^{-1}g_1)w\rangle_\sigma dg)dg_1$$
$$= \int_H f_s(1, g)\langle v|\mathbf{\Pi}(g)w\rangle_\sigma dg.$$

We note at this point for $g \in H$ the embedding $g \rightsquigarrow (1, g)$ is the same as taking $(1, g)$ in $M \times G$ (in $SO(W, \tilde{q})$)) or in $H \times H$. This implies that $g \rightsquigarrow f_s(1, g)$ is K_H finite on the left and right.

This unfolding is possible by making the following general assumption

$(*)$ Given $v \in \sigma, w \in \mathbf{\Pi}$ there exists s_0 large enough so that

$$|f_s(1, g)\langle v \mid \mathbf{\Pi}(g)w\rangle_\sigma|$$

is L^1 for all $s \in \mathbb{C}$ so that $Re(s) \geq s_0$.

In practice this assumption will be verified in several ways depending on the context of the calculation. For instance if $\mathbf{\Pi}$ is the local component of a global cuspidal representation $\mathbf{\Pi}$ and σ a component of an irreducible automorphic representation which is contained in the space of $A(M(\mathbb{A}))$ = slowly increasing automorphic forms then we define a M invariant local pairing between $\mathbf{\Pi}$ and σ as follows:

$$\langle v \mid w\rangle = \int\limits_{M(k)\backslash M(\mathbb{A})} f_v(m) g_w(m) dm$$

where $f_v \in \sigma(g_w \in \mathbf{\Pi})$ corresponds to v in σ (w in $\mathbf{\Pi}$) via an M embeddings of σ into (H embedding of $\mathbf{\Pi}$ into $\mathbf{\Pi}$. In any case since $\mathbf{\Pi}$ is cuspidal we have that

$$|\langle v \mid \mathbf{\Pi}(g)w\rangle_\sigma| \leq C$$

for all $g \in H$, i.e. that $\langle v \mid \mathbf{\Pi}(g)w\rangle_\sigma$ is a bounded function.

Another way to verify (*) is have an estimate of the form:

(**) Given $v \in \sigma, w \in \mathbf{\Pi}$ there exists a pair of representations $\tilde{\sigma} and \tilde{\mathbf{\Pi}}$ of M and H and vectors $\tilde{v} \in \tilde{\sigma}, \tilde{w} \in \tilde{\mathbf{\Pi}}$ and a M invariant pairing $\langle | \rangle_{\tilde{\sigma}}$ between $\tilde{\sigma}$ and $\tilde{\mathbf{\Pi}}$ so that

(i) $|\langle v \mid \mathbf{\Pi}(g)w\rangle_\sigma| \leq \tilde{v} \mid \tilde{\mathbf{\Pi}}(g)w\rangle_{\tilde{\sigma}}$

(ii) $\langle \tilde{v}, \tilde{\mathbf{\Pi}}(g)\tilde{w}\rangle_{\tilde{\sigma}} \geq 0$ for all $g \in H$.

In the case $\langle v \mid \mathbf{\Pi}(g)w\rangle_\sigma$ is bounded then (*) is satisfied since the function $g \rightsquigarrow f_s(1, g)$ lies in $L'(H)$ for s sufficiently large (see [PS-R]).

In the case (**) is satisfied then

$$|f_s(1,g)\langle v \mid \mathbf{\Pi}(g)w\rangle_\sigma| \leq |f_s(1,g)| \langle \tilde{v}, \tilde{\mathbf{\Pi}}(g)\tilde{w}\rangle_{\tilde{\sigma}}$$

The function $g \rightsquigarrow f_s(1, g)$ is K_H finite on the left and right. Thus there exists a compact open $K_f \leq K_H$ so that $f_s(1, kg) = f_s(1, g)$ or all $k \in K_f$ and all $g \in H$. This implies that

$$\int_H |f_s(1,g)| \langle \tilde{v}, \tilde{\mathbf{\Pi}}(g)\tilde{w}\rangle_{\tilde{\sigma}} dg = \int_{H \times K_f} |f_s(1,g)| \langle v, \tilde{\mathbf{\Pi}}(kg)\tilde{w}\rangle_{\tilde{\sigma}} dg dk$$

In any case the nonnegative function $m_{\mathbf{\Pi}}(g)$ given by

$$\int_{K_f} \langle v \mid \tilde{\mathbf{\Pi}}(kg)\tilde{w}\rangle_{\tilde{\sigma}} dg$$

is a H-matrix coefficient of the representation $\tilde{\mathbf{\Pi}}$. But we know that $|f_s(1, g) m_{\tilde{\mathbf{\Pi}}}(g)| \in L'(H)$ for $Re(s)$ sufficiently large. Indeed here we use the fact that the matrix coefficient $m_{\tilde{\pi}}(g) \leq \Xi^t(g)$ for some real t where $\Xi(g)$ is the spherical function defined by the integral

$$\Xi(g) = \int_{K_H} \delta_{B_H}^{1/2}(kg) dk$$

Thus Ξ^t (for real t) is a spherical function. But we know that $f_s(1, g)\Xi^t$ is L^1 for $Re(s)$ sufficiently large. Thus we have that (*) is satisfied in this case also!

Returning to the general setup; we note (by the same reasoning as above)

$$\int_H f_s(1,g)\mathbf{\Pi}(g)wdg =$$

$$\sum_{\xi' \in V_{\mathbf{\Pi}}} (\int_H f_s(1,g)\langle\hat{\xi'}|\mathbf{\Pi}(g)w\rangle dg)\xi'$$

with $\{\xi'\}(\{\hat{\xi'}\}$ resp.) a basis of $V_{\mathbf{\Pi}}$ (dual basis of V_π^ν). Thus again we deduce that

$$\langle v| \int_H f_s(1,g)\pi(g)wdg\rangle_\sigma =$$

$$\langle v| \sum_{\xi' \in V_\pi} (\int_H f_s(1,g)\langle\hat{\xi'}|\mathbf{\Pi}(g)w\rangle dg)\xi'\rangle_\sigma$$

$$= \int_H f_s(1,g)(\sum_{\xi' \in V_\pi} \langle v|\xi'\rangle_\sigma \langle\hat{\xi'}|\mathbf{\Pi}(g)w\rangle)dg$$

$$= \int_H f_s(1,g)\langle v|\mathbf{\Pi}(g)w\rangle_\sigma dg$$

Then we have determined the following formula

Lemma 1.1. *Let* $(*)$ *be satisfied. Let* $V_s(g) = \tilde{\Lambda}(f_s, v)(g)$ *defined above. Then*

$$\int_{(M\backslash H)} \langle V_s(g)|\mathbf{\Pi}(g)w\rangle_\sigma dg =$$

$$\langle v|(\int_H f_s(1,g)\mathbf{\Pi}(g)wdg)\rangle_\sigma$$

Remark 3. The equality just established in fact gives a proof of the rationality of the integral

$$\int_{(M\backslash H)} \langle V_s(g)|\mathbf{\Pi}(g)w\rangle_\sigma dg$$

The proof of **Lemma 1.1** shows that the integral

$$\int_H f_s(1,g)\langle v \mid \mathbf{\Pi}(g)w\rangle_\sigma dg$$

equals

$$\int_H f_s(1,g)\rho(g)dg$$

where ρ is a matrix coefficient of $\mathbf{\Pi}$. But this integral is a rational function in q^{-s}.

Indeed we use the invariance of $f_s(1,g)$ under K_f on the left. Then we have by change of variables

$$\int_{H\times K_f} f_s(1,kg)\langle v \mid \mathbf{\Pi}(g)w\rangle_\sigma dgdk = \int f_s(1,g)(\int_{K_f} \langle v, \mathbf{\Pi}(kg)w\rangle_\sigma dk)dg$$

The inner integral is a H-matrix coefficient ρ associated to $\mathbf{\Pi}$ but we know from [PS-R] that such an integral (the latter integral above) is rational in q^{-s}.

We note that also in the Archimedean case that the integral

$$\int_H f_s(1,g)\langle v|\pi(g)w\rangle_\sigma dg = \int_H f_s(1,g)\rho(g)dg$$

where ρ is a matrix coefficient of π. Indeed we assume here that f_s is K finite (for the group $SO(W,\tilde{q})$); we then can find a K_H finite function ϕ so that $f_s(1,g)*\phi = f_s(1,g)$. With this minor adaption then we can repeat the same arguments as in the p-adic case to deduce the above fact. In particular **Lemma 1.1** is valid in the Archimedean case! One subtle point that should be commented upon here is that each element of $I(\pi,s)$ which is $K_{0(V_1,q_{V_1})}$ finite can be obtained as a finite linear combination of $\Lambda(F_s,v)$ with the family of f_s and v suitably K finite! In the specific case where the function $\langle v|\pi(g)w\rangle_\sigma$ is *bounded* for all g (i.e. $|\langle v|\pi(g)w\rangle| \leq C$)

then the integral

$$\int_H f_s(1,g)\langle v|\pi(g)w\rangle_\sigma dg$$

is defined for all $f_s \in I(s)_\infty =$ smooth vectors in $I(s)$ with $Re(s) >> 0$. Also in this context the local integral above admits continuation to $s\in\mathbb{C}$ (for fixed $\langle v|\pi(g)w\rangle_\sigma$ we can use here exactly the same arguments as given in [K-R].

The essential point here is that $\langle v|\pi(g)w\rangle$ is bounded! Also we note that the analytic continuation defines a continuous linear functional on $I_\infty(s)$ (See [K-R]).

Corollary to Lemma 1.1. *Let the hypotheses of* **Theorem 1** *be in force. Let* $\mathbf{\Pi}$ *be a spherical representation of* H *with* T_0 *the unique (up to scalars) unramified vector.*

Let $V_s^0(g)$ *be the* K_G *fixed vector normalized so that* $V_s^0(e) = v_0$.

Then

$$\int_{(M\backslash H)} \langle V_s^0(h)|\mathbf{\Pi}(h)T_0\rangle_\sigma dh =$$

$$\left(\frac{L(\mathbf{\Pi},st,s+\frac{1}{2})}{L(\sigma,st,s+1)}\right)\left(\frac{d_V(s+\frac{1}{2})}{d_T(s)}\right) \qquad \langle v_0|T_0\rangle_\sigma$$

Remark 4. We note that

$$\frac{d_V(s+\frac{1}{2})}{d_T(s)} = \begin{cases} 1 & \text{if } (V,T) \text{ is (even, odd) pair} \\ \frac{1}{\zeta(2s+1)} & \text{if } (V,T) \text{ is (odd, even) pair} \end{cases}$$

Appendix to §1 The Local Spherical Uniqueness Principle (for any local field).

(1) We are given any pair of quadratic spaces $W_1 \subseteq W_2$ so that W_1 is nondegenerate and has codimension one in W_2. We let $O(W_1) \subseteq O(W_2)$ be the associated orthogonal groups. We let Π_1 and Π_2 be any pair of irreducible admissible representations of $O(W_1)$ and $O(W_2)$. The *Uniqueness Principle* asserts

$$\dim Hom_{O(W_1)}(\Pi_2, \Pi_1) \leq 1$$

for all Π_1, Π_2 as above.

We prove this statement in [R].

(2) The *generic Uniquenes Principle* is a refinement of (1). Namely we assume that Π_1 and Π_2 are spherical representations of $O(W_1)$ and $O(W_2)$. Then

$$\dim Hom_{O(W_1)}(\Pi_2, \Pi_1) \leq 1$$

for a *generic family* of pairs (Π_1, Π_2). Simply this means there is some open Zariski set in the data defining (Π_1, Π_2) where the above *Hom* statement is valid.

We note that (1) certainly implies (2).

(3) On the other hand if we replace $O(W_1)$ and $O(W_2)$ by $SO(W_1)$ and $SO(W_2)$ then the associated *Uniqueness Principle* becomes

$$[(\dim Hom_{SO(W_1)}(\Pi_2, \Pi_1^\nu)) \cdot (\dim Hom_{SO(W_1)}(\Pi_2^\nu, \Pi_1)] \leq 1$$

for all Π_1 and Π_2 admissible representations of $SO(W_1)$ and $SO(W_2)$. We prove *this statement* in [R] also.

(4) This version of the *Uniqueness Principle* is enough to prove the strong form of generic uniqueness for the pair $SO(W_1)$ and $SO(W_2)$. Indeed we assert the following **Proposition**.

Proposition (3) above implies

$$\dim Hom_{SO(W_1)}(\Pi_2, \Pi_1) \leq 1$$

for a generic family of pairs (Π_1, Π_2).

Proof. (sketch) From (3) above to prove the **Proposition** it suffices to show that for unramified pairs (Π_1, Π_2) of$SO(W_1) \times SO(W_2)$ if $Hom_{SO(W_1)}(\Pi_2, \Pi_1^\nu) \neq 0$ then

$Hom_{SO(W_1)}(\Pi_2^{\nu}, \Pi_1)$
$\neq 0$. However using **Theorem 7.1** of the text below we deduce that there is some polynomial condition between Π_1, Π_2 in order to insure nonvanishing of the $Hom_{SO(W_1)}(\Pi_2, \Pi_1) \neq 0$.

We emphasize that in the proof of **Theorem 7.1** *we have not assumed any uniqueness in any form. Specifically in* **Theorem 7.1** *we are specifically calculating the value of a possible nonzero Hom embedding.*

We note that the invariance of data in **Theorem 7.1** under the Weyl groups of the various groups in question implies the **Proposition** above /**QED**

We note that for all essential applications in this paper the only form of uniqueness required is that of the generic uniqueness of the **Proposition** above.

§2. Support Ideals

We can interpret **Lemma 1** in §1 in a geometric manner.

Namely we consider the Hecke algebra $\mathcal{H} = \mathcal{H}(H//K_H)$ = all compactly supported functions on H which are biinvariant under K_H. The structure of the algebra $\mathcal{H}$ (under convolution) is given by the Satake Isomorphism Theorem.

We in fact fatten up this Hecke algebra by considering $\mathbb{C}[X, X^{-1}] \otimes \mathcal{H} = A_H$. In fact A_H is isomorphic to the Hecke algebra $\mathcal{H}(k^x \times H//\mathcal{O}^x \times K_H)$.

The algebra A_H acts on the space $\tilde{I}(\sigma, s)$ in the following way. If $V_s \in \tilde{I}(\sigma, s)$ then

$$V_s * (X \otimes \varphi) = q^{-s} V_s * \varphi$$

where $\varphi \in \mathcal{H}$ and $V_s * \varphi$ is the left action on V_s via convolution.

We are now interested in determining $\tilde{I}(\sigma, s)$ as in H module. Geometrically this is similar to the problems considered in §1.

Specifically we are interested in determining the H orbits in the space $Y = \{$ the set of q_1 isotropic lines in $V_1\} \cong P_1 \backslash G$ where P_1 is the parabolic subgroup given by $GL_1 \times M \ltimes U_1$ (U_1 = the unipotent radical of P_1). Then the H orbits in Y consists of 2 sets : (i) $Y_0 = \{(\ell) \in Y | (\ell) \subseteq T\}$ (ii) $Y_1 = \{(\ell) \in Y | (\ell) \cap T = \{0\}\}$. Here (ℓ) = the line determined by the vector ℓ. We note here that the two sets Y_0 and Y_1 are $O(T, q_T)$ orbits (by use of Witt's Theorem). But in fact since the generic stabilizer in $O(T, q_T)$ of points in Y_0 and Y_1 contain some $\gamma \in O(T, q_T)$ with $det \gamma = -1$ it follows that $H = SO(T, q_T)$ operates transitively on Y_0 and Y_1.

Thus it follows that $\tilde{I}(\sigma, s)$ admits a certain composition series of H modules. Namely corresponding to the set Y_1 we have the H module

$$ind_M^H(\sigma)$$

and corresponding to the set Y_0 there is the H module

$$ind_{\tilde{P}}^H(|t|^{s+\frac{1}{2}} \otimes \sigma|_{\tilde{M}} \otimes 1)$$

where $\tilde{P}$ is a parabolic subgroup of H having the form $GL_1 \times \tilde{M} \times \tilde{U}$ (stabilizing an isotropic line). We note here that the shift by $\frac{1}{2}$ occurs in exponent of $|t|$ because of the comparison of the Jacobians of G and H relative to P_1 and $\tilde{P}$ respectively. Here $\tilde{M} \cong$ orthogonal group $O(V^{\#}, (q_T)|_{V^{\#}})$ where $V^{\#} \subset V$ is a codimension two space and $T|_{V^{\#}}$ is a hyperbolic plane relative to q_T. Here $\sigma|_{\tilde{M}}$ = the restriction of σ to the subgroup $\tilde{M}$. In any case we have that there exists an exact sequence of H modules

$$0 \to ind_M^H(\sigma) \to \tilde{I}(\sigma, s) \to ind_{\tilde{P}}^H|t|^{s+\frac{1}{2}} \otimes \sigma|_{\tilde{M}} \otimes 1) \to 0$$

Then we are interested in computing the support ideal of this exact sequence. Namely the sequence remains exact by considering in each space the set of K_H invariants.

In particular we want to determine the space

$$T_{supp} = \{f \in A_H | \tilde{I}(\sigma, s)^{K_H} * f \subseteq ind_M^H(\sigma)^{K_H}\}.$$

Then T_{supp} is an ideal in A_H. We note that $f \in T_{supp}$ if and only if

$$ind_{\tilde{P}}^H(|t|^{s+\frac{1}{2}} \otimes \sigma|_{\tilde{M}} \otimes 1)^{K_H} * f \equiv 0.$$

On the other hand every K_H invariant function in the space $ind_{\tilde{P}}^H(|t|^{s+\frac{1}{2}} \otimes \sigma|_{\tilde{M}} \otimes 1)$ is determined by its restriction to $\tilde{P}$. That is $F \in ind_{\tilde{P}}^H(|t|^{s+\frac{1}{2}} \otimes \sigma|_{\tilde{M}} \otimes 1)$ is determined by $F(t \cdot m \cdot u) = |t|^{s+\frac{1}{2}}\sigma(m)(F(e))$ where $F(e) \in \sigma^{K_{\tilde{M}}}$.

In particular we look at the $\tilde{M}$ cyclic module X in σ generated by $F(e)$. Then the original F lies in the space $\mathrm{ind}_{\tilde{P}}^H(|t|^{s+\frac{1}{2}} \otimes X \otimes 1)$. On the other hand we know that there is a $\tilde{M}$ surjective map from $S(\tilde{M})$ to X given by $\varphi_1 \rightsquigarrow \sigma(\varphi_1)(F(e))$. Here $\tilde{M}$ acts on $S(\tilde{M})$ via left action.

Thus to determine the $f \in A_H$ which annihilate the F above, it suffices to find the $f \in A_H$ which annihilate the space

$$\mathrm{ind}_{\tilde{P}}^H(|t|^{s+\frac{1}{2}} \otimes S(\tilde{M}) \otimes 1)^{K_H}.$$

We note that every irreducible admissible spherical representation ρ of $\tilde{M}$ (with $K_{\tilde{M}}$ fixed vector) can be realized as a quotient of $S(\tilde{M})$. That is there exists an $\tilde{M}$ intertwining map of $S(\tilde{M})$ to ρ.

Then we look at the space

$$T_{\mathrm{gen}} = \{f \in A_H \mid \mathrm{ind}_{\tilde{P}}^H(|t|^{s+\frac{1}{2}} \otimes \rho \otimes 1)^{K_H} * f \equiv 0 \text{ for all spherical } \rho\}.$$

We assert first $T_{\mathrm{gen}} \subseteq T_{\mathrm{supp}}$.

Indeed we assert that if $f \in T_{\mathrm{gen}}$, then f annihilates the space $\mathrm{ind}_{\tilde{P}}^H(|t|^{s+\frac{1}{2}} \otimes S(\tilde{M}) \otimes 1)^{K_H}$.

Thus assume there is a ξ in the above space so that $\xi * f \neq 0$. Thus for each s_0 where $\xi * f$ is nonzero, we know that $(\xi * f)(e)$ is a nonzero element in $S(\tilde{M})$. However we note that there exists an admissible irreducible module δ of $\tilde{M}$ and a nonzero intertwining map T of $S(\tilde{M}) \to \delta \otimes \check{\delta}$ so that $T(\xi * f(e)) \neq 0$. Here we use the $\tilde{M} \times \tilde{M}$ action on $S(\tilde{M})$ and $\delta \otimes \check{\delta}$ ($\check{\delta}$ = contra gradient of δ) respectively. In any case, since $\xi * f(e)$ is $K_{\tilde{M}}$ fixed (on the left) it follows that $T((\xi * f)(e))$ lies in the space $\delta^{K_{\tilde{M}}} \otimes \check{\delta}$. Hence δ and $\check{\delta}$ are $K_{\tilde{M}}$ spherical. In particular we can find a $\tilde{M}$ intertwining map $\tilde{T}$ of $S(\tilde{M})$ to δ so that $\tilde{T}((\xi * f)(e)) \neq 0$.

But now $\tilde{T}$ induces an H intertwining map $\tilde{T}'$ of $\mathrm{ind}_{\tilde{P}}^H(|t|^{s_0+\frac{1}{2}} \otimes S(\tilde{M}) \otimes 1)$ onto the space $\mathrm{ind}_{\tilde{P}}^H(|t|^{s_0+\frac{1}{2}} \otimes \delta \otimes 1)$ with $\tilde{T}'(\xi * f) \neq 0$. Then since $\tilde{T}'(\xi * f) \neq 0$ we have $\tilde{T}'(\xi) \neq 0$. But $\tilde{T}'(\xi)$ is a nonzero K_H fixed vector and we know that $\tilde{T}'(\xi) * f \equiv 0$.

Hence $\tilde{T}'(\xi * f) = 0$ (by the intertwining property of $\tilde{T}'$). This contradicts the fact $\tilde{T}'(\xi * f) \neq 0$ above.

Hence we have established that $T_{\text{gen}} \subseteq T_{\text{supp}}$.

We note that for any spherical ρ the space $ind_{\bar{P}}^{H}(|t|^{s} \otimes \rho \otimes 1)^{K_H}$ is one dimensional and that $f = \sum_k X^k \cdot c_k$ with $c_k \in \mathcal{H}$ has the property that f operates by a scalar on the space $ind_{\bar{P}}^{H}(|t|^{s+\frac{1}{2}} \otimes \rho \otimes 1)$ given by

$$\sum q^{-sk} trace_{(s,\rho)}(c_k).$$

where $trace_{(s,\rho)}(c_k)$ is the trace of the element $c_k \in \mathcal{H}$ acting on the space

$$ind_{\bar{P}}^{H}(|t|^{s+\frac{1}{2}} \otimes \rho \otimes 1)$$

Recalling that the group H is quasi-split we know that the Satake Isomorphism Theorem implies that the map $\varphi \in \mathcal{H} \rightsquigarrow trace_{s,\rho)}(\varphi)$ defines an isomorphism of $\mathcal{H}$ *onto* a Laurent ring of the form $\mathbb{C}[Z_1^{\pm 1}, \cdots, Z_r^{\pm 1}]^W$ where $W =$ Weyl group associated to a root system of type B, C or D. In any case the variable Z_1 corresponds to $q^{+s+\frac{1}{2}}$ under this isomorphism. Thus the element $f \in A_H$ can be expressed in the form

$$\sum X^k \hat{c}_k(Z_1, \cdots, Z_r)$$

We can rewrite this element in another way as

$$\sum_\nu \xi_\nu(X, Z_1)\xi_\nu^*(Z_2, \cdots, Z_r)$$

where ξ_ν and ξ_ν^* are appropriate Laurent polynomials in the variables (X, Z_1) and $(Z_2, \cdots, Z_r)$ respectively. Moreover we can assume that the sets $\{\xi_\nu\}$ and $\{\xi_\nu^*\}$ are linearly independent. Then under the action on A_H on $ind_{\bar{P}}^{H}(|t|^{s} \otimes \rho \otimes 1)$ above we deduce that

$$\sum_\nu \xi_\nu(q^{-s}, q^{s+\frac{1}{2}})\xi_\nu^*(\hat{\rho}) = 0.$$

where $\hat{\rho}$ represents an arbitrary point in the $r-1$ dimensional torus

$\prod_{r-1} (\mathbb{C}^x)$.

In particular if $f \in A_H$ vanishes for $(s, \hat{\rho})$ where s is arbitrary then the polynomial ξ_ν must be divisible by $(q^{-\frac{1}{2}} X Z_1 - 1)$ in the ring $\mathbb{C}[X^{\pm 1}, Z_1^{\pm 1}]$ since $XZ_1 - 1$ is a prime element in $\mathbb{C}[q^{-\frac{1}{2}} X, Z_1]$. Then using the invariance of f under W it follows that f is divisible by

$$L(X, Z) = \prod_{i=1}^{i=r} (q^{-\frac{1}{2}} X Z_i - 1)(q^{-\frac{1}{2}} X Z_i^{-1} - 1).$$

Remark 1. The significance of $L(X, Z)$ is the following. Indeed $L(q^{-\frac{1}{2}}q^{-s}, (q^{\beta_1}, \cdots, q^{\beta_t}))$ = the inverse of the standard L function of the unramified representation $\mathbf{\Pi} = \operatorname{ind}_B^H(||^{\beta_1} \otimes \cdots \otimes ||^{\beta_t})$ given by

$$(L_v(\mathbf{\Pi}, st, s + \frac{1}{2}))^{-1}\rho_v(s)$$

with

$$\rho_v(s) = \begin{cases} \zeta_v(2s+1)^{-1} & \text{if } H \text{ corresponds to} \\ & \text{an even nonsplit form} \\ 1 & \text{otherwise} \end{cases}$$

Thus we have established the following lemma

Lemma 2.1. *The ideal T_{supp} contains T_{gen}. T_{gen} is generated by $L(X, Z_1, \ldots, Z_r)$ given above.*

Proof. It is clear by the calculations above that $L(X, Z) \in A_H$/QED

We can characterize T_{gen} in another way. Here we extend the results of [PS-R] to the case where k is a general local field.

We let $(W, \tilde{q}) = (T, q_T) \oplus (T, -q_T)$.

Then we form $\tilde{I}(s)$ relative to $SO(W, \tilde{q})$. Then we embed H into $SO(W, \tilde{q})$ via the map

$$H \hookrightarrow \{e\} \times H \hookrightarrow H \times H \hookrightarrow SO(W, \tilde{q})$$

Then the action of A_H on $\tilde{I}(s)$ is given by

$$f_s * (X \otimes \varphi) = q^{-s} f_s * \varphi$$

where $f_s \in \tilde{I}(s)$ and $X \otimes \varphi \in A_H$ (see above).

The various $H \times H$ boundary components of $\tilde{I}(s)$ can be determined in a similar way as given in [K-R] and [PS-R]. In any case, we note two points here. First to determine T_{supp} for this specific example, it suffices to annihilate the functions in $\tilde{I}(s)$ which are supported on the $H \times H$ orbit in $P_S \backslash SO(W, \tilde{q})$ which corresponds to the next to largest orbit (recall $H^{\Delta} \backslash H \times H$ is the largest $H \times H$ orbit here). Secondly the determination of the annihilation property on this next to top orbit is equivalent to determining the ideal in A_H

$$\{f \in A_H \mid \operatorname{ind}_{\tilde{p}}^H(|t|^{s+\frac{1}{2}} \otimes S(\tilde{M}) \otimes 1)^{K_H} * f = 0\}.$$

But clearly by the arguments above we note that this ideal equals T_{gen}!

Lemma 2.2. *T_{gen} is the same as the space*

$$\{f \in A_H | \tilde{I}(s)^{K_H} * f \subseteq \mathcal{H}(H//K_H)\}.$$

We note here that the connection between T_{gen} and T_{supp} (given in **Lemma 2.1**) is approximately the following. Recall that T_{gen} is basically the ideal of A_H which annihilates the spherical vectors in the family of modules

$$\text{ind}_{\tilde{p}_1}^{H}(|t|^{s+\frac{1}{2}} \otimes \rho \otimes 1)$$

(for all spherical ρ of $\tilde{M}$). However, T_{supp} depends on $\sigma|_{\tilde{M}}$. In fact, we can roughly think of T_{supp} as being the ideal of A_H which annihilates the family of spherical modules above where ρ runs through all possible subquotients which occur in $\sigma|_{\tilde{M}}$. For instance if σ = identity representation, then T_{supp} also contains $1 \otimes I_{\{0\}}$ in $\mathbb{C}[X, X^{-1}] \otimes \mathcal{H}$ where $I_{\{0\}} = \{\varphi \in \mathcal{H} \mid \text{Trace}_{(s+\frac{1}{2}, \rho_{n-1})}(\varphi) \equiv 0\}$ with ρ_{n-1} a certain $n-1$ tuple.

Then as a consequence of the 2 **Lemmas** above we deduce the following support statement.

At this point we let

$$a_H(s) = \begin{cases} \zeta(2s) & \text{if } H \text{ corresponds to an even nonsplit form} \\ 1 & \text{otherwise} \end{cases}$$

Corollary to Lemma 2.2.

Let $\tilde{\Phi}_s$ be the unique $SO(W, \tilde{q})$ invariant vector in $\tilde{I}(s)$ normalized so that $\tilde{\Phi}_s(e) = 1$.

Then

$$\tilde{\Phi}_s * L(,) = d_T(s)^{-1} a_H(s + \frac{1}{2}) 1_{K_H}$$

Proof. From **Theorem 1.1** we note that

$$\int_H \tilde{\Phi}_s(1, h)\omega_\pi(h)dh = \frac{L(\pi, st, s + \frac{1}{2})}{d_T(s\ )}$$

where π is a spherical (unramified) representation of H. (ω_π is the unique spherical function associated to π). Thus we deduce that

$$\int_H (\tilde{\Phi}_s * L)(1, h)\omega_\pi(h)dh = (L(\pi, st, s + \frac{1}{2}))^{-1} a_H(s + \frac{1}{2})(\int_H \tilde{\Phi}_s(1, h)\omega_\pi(h)dh).$$

Moreover we note that $\tilde{\Phi}_s * L \in \mathcal{H}$ for each value of s. But it also follows that

$$\tilde{\Phi}_s * L = \sum c_\wedge(s) 1_{K_H \alpha_\wedge K_H};$$

the sum is a finite linear combination of characteristic functions of double coset $K_H \alpha_\wedge K_H$. Here $c_\wedge(s)$ is a polynomial in q^s and q^{-s}. However the integral also represents the Satake transform of the L^1 function (on H) $\tilde{\Phi}_s * L$ where s is sufficiently large.

There is one subtle point here. Namely the finiteness of the sum does not depend on s. Indeed, using the fact that $c_\Lambda(s)$ is a polynomial in q^s and q^{-s} and a standard Baire category argument we find that the terms $c_\Lambda(s) = 0$ for all Λ but a finite number.

Then comparing formulae above we deduce using again the Satake Isomorphism Theorem that $\alpha_\wedge = e$ and $c_\wedge(s) = \begin{cases} 0 & if \alpha_\wedge \neq e \\ (d_T(s))^{-1} a_H(s+\frac{1}{2}) & if \alpha_\wedge = e \end{cases}$.

Remark 3. It follows from **Corollary** to **Lemma 1.1** and **Corollary** to **Lemma 2.2** that

$$V_s^0 * L(h) = \frac{1_\sigma(h)}{L(\sigma_s st, s+1)} \begin{cases} 1 & \text{if } T \text{ odd, } V \text{ has an even split form} \\ \zeta(2s+1) & \text{if } T \text{ odd and } V \text{ has an even nonsplit form} \\ \frac{1}{\zeta(2s+1)} & \text{if } T \text{ even and } V \text{ odd.} \end{cases}$$

where $1_\sigma \in ind_M^H(\sigma)$ which is given by

$$1_\sigma(m\alpha k) = \begin{cases} \sigma(m)v_0 & \text{if } \alpha = 1 \\ 0 & \text{if } \alpha \in G \text{ so that } M\alpha K_H \cap MK_H = \emptyset. \end{cases}$$

§3. Certain Jacquet Functors

We are given a sequence of spaces

$$V \hookrightarrow T \hookrightarrow V_1 \hookrightarrow W_1$$

where V, T and V_1 are given in §1 and W_1 comes with a nondegenerate form q_{W_1} so that $W_1 = V_1 \oplus V_1^{\perp}$ (relative to q_{W_1}) and $(V_1^{\perp}, q_{W_1}) \cong H_{(r',r')}$ a split form in $2r$ variables.

We let $\mathbb{G} = SO(W_1)$.

We are to consider certain parabolic subgroups of $\mathbb{G}$.

Indeed we take a decomposition of V of the form $V = L \oplus V_0 \oplus L^*$ where L and L^* are q_V isotropic subspace with L and L^* being paired nonsingularly by q_V. Moreover V_0 is q_V anisotropic. Thus Witt index $(V, q_V) = \dim L$. Then it follows that (W_1, q_{W_1}) admits a compatible decomposition $W_1 = L_1 \oplus V_0 \oplus L_1^*$ where $L_1 \supset L$ and $L_1^* \supset L^*$.

We take any subspace $R \subset L_1$ and complementary subspace $R^* \subset L_1^*$ so that

$$W_1 = R \oplus W_1(R) \oplus R^*$$

with $W_1(R) = (R \oplus R^*)^{\perp}$ (relative to q_{W_1}).

We consider the parabolic subgroup $\mathbb{P}_R \subset \mathbb{G}$. Here $\mathbb{P}_R = \{g \in \mathbb{G} | g(R) = R\}$. It is possible now to describe the action of $\mathbb{P}_R$ on the space W_1.

First $\mathbb{P}_R = GL(R) \times SO(W_1(R)) \ltimes \mathbb{U}_R$. We know that $GL(R)$ acts on R^* through duality; that is, $g(v^*) \in R^*$ for $v^* \in R^*$ is defined via the formula $q_{w_1}(g(v^*), v) = q_{w_1}(v^*, g^{-1}(v))$ for all $v \in R$).

Let $\nu = \dim R$.

It remains to describe the action of $\mathbb{U}_R$. Indeed $\mathbb{U}_R$ is isomorphic to the space (with the group structure given below)

$$W_1(R) \otimes k^{\nu} \oplus Skew_{\nu}(k)$$

Then we choose bases $\{v_i\}, \{v_i^*\}$ of R and R^* so that $q_{W_1}(v_i, v_j^*) = \delta_{ij}$. Then the action of $\mathbb{U}_R \cong \{((\xi_1, \cdots, \xi_\nu), s_{ij}) | \xi_i \in W_1(R), s_{ij} \in Skew_{\nu}(k)\}$ is given as follows:

$$N((\xi, s))(v) = v \text{ for } v \in R$$

$$N((\xi, s))(v_i^*) = v_i^* + \xi_i - \frac{1}{2} \sum_{t=1}^{t=\nu} \{q_{W_1}(\xi_i, \xi_t) + s_{it}\} v_t$$

$$N((\xi, s))(Y) = Y - \sum_{i=1}^{i=\nu} q_{W_1}(Y, \xi_i) v_i \text{ for } Y \in W_1(R)$$

Then the usual law of composition on $\mathbb{U}_R$ is given as follows:

$$(\xi, s)(\xi', s') = (\xi + \xi', \{s_{i\ell} + s'_{i\ell}\} + \{q_{W_1}(\xi_i, \xi'_\ell) - q_{W_1}(\xi'_i, \xi_\ell)\})$$

Thus $\mathbb{U}_R$ is at most 2 step nilpotent group with center $Skew_\nu(k)$ if $\nu \geq 2$. If $\nu = 1$ then $\mathbb{U}_R$ is an abelian group isomorphic to $W_1(R)$.

At this point we consider a sequence of codimension subspaces $R_1 \subseteq R_2 \subseteq \cdots \subseteq R = L$. We denote $\mathbb{P}_i = \mathbb{R}_{R_i}$ and $\mathbb{U}_i = \mathbb{U}_{R_i}$. We say that R determines a split torus T_R which acts on R via a set of distinct characters on each R_i/R_{i-1}.

We are interested in determining the action of $\mathbb{P}_i$ on the space X_j of isotropic subspaces of W_1 of dimension j.

We note that the space $X_j \cong \mathbb{P}_j\backslash\mathbb{G}$ except in the case where W_1 is totally split and j = Witt index of W_1. Then X_j is a union of 2 distinct $\mathbb{G}$ orbits, X_j^+ and X_j^- where $X_j^+ = \mathbb{P}_j\backslash\mathbb{P}_j \cdot \mathbb{G}$ and $X_j^- = \mathbb{P}_j\backslash\mathbb{P}_j\sigma\mathbb{G}$ and $X_j^- = \mathbb{P}_j\backslash\mathbb{P}_j\sigma\mathbb{G}$ with $\sigma \in O(W_1) = \mathbb{G}$.

Let $S \in X_j$. We consider the flag in S given by

$$S \cap R_i \subseteq S \cap (R_i \oplus W(R_i)) \subseteq S$$

Let $X_j(\alpha, \beta) = \{S \in X_j | \dim S \cap R_i = \alpha, \dim S \cap (R_i \oplus W_1(R_i)) = \beta\}$.

Proposition 3.1. *The set $X_j(\alpha, \beta)$ (if non empty) is an orbit under $\mathbb{P}_i$ except in the case when $(W_1(R_i), q_{W_1})$ is split and $\beta - \alpha = 1/2 \dim W_1(R_i)$. In the exceptional case there exist 2 $\mathbb{P}_i$ orbits in $X_j(\alpha, \beta)$ (except if $i = j$ = Witt index of W_1 when there is one $\mathbb{P}_i$ orbit).*

Proof. The proof consists of three steps.

If S and $S' \in X_j(\alpha, \beta)$ then dim $S \cap R_i$ = dim $S' \cap R_i$. This implies that there exist $g \in GL(R)$ so that $g(S \cap R_i) = S' \cap R_i$. Thus we may assume that $S \cap R_i = S' \cap R_i$.

We choose a basis $\{L_1, \cdots, L_\beta\}$ of $S \cap (R_i \oplus W_1(R_i))$ so that $\{L_1, \cdots, L_\alpha\}$ span $S \cap R_i$. We consider the components $(L_j)_0$ in $W_1(R_i)$ of the elements L_j (relative to the decomposition $R_i \oplus W_1(R_i)$). Then the set of vectors $\{(L_{\alpha+1})_0, \cdots (L_\beta)_0\}$ is a linearly independent set of vectors. Moreover there clearly exist vectors $\xi_1, \cdots, \xi_i \in W_1(R_i)$ so that $q_{W_1}(\xi_\ell, (L_\rho)_0)$ = the coefficient of v_ℓ in the term $(L_\rho)_+$ (the projection of L_ρ onto $S \cap R_i$). Thus there exists an element $N_i((\xi, s))$ with $(\xi_1, \cdots, \xi_i) = \xi$ as above so that $[N_i((\xi, s))]^{-1}(S \cap (R_i \oplus W_1(R_i)))$ is a direct sum of the form $S \cap R_i \oplus S \cap W_1(R_i)$.

Then we write $S \cap (R_i \oplus W_1(R_i)) = (S \cap R_i) \oplus U$ and $S' \cap (R_i \oplus W_1(R_i)) = S' \cap R_i \oplus U'$ where U and U' are isotropic subspace of the same dimension in $W_1(R_i)$. In any case there exists $\gamma \in O(W_1(R_i), q_{W_1})$ so that $\gamma(U) = U'$. In fact if we are not in the exceptional case above the we may choose $\gamma \in SO(W_1(R_i), q_{W_1})$.

Here we use the fact that the special orthogonal group operates transitively on the set of isotropic subspaces of a fixed dimension (except in the case when the quadratic form is split and the dimension equals the Witt index of the form).

Thus we can assume that $S \cap R_i = S' \cap R_i$ and that $S \cap (R_i \oplus W_1(R_i)) = S' \cap (R_i \oplus W_1(R_i))$.

Then we choose complements S_1 and S'_1 to S and S' so that $S = S \cap (R_i \oplus W_1(R_i)) \oplus S_1$ and $S' = S' \cap (R_i \oplus W_1(R_i)) \oplus S'_1$. We consider an element $A \in S$ and decompose $A = Z_+ + Z_0 + Z_-$ where $Z_+ \in R_i, Z_0 \subset W_1(R_i)$ and $Z_- \in R_i^*$. Moreover if $\{Z_1, \cdots Z_\ell\}$ is a basis of S_1 then the set $\{(Z_1)_-, \cdots (Z_\ell)_-\}$ is a linearly independent subset in R_i^*. Similarly if $\{Z'_1, \cdots, Z'_\ell\}$ is a basis of S'_1 then the set $\{(Z'_1)_-, \cdots, (Z'_\ell)_-\}$ is a linearly independent set in R_i^*. Also we note the sets $\{(Z_1)_-, \cdots, (Z_\ell)_-\}$ ($\{(Z'_1)_-, \cdots, (Z'_\ell)_-\}$ resp.) are perpendicular to $S \cap R_i$ relative to q_{W_1}. This implies that there exists $g \in GL(R_i)$ which is 1 on $S \cap R_i$ and satisfies $g((Z_i)_-) = (Z'_i)_-$ for $i = 1, \cdots, \ell$. Hence the space $S^\# = S \cap (R_i \oplus W_1(R_i)) \oplus Span\{(Z_1)_-, \cdots, (Z_\ell)_-\}$ is $\mathbb{P}_i$ conjugate to $(S')^\# = S' \cap (R_i \oplus W_1(R_i)) \oplus Span\{(Z'_1)_-, \cdots, (Z'_\ell)_-\}$.

Thus it suffices to show that $S(S'$ resp.) is $\mathbb{P}_i$ conjugate to$S^\#((S')^\#$ resp.).

Since $q_{W_1}(Z_i, Z_j) = 0$ we deduce that

$$q_{W_1}((Z_i)_0, (Z_j)_0) = -q_{W_1}((Z_i)_+, (Z_j)_-) - q_{W_1}((Z_i)_-, (Z_j)_+).$$

We use the above fact to construct an element $u \in \mathbb{U}_i$ so that u conjugates $S^\#$ to S.

For this we let $\{\nabla_1, \cdots \nabla_i\}$ be a basis of R_i. Let $\{(Z_1)_-, \cdots, (Z_\ell)_-\}$ be completed to a basis of R_i^* given by $\{\nabla_1^* \cdots, \nabla_i^*\}$ so that $\nabla_i^* = (Z_i)_-(i = 1, \cdots, \ell)$ and $q_{W_1}(\nabla_i, \nabla_j^*) = \delta_{ij}$.

We construct $u = N((\xi, s))$ via the formula above.

Namely let $\xi_t = (Z_t)_0(t = 1, \cdots \ell)$ and $\xi_t = 0$ if $t > \ell$. Moreover the skew symmetric matrix $s = (s_{ij})$ is given by the following prescription: (with $i < j$)

$$s_{ij} = \begin{cases} q_{W_1}((Z_j)_+, (Z_i)_-) - q_{W_1}((Z_i)_+, (Z_j)_-) & \text{for } 1 \le i < j \le \ell \\ (-2)q_{W_1}((Z_i)_+, \nabla_j^*) & \text{for } 1 \le i < \ell < j \\ 0 & \text{otherwise} \end{cases}$$

By construction it is clear that u acts as identity on the space $S \cap (R_i \oplus W_1(R_i))$. Moreover

$$N((\xi, s))((Z_\nu)_-) = (Z_\nu)_- + (Z_\nu)_0 -$$
$$\frac{1}{2}\sum_{i=1}^{i=\ell}(q_{W_1}((Z_v)_0, (Z_i)_0) + q_{W_1}((Z_i)_+, (Z_\nu)_-) - q_{W_1}((Z_\nu)_+, (Z_i)_-)\nabla_i$$
$$+ \sum_{j>\ell} q_{W_1}((Z_\nu)_+, \nabla_j^*)\nabla_j$$

But by using the relations above we deduce that

$$N((\xi, s))((Z_\nu)_-) = Z_\nu.$$

In the exceptional case $X_j(\alpha, \beta)$ is a single orbit under $\mathbb{P}'_i = GL(R_i) \times O(W_1(R_i)) \ltimes \mathbb{U}_i$. In any case (from Proposition 2 below) the stabilizer in $\mathbb{P}'_i$ of a subspace has the form $V_1 \times V_2 \ltimes T$ where $V_1 \subset GL(R_i), V_2 \cong GL_{\alpha-\beta}\mathbb{U}_{\alpha-\beta} \subset O(W(R_i))$ and $T \subset \mathbb{U}_i$. Now $V_1 \times V_2 \ltimes T \subset \mathbb{P}_i$. This implies that $X_j(\alpha,\beta)$ cannot be a single orbit under $\mathbb{P}_i$; thus $X_j(\alpha,\beta)$ consists of $2\mathbb{P}_i$ orbits! /**QED**

We note that in the case where W_1 is totally split and j = Witt index of W_1, then $X_j(\alpha,\beta) \neq \emptyset$ if and only if $\beta - \alpha - \frac{1}{2}\dim W_1(R_i)$. If $i \neq j$, then $X_j(\alpha,\beta)$ is a union of two $\mathbb{P}_i$ orbits given by $X_j(\alpha,\beta) \cap X_j^+$ and $X_j(\alpha,\beta) \cap X_j^-$. If $i = j$, then $X_j(\alpha,\beta)$ lies in X_j^+ or in X_j^-. In any case we have the decomposition

$$X_j^+ = \cup X_j(\alpha,\beta) \cap X_j^+$$

into distinct $\mathbb{G}$ orbits.

Then we can determine the general structure of the stabilizer in $\mathbb{P}_i$ of a canonical subspace in $X_j(\alpha,\beta)$.

Namely, we take the subspace S in $X_j(\alpha,\beta)$ which has the following form:

i) $S = (S \cap R_i) \oplus (S \cap W_1(R_i)) \oplus (S \cap R_i^*)$
ii) $S \cap R_i^*$ has basis $\{v_1^*, \cdots, v_{\dim S-\beta}^*\}$
iii) $S \cap R_i$ has basis $\{v_{i-\alpha+1}, \cdots, v_i\}$

We note that $i \geq \dim S \cap R_i + \dim S \cap R_i^*$. This implies that $i - \alpha \geq \dim S - \beta$.

We let $V_{S\cap R-i}$ and $V_{S\cap R_i^*}$ be the parabolic subgroups of $GL(R_i)$ of the subspaces $S \cap R_i$ and $S \cap R_i^*$ (in this case $GL(R_i)$ acts by duality on R_i^* as specified above).

We let $P_{S\cap W_1(R_i)}^{SO(W_1(R_i))}$ = parabolic subgroup of $SO(W_1(R_i))$ stabilizing $S\cap W_1(R_i)$.

Proposition 3.2. *The stabilizer in $\mathbb{P}_i$ of the subspace S in $X_j(\alpha,\beta)$ is the group*

$$(V_{S\cap R_i} \cap V_{S\cap R_i^*}) \times \left(\mathbb{P}_{S\cap W_1(R_i)}^{SO(W_1(R_i))}\right) \ltimes$$
$$\{N((\xi,s)|\xi_j \in S \cap W_1(R_i), j = 1, \cdots \dim S - \beta,$$

$\xi_j \in (S \cap W_1(R_i))^\perp$ (in $W_1(R_i)$) for $j = \dim S - \beta + 1, \cdots i - \alpha$ and $s_{ij} = 0$ for $i \in \{1, \cdots \dim S - \beta\}, j \in \{1, \cdots, i - \alpha\}\}$

Proof. It is clear by direct calculation that each of the components above stabilize S. On the other hand we then can establish that relative to the decomposition $g = g_1 \cdot g_2 \cdot u(g_1 \in GL(R_i), g \in SO(W_1(R_i))$ and $u \in \mathbb{U}_i)$ if g stabilizes S then g_1, g_2 and u stabilize S. Indeed if g stabilizes S and R_i simultaneously then $g(S \cap R_i) = S \cap R_i$ Since g_2 and u acts as identity on R_i it follows that $g_1(S \cap R_i) = S \cap R_i$. On the other hand for $\xi \in S \cap R_i^*$ we have $g_1 g_2 u(\xi) = g_1 g_2(\xi_{R_i} + \xi_0 + \xi) = g_1(\xi_{R_i} + g_2(\xi_0) + \xi) = g_1(\xi_{R_i}) + g_2(\xi_0) + g_1(\xi)$ (here $\xi_{R_i} + \xi_0 + \xi = u(\xi)$ is the splitting relative to $S = S \cap R_i \oplus S \cap W_1(R_i) \oplus S \cap R_i^*$); thus $g_1(\xi) \in S \cap R_i^*$. This implies that $g_1 \in V_{S\cap R_i} \cap V_{S\cap R_i^*}$. Then for $\xi_0' \in S \cap W_1(R_i)$ $g_2 u(\xi_0') = g_2(\xi_{R_i} + \xi_0') = \xi_{R_i} + g_2(\xi_0')$ (decompose $u(\xi_0')$ in $S \cap R_i \oplus S \cap W_1(R_i)$). Hence $g_2(\xi_0') \in S \cap W_1(R_i)$; thus $g_2 \in P_{S\cap W_1(R_i)}^{LSO(W_1(R_i))}$. calculation u lies in the subgroup of $\mathbb{U}_i$ defined in the **Proposition** /QED

At this point we require a more detailed structural decomposition of the group described in **Proposition 3.2**

$$N_{(i,j)}^{(\alpha,\beta)} = \{N(\xi,s))|\xi_j \in S\cap W_1(R_i), j=1,\cdots,\dim S-\beta,$$

$\xi_j \in (S\cap W_1(R_i))^{\perp}, j=\dim S-\beta+1,\cdots,i-\alpha$ and $s_{ij}=0$ for $i\in\{1,\cdots,\dim S-\beta\}, j\in\{1,\cdots,i--\alpha\}\}$.

We are going to decompose the group $N_{(i,j)}^{(\alpha,\beta)}$ according to the Levi decomposition of $\mathbb{P}_S$.

We are going to choose a certain splitting of the space W_1 relative to S. First we note that $S^{\perp}$ (in W_1) is a direct sum

$$A\oplus B\oplus C$$

where $A=\{w\in R_i \mid w\perp S\cap R_i^*\} = Sp\{v_{\dim S-\beta+1},\cdots,v_1\}, B=\{v\in W_1(R_i) \mid v\perp S\cap W_1(R_i)\}$, and $C=\{w\in R_i^* \mid w\perp S\cap R_i\} = Sp\{v_1^*,\cdots,v_{i-\alpha}^*\}$. In particular $A\supseteq S\cap R_i, B\supseteq S\cap W_1(R_i)$ and $C\supsetneqq S\cap R_i^*$. On the other hand let $(S\cap W_1(R_i))^*$ be the isotropic subspace in $W_1(R_i)$ which is nonsingularly paired to $S\cap W_1(R_i)$. Then we choose $S_{\#}$ to be a nondegenerate space in $W_1(R_i)$ so that $W_1(R_i)=S\cap W_1(R_i)\oplus S_{\#}\oplus(S\cap W_1(R_i))^*$. Then $B=S\cap W_1(R_i)\oplus S_{\#}$. We then define an orthogonal splitting of W_1 relative to S as follows:

$$W_1 = S\oplus W_1(S)\oplus S^*$$

Here $S^*=Sp\{v_1,\cdots,v_{\dim S-\beta}\}\oplus(S\cap W_1(R_i))^*\oplus Sp\{v_{i-\alpha+1}^*,\cdots,v_i^*\}$ and $W_1(S)=Sp\{v_{\dim S-\beta+1},\cdots,v_{i-\alpha}\}\oplus S_{\#}\oplus Sp\{v_{\dim S-\beta+1},\cdots,v_{i-\alpha}^*\}$. Here we assume $i-\alpha>\dim S-\beta$; otherwise $W_1(S)=S_{\#}$. We note here that $R_i=R_i\cap S\oplus R_i\cap W_1(S)\oplus R_i\cap S^*$

Then the group $\mathbb{P}_S\cong GL(S)\times SO(W_1(S))\ltimes\mathbb{U}_S$. For notation we decompose an element $N_i((\xi,s))$ in the form $\xi=(Z_1,Z_2,Z_3)$ where $Z_1\in W_1(R_i)\otimes k^{\dim S-\beta}$, $Z_2\in W_1(R_i)\otimes k^{(i-\alpha)-(\dim S-\beta)}$, and $Z_3\in W_1(R_i)\otimes k^{\alpha}$ and

$$s=\begin{bmatrix} s_{11} & s_{12} & s_{13} \\ --^t s_{12} & s_{22} & s_{23} \\ --^t s_{13} & -^t s_{23} & s_{33}\end{bmatrix} \text{ where}$$

$$s_{11}\in Skew_{\dim S-\beta}(k), s_{22}\in Skew_{(i-\alpha)-(\dim S-\beta)}(k)$$
$$s_{33}\in Skew_{\alpha}(k), s_{12}\in M_{\dim S-\beta,(i-\alpha)-(\dim S-\beta)}(k),$$
$$s_{13}\in M_{\dim S-\beta,\alpha}(k) \text{ and } s_{23}\in M_{(i-\alpha)-(\dim S-\beta),\alpha}(k).$$

Remark 1. By the same reasoning as used in the proof of **Proposition 3.2** we verify that if $g\in\mathbb{P}_S\cap\mathbb{P}_i$ then when we write $g=g^1g^2u'$ with $g'\in GL(S), g^2\in SO(W_1(S))$ and $u'\in\mathbb{U}_S$ we have that g^1,g^2 and u' fix simultaneously S and R_i. Thus in fact **Proposition 3.2** shows that $\mathbb{P}_S\cap\mathbb{P}_i=(\mathbb{P}_S\cap GL(R_i))\cdot(\mathbb{P}_S\cap SO(W_1(R_i)))\cdot(\mathbb{P}_S\cap\mathbb{U}_i)=(\mathbb{P}_i\cap GL(S))\cdot(\mathbb{P}_i\cap SO(W_1(S)))(\mathbb{P}_i\cap\mathbb{U}_S)$; hence $N_{(i,j)}^{(\alpha,\beta)}=\mathbb{P}_S\cap\mathbb{U}_i$.

Lemma 3.1. *The group* $N_{(i,j)}^{(\alpha,\beta)}$ *decomposes as*

$$N_{(i,j)}^{(\alpha,\beta)} = (N_{(i,j)}^{(\alpha,\beta)} \cap GL(S)) \cdot (N_{(i,j)}^{(\alpha,\beta)} \cap SO(W_1(S))) \ltimes (N_{(i,j)}^{(\alpha,\beta)} \cap \mathbb{U}_S)$$

Moreover

$$\begin{aligned} &N_{(i,j)}^{(\alpha,\beta)} \cap GL(S) \\ &= \{N((\xi,s))|Z_1 \in S \cap W_1(R_i), Z_2 = 0, Z_3 \in (S \cap W_1(R_i))^*, \\ &s_{11} = s_{12} = s_{22} = s_{23} = s_{33} = 0\} \\ &N_{(i,j)}^{(\alpha,\beta)} \cap SO(W_1(S)) \\ &= \{N((\xi,s))|Z_1 = 0, Z_3 = 0, Z_2 \in (S \cap W_1(R_i) + (S \cap W_1(R_i))^*)^\perp \text{ in } W_1(R_i), \\ &s_{11} = s_{12} = s_{13} = s_{23} = s_{33} = 0\} \\ &N_{(i,j)}^{(\alpha,\beta)} \cap \mathbb{U}_S \\ &= \{N((\xi,s))|Z_1 = 0, Z_2 \in S \cap W_1(R_i), Z_3 \in (S \cap W_1(R_i))^\perp \text{ in } W_1(R_i), \\ &s_{11} = s_{12} = s_{13} = s_{22} = 0\}. \end{aligned}$$

(we use the shorthand notation $Z_i \in \Omega$*, a subspace of* $W_1(R_i)$ *to mean* $Z_i \in \Omega \otimes k^\nu$ *with* ν *the appropriate integer associated to* i *above).*

Proof. We note that $\mathbb{U}_S = \{p \in \mathbb{P}_S | p|_S = 1 \text{ and } p(x) \equiv x \mod S \text{ for all } x \in S^\perp\}$. With this characterization of $\mathbb{U}_S$ we can compute $N_{(i,j)}^{(\alpha,\beta)} \cap \mathbb{U}_S$.

First we compute $\{n \in N_{(i,j)}^{(\alpha,\beta)} | n|_S = 1\}$. By direct calculation this group equals

$$N(((Z_1, Z_2, Z_3), s))|Z_1 = 0, Z_2 \text{ and } Z_3 \in (S \cap W_1(R_i))^\perp, s_{11} = s_{12} = s_{13} = 0\}.$$

Thus using this decomposition we then determine $N_{(i,j)}^{(\alpha,\beta)} \cap \mathbb{U}_S$ as given in the **Lemma**.

On the other hand the quotient $\{p \in \mathbb{P}_S | p|_S = 1\}/\mathbb{U}_S \cong SO(W_1(S))$. Moreover the subgroup

$$\begin{aligned} (*) = \{&N((\xi,s))|Z_1 = 0, Z_2 = 0 \text{ and } Z_3 \in (S \cap W_1(R_i) + \\ &(S \cap W_1(R_i))^*)^\perp \text{ in } S \cap W_1(R_i), s_{11} = s_{12} = s_{13} = s_{23} = s_{33} = 0\} \end{aligned}$$

forms a complement to $N_{(i,j)}^{(\alpha,\beta)} \cap \mathbb{U}_S$ in the group $\{n \in N_{(i,j)}^{(\alpha,\beta)} | n|_S = 1\}$. Moreover we note that the root subgroups (relative to the torus T_R associated to R) that appear in $N_{(i,j)}^{(\alpha,\beta)} \cap \mathbb{U}_S$ and $(*)$ above are complementary (i.e. disjoint from each other). Also we note that the torus T_R separates the root subgroups in $\mathbb{U}_S$ and $SO(W_1(S))$. Thus we deduce that $(*) = N_{(i,j)}^{(\alpha,\beta)} \cap SO(W_1(S))$.

In a similar fashion we know that the quotient $\mathbb{P}_S/\{p \in P_S | p_S = 1\} \cong GL(S)$. Again the subgroup

$$\begin{aligned} (**) = \{&N((\xi,s))|Z_1 \in S \cap W_1(R_i), Z_2 = 0, Z_3 \in (S \cap W_1(R_i))^* \\ &s_{11} = s_{12} = s_{22} = s_{23} = s_{33} = 0\} \end{aligned}$$

is complementary to $\{n \in N_{(i,j)}^{(\alpha,\beta)} | n|_S = 1\}$ in $N_{(i,j)}^{(\alpha,\beta)}$. Then by similar reasoning as above $(**) = N_{(i,j)}^{(\alpha,\beta)} \cap GL(S)$. /QED.

Moreover we can also characterize $N_{(i,j)}^{(\alpha,\beta)} \cap GL(X)$ and $N_{(i,j)}^{(\alpha,\beta)} \cap SO(W_1(S))$ as subgroups of $GL(S)$ and $SO(W_1(S))$.

Corollary to Lemma 3.1. *The group $N_{(i,j)}^{(\alpha,\beta)} \cap GL(S)$ is the unipotent radical of the parabolic subgroups of $GL(S)$ which stabilizes the flag of subspaces*

$$S \cap R_i \subseteq S \cap (R_i \oplus W_1(R_i)) \subseteq S.$$

The group $N_{(i,j)}^{(\alpha,\beta)} \cap SO(W_1(S))$ is the unipotent radical of the parabolic subgroup of $SO(W_1(S))$ which stabilizes the isotropic subspace $R_i \cap W_1(S)$.

Proof. It follows by definition that $N_{(i,j)}^{(\alpha,\beta)} \cap GL(S)$ is contained in the appropriate parabolic subgroup of $GL(S)$. It follows (by dimension reasons and use of **Lemma 1**) that we have equality when the field is algebraically closed. This equality is preserved by taking k-rational points (for k local or global of characteristic $= 0$) since all groups in question are defined over k.

A similar argument works for the case of $N_{(i,j)}^{(\alpha,\beta)} \cap SO(W_1(S))$./Q.E.D.

Remark 2. In the case that $\dim S - \beta = i - \alpha$ then $N_{(i,j)}^{(\alpha,\beta)} \cap SO(W_1(S)) = \{e\}$ and $R_i \cap W_1(S) = \{0\}$.

We now determine similar structural decompositions of $V_{S\cap R} \cap V_{S\cap R_i^*}$ and $\mathbb{P}_{S\cap W_1(R_i)}^{SO(W_1(R_i))}$ relative to the embedding into $\mathbb{P}_S$

Lemma 3.2. *The group $V_{S\cap R_i} \cap V_{S\cap R_i^*}$ is the parabolic subgroup of $GL(R_i)$ which stabilizes the flag*

$$S \cap R_i \subset S^{\perp} \cap R_i \subseteq R_i$$

Then

$$\begin{aligned} &V_{S\cap R_i} \cap V_{S\cap R_i^*} \\ &= (V_{S\cap R_i} \cap V_{S\cap R_i^*}) \cap GL(S) \cdot (V_{S\cap R_i} \cap V_{S\cap R_i^*} \cap SO(W_1(S))) \\ &\ltimes V_{S\cap R_i} \cap V_{S\cap R_i^*} \cap \mathbb{U}_S \end{aligned}$$

where

$$\begin{aligned} &V_{S\cap R_i} \cap V_{S\cap R_i^*} \cap GL(S) = GL(S \cap R_i) \cdot GL(S \cap R_i^*) \\ &V_{S\cap R_i} \cap V_{S\cap R_i^*} \cap SO(W_1(S)) = GL(R_i \cap W_1(S)) \\ &V_{S\cap R_i} \cap V_{S\cap R_i^*} \cap \mathbb{U}_S = \textit{ the unipotent radical of } V_{S\cap R_i} \cap V_{S\cap R_i^*} \end{aligned}$$

The group $\mathbb{P}_{S\cap W_1(R_i)}^{SO(W_1(R_i))} = GL(S \cap W_1(R_i)) \cdot$

$$SO(W_1(S) \cap W_1(R_i)) \ltimes \mathbb{U}_{S\cap W_1(R_i)}^{SO(W_1(R_i))}$$

where

$$\mathbb{U}_{S\cap W_1(R_i)}^{SO(W_1(R_i))} \subseteq \mathbb{U}_S.$$

Proof. Since $V_{S\cap R_i} \cap V_{S\cap R_i^*}$ preserves S it follows that $V_{S\cap R_i} \cap V_{S\cap R_i^*}$ preserves the flag in question . On the other hand by the very definition of $V_{S\cap R_i}$ and $V_{S\cap R_i^*}$ (and the disjointness properties of $S\cap R_i$ with $S\cap R_i^*$) we deduce that $V_{S\cap R_i} \cap V_{S\cap R_i^*}$ is a parabolic subgroup in $GL(R_i)$ which preserves a certain fixed flag of subspaces in R_i of the form

$$S\cap R_i \subseteq V \subseteq R_i$$

where V is a subspace of R_i with $\dim(V/S\cap R_i) = \dim(S^\perp \cap R_i/S\cap R_i)$. This condition forces that $V_{S\cap R_i} \cap V_{S\cap R_i^*}$ is the required parabolic subgroup.

Now if $u \in$ unipotent radical of $V_{S\cap R_i} \cap V_{S\cap R_i^*}$ then u fixes pointwise $S\cap R_i, S\cap R_i^*$ and $S\cap W_1(R_i)$ respectively. Thus u fixes S pointwise.

On the other hand using the decomposition of $S^\perp$ given above it is clear that the unipotent radical of the parabolic in $GL(R_i)$ stabilizing the flag $S\cap R_i \subseteq S^\perp \cap R_i \subseteq R_i$ operates as identity on $S^\perp$ modulo S. This implies that the unipotent radical of $V_{S\cap R_i} \cap V_{S\cap R_i^*} = V_{S\cap R_i} \cap V_{S\cap R_i^*} \cap \mathbb{U}_S$.

Now the Levi factor of $V_{S\cap R_i} \cap V_{S\cap R_i^*}$ is given by $GL(S\cap R_i) \times GL(S\cap R_i^*) \times GL(S\cap R_i\backslash V)$. It is clear that $GL(S\cap R_i\backslash V)$ lies in $SO(W_1(S))$. Indeed the isotropic space $(S\cap R_i\backslash V)$ is contained in $W_1(S)$. Thus we deduce the relevant statements about $V_{S\cap R_i} \cap V_{S\cap R_i^*}$ in the Lemma.

Finally we note that the space $(S\cap W_1(R_i)\backslash(S\cap W_1(R_i))^\perp$ in $W_1(R_i)$ equals $W_1(S)\cap W_1(R_i)$. From this we deduce the structural statement about $\mathbb{P}_{S\cap W_1(R_i)}^{SP(W_1(R_i)}$. / **Q.E.D.**

With the calculations of **Propositions 1** and **2**, **Lemma 1** and **2** can now compute (in general) a certain Jacquet functor.

We assume here that k is a local nonarchimedian field.

We consider a representation of $\mathbb{P}_j = GL_j(R_j) \times SO(W_1(R_j)) \ltimes \mathbb{U}_j \cong GL_j \times SO(W_1(R_j))$
$\ltimes \mathbb{U}_j \rightsquigarrow \tau \otimes \sigma \otimes 1$ where τ and σ are admissible representations of GL_j and $SO(W_1(R_j))$. Then we consider the representation

$$\operatorname{ind}_{\mathbb{P}_j}^{\mathbb{G}}(\tau\otimes\sigma\otimes 1)$$

A $\mathbb{P}_i$ in orbit $X_j(\alpha,\beta)$ in $\mathbb{P}_j\backslash\mathbb{G}$ determines the $\mathbb{P}_i$ module given by the following data. There is a injective homomorphism Γ of $P_i^j(\alpha,\beta) = (V_{S\cap R_i} \cap V_{S\cap R_i^*}) \cdot \mathbb{P}_{S\cap W_1(R_i)}^{S(W_1(R_i)} \ltimes N_{(i,j)}^{(\alpha,\beta)}$ into $GL(S)\cdot SO(W_1(S))\cdot \mathbb{U}_S$ defined in the following manner (from the facts above):

$$GL(S\cap R_i)\cdot GL(S\cap W_1(R_i))\cdot GL(S\cap R_i^*) \ltimes (N_{(i,j)}^{(\alpha,\beta)} \cap GL(S)) \hookrightarrow GL(S)$$

$$GL(R_i\cap W_1(S))\cdot SO(W_1(R_i)\cap W_1(S)) \ltimes (N_{(i,j)}^{(\alpha,\beta)} \cap SO(W_1(S)) \hookrightarrow SO(W_1(S))$$

$$N_{(i,j)}^{(\alpha,\beta)} \cap \mathbb{U}_S \cdot (V_{S\cap R_i} \cap V_{S\cap R_i^*} \cap \mathbb{U}_S)\cdot \mathbb{U}_{S\cap W_1(R_i)}^{SO(W_1(R_i))} \hookrightarrow \mathbb{U}_S$$

Then we consider the composite map

$$w_S \circ \Gamma$$

where $w_S \in \mathbb{G}$ so that $w_S(S) = R_j$ as subspaces. The element w_S is defined more precisely in §5 (after **Remark 3**).

Now the $\mathbb{P}_i$ module which corresponds to $X_j(\alpha, \beta)$ is

$$\mathrm{ind}_{\mathbb{P}_i^j(\alpha,\beta)}^{\mathbb{P}_i}(w_S \circ \Gamma|_{\tau \otimes \sigma \otimes 1}).$$

Here $w_S \circ \Gamma$ represents conjugating the element $Z \in \mathbb{P}_i^j(\alpha, \beta)$ by w_S, i.e. $w_S Z w_S^{-1}$.

We are interested in determining a certain Jacquet functor of the above module.

We consider the character on the group $\mathbb{U}_i$ given by the following formula

$$N((\xi, s)) \underset{\psi_Z}{\rightsquigarrow} \psi(\Sigma q_{W_1}(\xi_i, Z_i))$$

Here $Z \in W_1(R_i) \otimes k^i$ and ψ is an additive character on the field k which has order 0. We note that if $\mathbb{U}_i \neq Skew_i(k)$ then the commutator of $\mathbb{U}_i$ equals $Skew_i(k)$; thus in that case ψ_Z represents the most general character on $\mathbb{U}_i$.

We consider the specific case where $Z = (\xi_0|0|\cdots|0)$ with $q_{W_1}(\xi_0) \neq 0$. For (shorthand notation) we denote $\psi_Z = \psi_{\xi_0}$.

We assume that the space R_i is identified with the row vector space k^i in the following way. The vector v_ℓ corresponds to the row vector $(0, \cdots, 1, \cdots, 0)$ (which is 1 in the ℓ-th slot and zero elsewhere). Then we let

$$\text{matrix } (X) = \{X_{ij}\}$$

which is defined by $X(v_\ell) = \Sigma X_{\ell j} v_j$. Then we know that the map $\Lambda_i : GL(R_i) \to GL(k^i)$ given by $\Lambda_i(X) = [\text{matrix } (X)]^{-1}$ is an isomorphism of groups.

Remark 3. Relative to the map Λ_i above we see that $\Lambda_i(V_{S\cap R_i} \cap V_{S \cap R_i^*}) =$

$$\left\{ \begin{bmatrix} * & * & * \\ 0 & * & * \\ 0 & 0 & * \end{bmatrix} \begin{matrix} \}\dim S - \beta \\ \}i - (\alpha + \dim S - \beta) \\ \}\alpha \end{matrix} \right.$$

Here $\Lambda_i(GL(S \cap R_i)) =$

$$\left\{ \begin{bmatrix} I & 0 & 0 \\ 0 & I & 0 \\ 0 & 0 & X \end{bmatrix} \middle| X \in GL_\alpha(k) \right\},$$

$\Lambda_i(GL(S \cap R_i^*)) =$

$$\left\{ \begin{bmatrix} Z & 0 & 0 \\ 0 & I & 0 \\ 0 & 0 & I \end{bmatrix} \middle| Z \in GL_{\dim S - \beta}(k) \right\}.$$

$\Lambda_i(GL(S \cap W_1(R_i))) =$

$$\left\{ \begin{bmatrix} I & 0 & 0 \\ 0 & T & 0 \\ 0 & 0 & I \end{bmatrix} \middle| T \in GL_{i-(\alpha+\dim S-\beta)}(k) \right\},$$

$\Lambda_i(V_{S\cap R_i} \cap V_{S\cap R_i^*} \cap \mathbb{U}_S) =$

$$\left\{ \begin{bmatrix} I & X & Y \\ 0 & I & Z \\ 0 & 0 & I \end{bmatrix} \middle| X, Y, Z \text{ arbitrary} \right\}.$$

The group $GL(R_i) \times SO(W_1(R_i))$ acts on $\mathbb{U}_i$ by means of the following formula:

$$\begin{aligned} Ad(g_1, g_2)(N_i((\xi, s))) &= (g_1, g_2)N_i((\xi, s))(g_1, g_2)^{-1} \\ &= N_i((g_2\xi\Lambda_i(g_1)^{-1}, {}^T\Lambda_i(g_1)s\Lambda_i(g_i))) \end{aligned}$$

Here the action $g_2\xi\Lambda_1(g_1)^{-1} =$

$$(g_2\xi_1|\cdots|g_2\xi_i) \cdot \text{matrix } (g_1)$$

Then the stabilizer of the character ψ_{ξ_0} in the group $GL(R_i)\times SO(W_1(R_i))$ equals

$$\{(g_1, g_2)|\Lambda_i(g_1) = \begin{bmatrix} \nu(g_2) & ** \\ 0 & \\ \vdots & ** \\ 0 & \end{bmatrix} \quad \text{with} \quad g_2\xi_0 = \nu(g_2)\xi_0\}.$$

We note here that $\nu(g_2) = \pm 1$. Thus the stabilizer of $\psi_{(\xi_0|0|\cdots|0)}$ contains the subgroup

$$\{(g_1, g_2)|\Lambda_i(g_1) = \begin{bmatrix} 1 & & & * \\ & 1 & & \\ & & \ddots & \\ & 0 & & 1 \end{bmatrix}, g_2(\xi_0) = \xi_0\}$$

The above group (in shorthand notation) is

$$N^i \cdot SO((\xi_0)^\perp, q_{W_1}).$$

Then we consider a fixed generic character on N^i. That is, we let ψ_i be the unitary character on N^i which is trivial on each non simple root subgroup of N^i and satisfies

$$\psi_i : N^i/[N^i, N^i] \cong k \oplus \cdots \oplus k\psi \overrightarrow{\otimes \cdots} \otimes \psi\mathbb{C}$$

where ψ is the usual character on k which is trivial on $\mathcal{O}_v$ but nontrivial on $\pi\mathcal{O}_v$ (see introduction). (Each summand above corresponds to a simple positive root subgroup).

Then we define the character Λ_{ξ_0} on $N^i \times \mathbb{U}_i$ by

$$\Lambda_{\xi_0}(nu) = \psi_i(n)\psi_{\xi_0}(u)$$

Then we determine the Jacquet module

$$\text{ind}_{\mathbb{P}_i^j(\alpha,\beta)}^{\mathbb{P}_i} (w_S \circ \Gamma|_{\tau\otimes\sigma\otimes 1})_{(N^i\mathbb{U}_i, \Lambda_{\xi_0})}$$

We first have a simple vanishing criteria

Lemma 3.3. *Let $\alpha > 0$. Then*

$$ind_{\mathbb{P}_i^j(\alpha,\beta)}^{\mathbb{P}_i}(w_S \circ \Gamma|_{\tau\otimes\sigma\otimes 1})_{(N^i\mathbb{U}_i,\Lambda_{\xi_0})} \equiv 0.$$

Proof. We are going to show that each orbit of $N^i \times \mathbb{U}_i$ in $\mathbb{P}_i^j(\alpha,\beta)\backslash\mathbb{P}_i$ cannot carry a functional of the form Λ_{ξ_0} given above.

We note that elements of the form

$$\{(w,g)|w \in W_{GL_i}, g \subset SO(W_1(R_i))\}$$

form a set of double coset representatives for

$$\mathbb{P}_i^j(\alpha,\beta)\backslash\mathbb{P}_i/N^i \times \mathbb{U}_i.$$

Here W_{GL_i} is the Weyl group of $GL(R_i)$ realized as permutation matrices in $GL(R_i)$. Then to determine the $N^i \times \mathbb{U}_i$ orbit through $\mathbb{P}_i^j(\alpha,\beta)\cdot(x,g)$ we first determine the stabilizer $S_{(x,g)}$ of $\mathbb{P}_i^j(\alpha,\beta)(x,g)$ in the group $N^i \times \mathbb{U}_i$. This group equals

$$\begin{aligned}&\{n\cdot u|(x,g)(n\cdot u)(x,g)^{-1} \in \mathbb{P}_i^j(\alpha,\beta)\} = \\ &(N^i \cap x^{-1}(V_{S\cap R_i} \cap V_{S\cap R_i^*})x)\cdot(xg)^{-1}N_{(i,j)}^{(\alpha,\beta)}(xg).\end{aligned}$$

We note here that $(xg)^{-1}N_{(i,j)}^{(\alpha,\beta)}xg \subseteq \mathbb{U}_i$. Then we note that (by Frobenius reciprocity)

$$\begin{aligned}&Hom_{N^i\times\mathbb{U}_i}(\text{ind}_{S_{(x,g)}}^{N^i\times\mathbb{U}_i}(w_S\circ\Gamma|\cdots)^w,\Lambda_{\xi_0})\\ =&Hom_{S_{(x,g)}}((w_S\circ\Gamma|\cdots)^w,\Lambda_{\xi_0})\\ \cong&Hom_S(w_S\circ\Gamma,\psi_i^{x^{-1}}\otimes\psi_{g(\xi_0|0|\cdots|0)^T\Lambda_i(x)})\end{aligned}$$

Here $S = (x,g)S_{(x,g)}(x,g)^{-1} = (xN^ix^{-1}\cap V_{S\cap R_i}\cap V_{S\cap R_i^*})\cdot N_{(i,j)}^{(\alpha,\beta)}$. Moreover $\psi_i^{x^{-1}}$ is the twisted character on $xN^ix^{-1}\cap V_{S\cap R_i}\cap V_{S\cap R_i^*}$: $\xi \rightsquigarrow x^{-1}\xi x \to \psi_i(x^{-1}\xi x)$. The representation on the $N_{(i,j)}^{(\alpha,\beta)}$ subgroup is explicitly given by

$$N((\xi_1,\cdots,\xi_i),s) \rightsquigarrow \psi(\sum x_{1\ell}^* q_{W_1}(g\xi_0,\xi_\ell))$$

where ${}^T\Lambda_i(X) = {}^T[\text{matrix}(X)]^{-1} = (x_{ij}^*)$. Also we note here that $(w_S\circ\Gamma|\cdots)^w$ is the representation of $S_{(x,g)}$ given by the composition of maps (1) and (2). First (2) is the map which conjugates $S_{(x,g)}$ by (x,g). The (1) is the map $w_S\circ\Gamma$.

We first examine the subgroup $N_{(i,j)}^{(\alpha,\beta)}\cap\mathbb{U}_S$ given in **Lemma 1**. Since $w_S\circ\Gamma|$ is trivial on $N_{(i,j)}^{(\alpha,\beta)}\cap\mathbb{U}_S$ it follows that for the space $Hom_S(\cdots)$ above to be nonzero we require that $\psi_{g(\xi_0|0\cdots|0)^\perp\Lambda_1(x)}$ is trivial on $N(((0|Z_2|Z_3),s)|Z_2 \in S\cap W_1(R_i), Z_3 \in (S\cap W_1(R_i))^\perp$ in $W(R_i), s_{11} = s_{12} = s_{13} = s_{22} = 0\}$.

This latter condition translates simply to ($x^* = {}^T\Lambda_i(x)$ here)

$$\sum_{\ell \geq \dim S - \beta + 1} x^*_{1\ell} Z_\ell \perp g(\xi_0)$$

(relative to q_{W_1}) with $(0|Z_2|Z_3)$ satisfying the conditions above.

The hypothesis that $\alpha > 0$ implies that $Z_3 \neq 0$. Thus if $x^*_{1\ell} \neq 0$ for $\ell > i - \alpha$ we deduce that

$$g(\xi_0) \perp \{Z \in W_1(R_i) | Z \in (S \cap W_1(R))^\perp\}$$

This implies that $g(\xi_0) \in S \cap W_1(R_i)$, which is a contradiction since $q_{W_1}(\xi_0) \neq 0$.

Thus we may assume that $x \in GL(R_i)$ has the property that $x^*_{1\ell} = 0$ for $\ell > i - \alpha$ (again $\alpha > 0$). Then $\{x \in GL(R_i) | x^*_{1\ell} = 0 \text{ for } \ell > i - \alpha\}$ is stable by multiplication on the left by $V_{S \cap R_i} \cap V_{S \cap R_i^*}$ and on the right by N^i. Thus $\{x \in GL(R_i) | x^*_{1\ell} = 0, \ell > i - \alpha\} = \bigcup_{\Omega} (V_{S \cap R_i} \cap V_{S \cap R_i^*}) w N^i$ where w ranges over a subset Ω of W_{GL_i} (the Weyl group of GL_i).

On the other hand we note the usual Bruhat decomposition of $GL(R_i)$ is given by

$$GL(R_i) = \bigcup_{w \in \Omega'} (V_{S \cap R_i} \cap V_{S \cap R_i^*}) w N^i$$

where $w \in \Omega' = \{$ the set of elements w in W_{GL_i} which have the property w^{-1}(positive roots of Levi factor of $V_{S \cap R_i} \cap V_{S \cap R_i^*})w \subseteq$ positive roots $\}$.

Combinatorially such $w \in \Omega^1$, can be described as follows. The element w^{-1} has the form

$$\begin{array}{ccc} (1, \cdots, \dim S - \beta) & (\dim S - \beta + 1, \cdots, i - \alpha) & (i - \alpha + 1, \cdots, i) \\ \downarrow & \downarrow & \downarrow \\ (\ell_1, \ell_2, \cdots, \ell_{\dim S - \beta}) & (\ell_{\dim S - \beta + 1}, \cdots, \ell_{i - \alpha}) & (\ell_{i - \alpha + 1}, \cdots \ell_i) \end{array}$$

where in each $(\cdots)$ we have an increasing sequence of integers chosen from the set $\{1, \cdots, i\}$.

Now if the tuple $(\ell_{i-\alpha+1}, \cdots \ell_i) \neq (1, \cdots, \alpha)$ we note that there exists a ℓ_ν in one of the first 2 tuples so that $\ell_\nu = \ell_j - 1$ for ℓ_j in the last (or third) tuple.

Thus such a w has the property that w^{-1} (unipotent radical of $V_{S \cap R_i} \cap V_{S \cap R_i^*})w$ contains a simple positive root subgroup. Hence the group $N^i \cap w^{-1}$ (unipotent radical of $V_{S \cap R_i} \cap V_{S \cap R_i^*})w$ contains a simple positive root subgroup.

Then as a consequence of the above statement (*for such* w)

$$Hom_{S_{(w,g)}}(w_S \circ \Gamma | \cdots, \Lambda_{\xi_0}) = \{0\}.$$

Then we can assume $(\ell_{i-\alpha+1}, \cdots \ell_i) = (1 \cdots \alpha)$. Again by the same reasoning if $(\ell_{\dim S - \beta + 1}, \cdots \ell_{i-\alpha+1}) \neq$ (a consecutive tuple starting with $\alpha + 1$) we deduce (for such w)

$$Hom_{S_{(x,g)}}(w_S \circ \Gamma | \cdots, \Lambda_{\xi_0}) = \{0\}.$$

Thus we finally deduce that the remaining $w \in \Omega'$ is uniquely determined by the property that w^{-1} (unipotent radical of $V_{S\cap R_i} \cap V_{S\cap R_i^*})w \subset N_-^i$ (= the lower unipotent group of $GL(R_i)$). On the other hand we know $(V_{S\cap R_i} \cap V_{S\cap R_i^*})wN^i$ is the open Bruhat cell in $GL(R_i)$ relative to the double cosets $V_{S\cap R_i} \cap V_{S\cap R_i^*} \backslash GL(R_i)/N^i$. (This follows by the above fact and the choice of the set Ω'). But we note that the open cell is clearly disjoint from $\{x \in GL(R_i) | x_{1\ell}^* = 0, \ell > i - \alpha\}$.

In particular we see directly that $w \in \Omega'$, the long element does not lie in $\{x \in GL(R_i) | x_\ell^* = 0, \ell > i - \alpha\}$.

Thus we have shown that there is no $N^i \times \mathbb{U}_i$ orbit in $\mathbb{P}_i^j(\alpha,\beta)\backslash\mathbb{P}_i$ which can determine a nonzero Λ_{ξ_0} functional ! /Q.E.D.

Remark 4. The proof above shows in fact that for any (α,β) with $\dim S - \beta < i$ the only possible pairs (w,g) that could possibly contribute a double coset in $\mathbb{P}_i^j(\alpha,\beta)\backslash\mathbb{P}_i/N^i \times \mathbb{U}_i$ (that determines a nonzero Λ_{ξ_0} functional) is when $w = w_\ell$ (the long element which has the property that w_ℓ^{-1} (unipotent radical of $V_{S\cap R_i} \cap V_{S\cap R_i^*})w_\ell \subseteq N_-^i$) and g satisfies the condition $g(\xi_0) \perp S \cap W_1(R_i)$. We note that the set of such g satisfying this condition is stable by multiplication on the left by $\mathbb{P}_{S\cap W_1(R_i)}^{SO(W_1(R_i))}$ and on the right by $SO((\xi_0)^\perp, q_{W_1})$. Thus g belongs to a certain number of double cosets of $(\mathbb{P}_{S\cap W_1(R_i)}^{SO(W_1(R_i))}, SO((\xi_0)^\perp, q_{W_1}))$ in $SO(W_1(R_i))$. Moreover the condition $\xi_0 \perp g^{-1}(S\cap W_1(R_i))$ implies that $g^{-1}(S\cap W_1(R_i)) \subseteq (\xi_0)^\perp$. Thus there exist either one or two double cosets in $\mathbb{P}_{S\cap W_1(R_i)}^{SO(W_1((R_i))}\backslash SO(W_1(R_i))/SO((\xi_0)^\perp, q_{W_1})$ which carry such g (two if $\dim S\cap W_1(R_i) = 1/2 \dim(\xi_0)^\perp$ and otherwise one; this is given by an extension of the type of argument used in the proof of **Proposition 1**). In particular these $SO((\xi_0)^\perp, q_{W_1})$ orbits have the form $Q_{S\cap W_1(R_i)}\backslash SO((\xi_0)^\perp, q_{W_1})$ where $Q_{S\cap W_i(R_i)}$ is a maximal parabolic subgroup of $SO((\xi_0)^\perp, q_{W_1})$ which stabilizes an isotropic subspace $\eta^{-1}(S\cap W_1(R_i)) \subseteq (\xi_0)^\perp$ where $\eta \in SO(W_1(R_i), q_{W_1(R_i)})$. We note in fact that η can be chosen so that $\eta^{-1}(\xi_0)$ lies in the space $W_1(R_i) \cap W_1(S)$(= $S_\#$ defined above). Indeed if it is possible to find an $z \in SO(W_1(R_i))$ so that $z(\xi_0) \perp S\cap W_1(R_i)$ (or $z(\xi_0) \in (S\cap W_1(R_i))^\perp = S\cap W_1(R_i) \oplus W_1(R_i) \cap W_1(S)$ in $W_1(R_i)$); in particular there then exists $g' \in SO(W_1(R_i))$ so that $g'z(\xi_0)$ lies in $W_1(R_i) \cap W_1(S)$.

Then there is a decomposition

$$(\xi_0)^\perp = \eta^{-1}(S \cap W_1(R_i)) \oplus \mathcal{M} \oplus \eta^{-1}(S \cap W_1(R_i))^*$$

where $\eta^{-1}(S\cap W_1(R_i))$ and $\eta^{-1}(S\cap W_1(R_i))^*$ are isotropic subspaces (paired non-singularly by q_{W_1}) and $\mathcal{M}$ is the q_{W_1} orthogonal complement to $\eta^{-1}(S\cap W_1(R_i)) \oplus \eta^{-1}(S \cap W_1(R_i))^*$ in $(\xi_0)^\perp$. In particular

$$Q_{S\cap W_1(R_i)} = (GL(\eta^{-1}(S \cap W_1(R_i)) \times SO(\mathcal{M}, q_{\mathcal{M}})) \ltimes \tilde{U}_{\eta^{-1}(S\cap W_1(R_i)}$$

Very concretely we note that

$$\eta Q_{S\cap W_1(R_i)}\eta^{-1} = GL(S \cap W_1(R_i)) \times SO(\eta(\mathcal{M}), q_{\eta(\mathcal{M})})\tilde{U}_{S\cap W_1(R_i)}$$

Relative to this decomposition $SO(\eta(\mathcal{M}), q_{\eta(\mathcal{M})}) \subseteq SO(W_1(R_i) \cap W_1(S))$ and $\tilde{U}_{S\cap W_1(R_i)} \subseteqq \mathbb{U}^{SO(W_1(R_i))}_{S\cap W_1(R_i)}$ = the unipotent radical of $\mathbb{P}^{SO(W_1(R_i))}_{S\cap W_1(R_i)}$.

We also recall the construction of "generalized derivative " of representations of a classical reductive group G_v. Indeed let $P_{\ell,v}$ be a maximal parabolic subgroup which has the form $GL_\ell \times M_\ell \ltimes U$. Given a smooth G_v module $\mathbf{\Pi}_v$ we consider the usual (unnormalized) Jacquet function $(\mathbf{\Pi}_v)_U$. Then we consider the *Whittaker Jacquet functor* of $(\mathbf{\Pi}_v)_U$ relative to a maximal unipotent subgroup $U_\ell \subset GL_\ell$ and a fixed generic character ψ_ℓ of U_ℓ. That is, we form $((\mathbf{\Pi}_v)_U)_{(U_\ell,\psi_\ell)} = \mathbf{\Pi}_v^{(\ell)}$. We call $\mathbf{\Pi}_v^{(\ell)}$ the ℓ-th *derivative of* $\mathbf{\Pi}_v$. We note that $\mathbf{\Pi}_v^{(\ell)}$ is an admissible M_ℓ module of finite length. In any case $\mathbf{\Pi}_v^{(\ell)}$ is a module for M_ℓ in the obvious way.

Remark 5. In the case where $G_v = GL_N$ then $\mathbf{\Pi}_v^{(\ell)}$ is "essentially" the ℓ-th derivative of $\mathbf{\Pi}_v$ is taken in the following manner. Start with a standard parabolic $D_{\ell'} = GL_{\ell'} \times GL_{N-\ell'} \ltimes U_{\ell'}$ of $GL(N)$. Then the $N - \ell'$ derivative in [B-Z] is given as the composite Jacquet functor:

$$\mathbf{\Pi}_v^{\{N-\ell'\}} = ((\mathbf{\Pi}_v)_{(U_{\ell'}\mathbb{1})})_{(N_{N-\ell'},\psi)}$$

Here $N_{N-\ell'}$ is a standard maximal unipotent subgroup of $GL(N - \ell')$ and ψ the usual generic character. Then associated to $\mathbf{\Pi}_v$ we construct the associated representation $\underset{\sim}{\mathbf{\Pi}}_v$ given by

$$\underset{\sim}{\mathbf{\Pi}}_v(g) = \mathbf{\Pi}_v(w_N{}^t g^{-1} w_N)$$

Here $w_N =$

$$\begin{bmatrix} 0 & & 1 \\ & \ddots & \\ 1 & & 0 \end{bmatrix}$$

and thus $\underset{\sim}{\mathbf{\Pi}}_v$ is equivalent to the contragredient $\mathbf{\Pi}_v^{\nu}$ of $\mathbf{\Pi}_v$.

Thus the relation between the "derivative" $\mathbf{\Pi}_v^{(\ell)}$ defined above and derivative $\mathbf{\Pi}_v^{\{\ell'\}}$ given in [BZ] is as follows. Starting with $a \in GL_\ell$ we let $a^* = w_\ell{}^t a^{-1} w_\ell$. Then there exists an isomorphism J_ℓ between the spaces $\mathbf{\Pi}_v^{(\ell)}$ and $(\underset{\sim}{\mathbf{\Pi}}_v)^{\{\ell\}}$ in such a way that $J_\ell(\mathbf{\Pi}_v^{(\ell)}(a)v) = (\underset{\sim}{\mathbf{\Pi}}_v)^{\{\ell\}}(a^*)J_\ell(v)$ with v some arbitrary vector in the space of the representation $\mathbf{\Pi}_v^{(\ell)}$. For example we deduce from this that if $\mathbf{\Pi}_v = \mathrm{ind}_{B_N}^{GL_N}(\chi_1 \otimes \chi_2 \otimes \cdots \otimes \chi_N) =$ "$\chi_1 \otimes \cdots \otimes \chi_N$", then $\mathbf{\Pi}_v^{(\ell)}$ (after semi simplification) as a $GL_{N-\ell}$ module

$$\oplus \text{``}\chi_{z_1} \otimes \cdots \otimes \chi x_{N-\ell}\text{''} \otimes |\ |^{-\frac{\ell}{2}}$$

where the direct sum is taken over all tuples $(z_1, z_2, \cdots, z_{N-\ell})$ with $1 \leq z_1 < z_2 < \cdots < z_{N-\ell} \leq N$.

We also consider another type of derivative. Namely we consider a nontrivial character $\Psi_{\mathbb{U}_{Dec}}$ which has the property that $Stab_{GL_\ell \times M_\ell}(\Psi_{U_{Dec}})$ contains a subgroup of $GL_\ell \times M_\ell$ which fixes pointwise a "decomposable" vector $v \otimes w$ in the standard representation of $GL_\ell \times M_\ell$. *The above representation of* $\mathbb{U}_i$ *given by* $\psi_2 = \psi_{\xi_0}$ *is exactly of this type.* Then there is a maximal unipotent group $U_\ell \subseteq GL_\ell$ which fixes $\Psi_{U_{Dec}}$. We form the twisted Jacquet module:

$$\mathbf{\Pi}_v^{\langle \ell \rangle} = ((\mathbf{\Pi}_v)_{(U,\Psi_{U_{Dec}})})_{(U_\ell,\psi_\ell)}.$$

This space is the "ℓ-th twisted" derivative. In fact it is a module for $M_\ell^w = \{g \in M_\ell | g$ *fixes* w in the standard module of $M_\ell\}$.

For emphasis and notation here we note that the parabolic subgroup of $GL(S)$ which stabilizes $S \cap W_1(R_i)$ admits a Langlands decomposition as $GL(S \cap R_i^*) \cdot GL(S \cap W_1(R_i)) \ltimes (N_{(i,j)}^{(0,\beta)} \cap GL(S))$. Moreover the parabolic subgroup of $SO(W_1(S))$ which stabilizes the

isotropic subspace $R_i \cap W_1(S)$ admits a Langlands decomposition as $GL(R_i \cap W(S)) \cdot SO(W_1(S) \cap W_1(R_i)) \ltimes (N_{(i,j)}^{(0,\beta)} \cap SO(W_1(S))$.

Lemma 3.4. *Let* $\alpha = 0$ *and* $j - \beta < i$ *(here* $j = \dim S$*). Then*

$$ind_{\mathbb{P}_i^j(0,\beta)}^{\mathbb{P}_i} (w_S \circ \Gamma|_{\tau\otimes\sigma\otimes 1})_{(N^i\mathbb{U}_i,\Lambda_{\xi_0})}$$

is equivalent as a $SO((\xi_0)^\perp, q_{W_1})$ *module to (with unnormalized induction) the possible direct sum*

$$\bigoplus_\eta ind_{GL(\eta^{-1}(S\cap W_1(R_i))\times SO(\mathcal{M},q_{\mathcal{M}})\ltimes \tilde{U}_{\eta^{-1}(S\cap W_1(R_i))}}^{SO((\xi_0)^\perp,q_{W_1})} (|detg_1|^{-i}\otimes$$
$$(\tau^{w_S})^{(j-\beta)}(\eta g_1 \eta^{-1}) \otimes (\sigma^{w_S})^{\langle i-(j-\beta)\rangle}(\eta g_2 \eta^{-1}) \otimes 1(u)))$$

Here $(g_1, g_2, u) \in GL(\eta^{-1}(S \cap W_1(R_i)) \times SO(\mathcal{M}, q_{\mathcal{M}}) \ltimes \tilde{U}_{\eta^{-1}(S\cap W_1(R_i))}$. *Moreover* τ^{w_S} *(*σ^{w_S} *resp.) is the representation of* $GL(S)$ *given by* $\tau^{w_S}(x) = \tau(w_S x w_S^{-1})$ *for* $x \in GL(S)$ *(*$\sigma^{w_S}(y) = \sigma(w_S y w_S^{-1})$ *for* $y \in SO(W_1(S))$ *resp.) Moreover the* $(j-\beta)$ *derivative* $(\tau^{w_S})^{(j-\beta)}$ *is taken relative to the parabolic subgroup of* $GL(S)$ *which stabilizes the subspace* $S \cap W_1(R_i)$*. The* $i-(j-\beta)$ *twisted derivative* $(\sigma^{w_S})^{\langle i-(j-\beta)\rangle}$ *is taken relative to the parabolic subgroup of* $SO(W_1(S))$ *which stabilizes the isotropic subspace* $R_i \cap W_1(S)$*.*

Here η *ranges over either one or two elements (depending on whether* $\dim S \cap W_2(R_i) \neq \frac{1}{2}\dim(\xi_0)^\perp$ *or* $= \frac{1}{2}\dim(\xi_0)^\perp$*.)*

Proof. We let $G_{\xi_0} = N^i \mathbb{U}_i SO((\xi_0)^\perp, q_{W_1})$. By the comments above the space (as a $SO((\xi_0)^\perp, q_{W_1})$ module)

$$\mathrm{ind}_{\mathbb{P}_i^j(0,\beta)}^{\mathbb{P}_i} (w_S \circ \Gamma|_{\tau\otimes\sigma\otimes 1})_{(N^i\mathbb{U}_i,\Lambda_{\xi_0})}$$

is equivalent to

$$\mathrm{ind}_{(w_\ell\eta)^{-1}\mathbb{P}_i^j(0,\beta)(w_\ell\eta)\cap G_{\xi_0}}^{G_{\xi_0}} (w_S \circ \Gamma \circ w_\ell\eta|_{\tau\otimes\sigma\otimes 1})_{(N^i\mathbb{U}_i,\Lambda_{\xi_0})}.$$

Here $w_S \circ \Gamma \circ w_\ell \eta$ means mapping the element $g \in (w_\ell \eta)^{-1} \mathbb{P}_i^j(0,\beta)(w_\ell \eta) \cap G_{\xi_0}$ to

$$(\pi \otimes \sigma \otimes 1)((w_S(w_\ell \eta) g (w_\ell \eta)^{-1}(w_S)).$$

In explicit terms we conjugate $g \in (w_\ell \eta)^{-1} P_i^j(0,\beta) w_\ell \eta \cap G_{\xi_0}$ to $(w_\ell \eta) g (w_\ell \eta)^{-1}$. Then Γ separates $(w_\ell \eta) g (w_\ell \eta)^{-1}$ into its $\mathbb{P}_S$ components as above and then we conjugate each such component by w_S and apply the $\tau \otimes \sigma \otimes \mathbb{1}$ representation

On the other hand we know from above that

$$(w_\ell \eta)^{-1} \mathbb{P}_i^j(0,\beta)(w_\ell \eta) \cap G_{\xi_0} = \hat{T}_i \cdot Q_{S \cap W_1(R_i)} \cdot N_\beta$$

where

$$\hat{T}_i = w_\ell^{-1} V_{S \cap R_i^*} w_\ell \cap N^i = \left\{ g \in GL(R_i) | \Lambda_i(g) = \begin{bmatrix} 1 & & * & & & \\ & \ddots & & & 0 & \\ 0 & & 1 & & & \\ & & & 1 & & * \\ & 0 & & & \ddots & \\ & & & 0 & & 1 \end{bmatrix} \right\}$$

Recall that the matrix above is determined by the map Λ_i prescribed above. Here the first block has size $i - (\dim S - \beta)$ and the second block has size $\dim S - \beta$. Conjugation by w_ℓ permutes the two blocks. Also

$$N_\beta = (w_\ell \eta)^{-1} N_{(i,j)}^{(0,\beta)} (w_\ell \eta)$$

where $N_{(i,j)}^{(0,\beta)}$ is given as above. We note here that N_β is a subgroup of N^i.

Then we recall that (from **Corollary** to **Lemma 3.1**)

$$N_{(i,j)}^{(0,\beta)} = (N_{(i,j)}^{(0,\beta)} \cap GL(S))(N_{(i,j)}^{(0,\beta)} \cap SO(W_1(S)))(N_{(i,j)}^{(O,\beta)} \cap \mathbb{U}_S)$$

where the first and second components (in the decomposition above) are the unipotent radicals of parabolic subgroups of $GL(S)$ and $SO(W_1(S))$ stabilizing the isotropic subspaces $S \cap W_1(R_i)$ and $R_i \cap W_1(S)$.

Then we apply $w_S \circ \Gamma \circ (w_\ell \eta)$ to an element of $\hat{T}_i \times Q_{S \cap W_1(R_i)} \times N_\beta$. Using the parametrizations above the element

$$\Lambda_i^{-1} \begin{bmatrix} A & 0 \\ 0 & B \end{bmatrix} \times (g, z, u) \times (w_\ell \eta)^{-1} (n_1 \cdot n_2 \cdot n_3)(w_\ell \eta)$$

is sent to (with one component arbitrary and the rest $= e$)

$$(\tau^{w_S}) \left(\Lambda_S^{-1} \begin{bmatrix} {}^T B & \tilde{n}_1 \\ 0 & \Lambda_{r_1}(g) \end{bmatrix} \right) \otimes (\sigma^{w_S})(w_\ell^{-1} \Lambda_{r_2}^{-1}(A) w_\ell, \eta z \eta^{-1}, n_2) \otimes \mathbb{1}(u n_3)$$

Here τ^{w_S} and σ^{w_S} represent the twisting of τ and σ by w_S, i.e. $\tau^{w_S}(x) = \tau(w_S x w_S^{-1})$ and $\sigma^{w_S}(y) = \sigma(w_S y w_S^{-1})$.

Here Λ_S represents an embedding of S into k^j given by the choice of basis of S used above, i.e. $S = Sp\{v_1^*, \cdots, v_{\dim S-\beta}^*, z_1, \cdots, z_\beta\}$ with $\{z_1, \cdots, z_\beta\}$ some basis of $S \cap W_1(R_i)$. Similarly Λ_{r_1} and Λ_{r_2} represent embeddings of $\eta^{-1}(S \cap W_1(R_i))$ to $k^\beta(Sp\{v_1, \cdots, v_{i-(\dim S-\beta)}\}$ to $k^{i-(\dim S-\beta)}$ resp.) relative to the choice of basis $\eta^{-1}\{z_1, \cdots, z_\beta\}(\{v_1 \cdots, v_{i-(\dim S-\beta)}\}$ resp.). First ${}^TB$ is the transpose of the matrix B where ${}^TB$ is taken relative to the basis $\{v_1^*, \cdots, v_{\dim S-\beta}^*\}$. Moreover $\tilde{n}_1$ represents $\Lambda_S(n)$, which runs over the subgroup

$$\left\{ \begin{bmatrix} I & Z \\ 0 & I \end{bmatrix} \middle| Z \in M_{\dim S-\beta, i-(\dim S-\beta)}(k) \right\}.$$

The element $w_\ell^{-1}\Lambda_{r_2}(A)w_\ell \times \eta z \eta^{-1}$ lies in the Levi factor of the parabolic subgroup of $SO(W_1(S))$ associated to the isotropic subspace $R_i \cap W_1(S)$. In fact $\{w_\ell^{-1}\Lambda_{r_2}^{-1}(A)w_\ell | A \cdots\}$ ranges over a maximal unipotent radical of $GL(R_i \cap W_1(S))$; in fact $w_\ell^{-1}\Lambda_{r_2}^{-1}(A)w_\ell$ lies in $N^i_{GL(R)}$ = the commutator subgroup of $GL(R_i \cap W_1(S))$ which stabilizes the flag $\{v_{j-\beta+1}, \cdots, v_i\} \supseteq \cdots \supseteq \{v_i\}$!

Thus we have reduced to a problem of the following type.

Let $H = S \cdot T$ be a semidirect product where T is the unipotent radical and S is reductive. Let Ψ_T be a character on T so that S stabilizes Ψ_T. Let R be a closed subgroup of H with $R = (R \cap S) \cdot (R \cap T)$. Let σ be a smooth representation of R.

Then relative to the S action (see **Appendix**)

$$(\operatorname{ind}_R^H(\sigma))_{(T,\Psi_T)} \cong \operatorname{ind}_{R\cap S}^S(| \, | \otimes \sigma_{(R\cap T, \Psi_T)})$$

where $| \, | = |\det_{R\cap T \backslash T}(\ )|$. Induction in the second factor is taken as unnormalized to emphasize the dependence on the $| \, |$ factor.

We apply the above considerations to our specific case and deduce the statement of the Lemma. We note here that the character ψ_Z (defined above where $Z = \{\xi_0, 0, \cdots, 0)$ with $q_{W_1}(\xi_0) \neq 0$) can be explicitly determined on the subgroup N_β. Indeed we note that $N_{(i,j)}^{(0,\beta)} = \{N_i((Z_1|Z_2), s) | Z_1 \in S \cap W(R_i) \otimes k^{\dim S-\beta}$ and $Z_2 \in (S \cap W(R_i))^\perp \otimes k^{i-(\dim S-\beta)}$ and $s_{ij} = 0$ for $r \in \{1, \cdots, \dim S - \beta\}$ and $j \in \{1, \cdots, i\}\}$. Then we have that

$$(w_\ell \eta)^{-1} N_{(i,j)}^{(0,\beta)} (w_\ell \eta) = N_i((\eta^{-1}Z_1 | \eta^{-1}Z_2)\Lambda_i(w_\ell), {}^T\Lambda_i(w_\ell) s \Lambda_i(w_\ell))$$

Then we apply the character ψ_Z to N_β. We obtain

$$\psi(q_{W_1}(\eta^{-1}Z_2^{(1)}, \xi_0)) = \psi(q_{W_1}(Z_2^{(1)}, \eta\xi_0))$$

Here $Z_2^{(1)} \in (S \cap W_1(R_i))^\perp$ represents the first component of the Z_2 part (in N_i) defined above. Recalling the explicit form of $N_{(i,j)}^{(0,\beta)} \cap GL(S)$ and $N_{(i,j)}^{(0,\beta)} \cap \mathbb{U}_S$ it follows that ψ_Z is the identity on *these groups*. However since $N_{(i,j)}^{(0,\beta)} \cap SO(W_1(S)) = \{N_i((0|Z_2), s) | Z_2 \in (S \cap W_1(R_i) + (S \cap W_1(R_i))^*)^\perp$ in $W_1(R_i), s_{ij} = 0$ for $i \in$

$\{1, \cdots, \dim S - \beta\}, j \in \{1, \cdots, i\}\}$, then ψ_Z is clearly *nontrivial* on this group. On the other hand we consider the space $W_1(S)$ containing the isotropic subspace $R_i \cap W_1(S)$. We note that the corresponding unipotent group $\mathbb{U}^{SO(W_1(S))}_{R_i \cap W_1(S)} = N^{(0,\beta)}_{(i,j)} \cap SO(W_1(S))$ has the form as given in the beginning of this section. In fact by choosing as a basis of $W_1(S)$ the set $\{v_{\dim S - \beta + 1}, \cdots, v_i, v^*_{\dim S - \beta + 1}, \cdots, v^*_i\} \cup$ $\{$a basis of $W_1(S) \cap W_1(R_i)\}$, it is possible to put $\mathbb{U}^{SO(W_1(S))}_{R_i \cap W_1(S)}$ into the form

$$N_\nu((T_1, \cdots, T_\nu), s)$$

where $T_i \in [R_i \cap W_1(S) \oplus (R_i \cap W_1(S))^*]^\perp$ in $W_1(S)$ and $s \in Skew_\nu(k)(\nu = i - (\dim S - \beta))$. Moreover it is then easy to compute directly that

$$N_i(((0, Z_2), \begin{bmatrix} 0 & 0 \\ 0 & s \end{bmatrix})) \equiv N_\nu((Z_2, s))$$

(Again we *emphasize* the identity is comparing the same element in $N_i \subseteq \mathbb{U}_i$ and $N_\nu \subseteq \mathbb{U}^{SO(W_1(S))}_{R_i \cap W_1(S))}$ relative to the suitable choice of bases given above.) Thus in fact the character ψ_Z restricted to $(w_\ell \eta)^{-1}(N^{(0,\beta)}_{(i,j)} \cap SO(W_1(S)))(w_\ell \eta)$ equals the character on $\mathbb{U}^{SO(W_1(S))}_{R_i \cap W_1(S)}$ given by $\psi_{\eta \xi_0}$ (defined relative to the group N_ν). Thus we deduce in the usual way that $N^i_{R_i \cap W_1(S)} (\subseteq GL(R_i \cap W_1(S)))$ stabilizes the character $\psi_{\eta \xi_0}$! Thus in the determination of the Jacquet functor of $\mathrm{ind}^{G_{\xi_0}}_{(w_\ell \eta)^{-1} \mathbb{P}^j_i(0,\beta)(w_\ell \eta) \cap G_{\xi_0}} (w_S \circ \Gamma \circ w_\ell \eta|_{\tau \otimes \sigma \otimes 1})_{(N^i \mathbb{U}_i, \Lambda_{\xi_0})}$ we see that the term

$$(\sigma^{w_S})_{(N^i_{R_i \cap W_1(S)} \mathbb{U}^{SO(W_1(S))}_{R_i \cap W_1(S)}, \psi_i \otimes \psi_{\eta \xi_0})}$$

gives the twisted derivative of the form

$$(\sigma^{w_S})^{\langle i-(j-\beta) \rangle}$$

We note in fact that this twisted derivative comes from exactly the type of character $\psi_Z = \psi_{\xi_0}$ defined in $\mathbb{U}_i$ above. Here $\mathbb{U}_i$ is replaced by $\mathbb{U}^{SO(W_1(S))}_{R_i \cap W_1(S)}$ and N^i is replaced by $N^i_{R_i \cap W_1(S)}$. Moreover the group $\eta SO(\mathcal{M}, q_\mathcal{M}) \eta^{-1}$ acts on the space since

$$SO(\mathcal{M}, q_\mathcal{M}) \subseteq SO((\xi_0^\perp, q_{W_1}).$$

Then we compute the coinvariants of the group

$$\left\{ \Lambda_i \left(\begin{bmatrix} I & 0 \\ 0 & B \end{bmatrix} \right) \middle| B = \begin{bmatrix} 1 & * \\ 0 & 1 \end{bmatrix} \right\} \times (w_\ell \eta)^{-1} N^{(0,\beta)}_{i,j)} (w_\ell \eta)$$

acting through

$$\tau^{w_S} \left[\Lambda_S \begin{bmatrix} {}^T B & \eta_2 \\ 0 & I \end{bmatrix} \right]$$

relative to the trivial character in η_2 and the generic Whittaker character on B. In any case we obtain

$$(\tau^{ws})^{j-\beta)}$$

considered as $GL(\eta^{-1}(S \cap W_1(R_I))$ module.

This completes the proof of the **Lemma.**/Q.E.D.

Finally we consider the case when $\alpha = 0$ and $\dim S - \beta = i$. This implies that $\dim S \geq \dim R_i$ and in fact $S \supseteq R_i^*$. In fact we have in this case that the space $W_1(S) \subsetneqq W_1(R_i)$. In this case $V_{S\cap R_i^*} = V_{S\cap R_i} = GL(R_i)$. Thus the orbits of $N^i \times \mathbb{U}_i$ in $\mathbb{P}_i^j(0,j)\backslash\mathbb{P}_i$ are parametrized by the elements in $\mathbb{P}_{S\cap W_1(R_i)}^{SO(W_1(R_I))}\backslash SO(W_1(R_i))$.

In fact $SO((\xi_0)^\perp, q_{W_1})$ admits at most 3 orbits in the space $\mathbb{P}_{S\cap W_1(R_i)}^{SO(W_1(R_i))}\backslash SO(W_1(R_i))$ provided $S \cap W_1(R_i) \neq 0$. The possible 3 orbits are contained in the sets (i) $\{L \text{ isotropic } |L \subseteq (\xi_0)^\perp\}$ (In fact there is exactly one orbit unless $\dim(\xi_0)^\perp = 2\dim L$ in which case there are 2 orbits). (ii) $\{L$ isotropic $|L \cap (\xi_0)^\perp$ has codimension 1 in $L\}$. By the arguments above it is possible to choose $\eta \in SO(W_1(R_i))$ so that $\eta(S \cap W_1(R_i)) \subseteq (\xi_0)^\perp$. Thus we let *in case (i)* $Q_{S\cap W_1(R)}$ be the stabilizer of $(S \cap W_1(R_i))$ in $SO((\xi_0)^\perp, q_{W_1})$. The group admits the same structure as before. Thus $(\eta)^{-1}\mathbb{P}_{(i,j)}^{(0,i)}\eta \cap G_{\xi_0} = N^i \cdot Q_{S\cap W_1(R_i)} \cdots N_\beta(N_\beta$ defined above).

In case (ii) above the stabilizer $Q^*_{S\cap W_1(R)}$ of the subspace $S \cap W_1(R_i)$ in $SO(\xi_0)^\perp, q_{W_1})$ can be determined in the following manner. First (except for the case where W_1 is totally split and $j =$ Witt index of W_1 and i odd) we can choose a basis $\{z_i\}$ of $S \cap W_1(R_i)$ so that $\{z_2, \cdots z_l\} \subseteq (\xi_0)^\perp$ and $z_1 = z_1^+ + z_1^-(z_1^+ \in (\xi_0), z_1^- \in (\xi_0)^\perp$ both nonzero q_{W_1} length). Then it is possible to complete this basis to one of $W(R_i)$, where the remaining set consists of $\{z_i^* | i \geq 2\} \cup \{y_t | \cdots\}$. Then $Span\{z_1, \cdots, z_\ell\} \oplus Span\{y_t | \cdots\} \oplus Span\{z_1^*, \cdots z_\ell^*\} = W_1(R_i)$. Here $Span\{z_1, \cdots, z_\ell\}$ and $Span\{z_1^*, \cdots z_1^*\}$ are isotropic subspaces nonsingularly paired by q_{W_1} and $Span\{y_t | \cdots\}$, the perpendicular complement to $Span\{z_1, \cdots, z_\ell\}$ $\oplus Span\{z_1^*, \cdots, z_\ell^*\}$. The

$$(\xi_0)^\perp = Span\{z_2, \cdots, z_\ell\} \oplus Span\{z_1^-, y_t \cdots\} \oplus Span\{z_2^* \cdots, z_\ell^*\}.$$

Moreover we choose the basis $\{y_t | \cdots\}$ so that $Span\{y_t | \cdots\} = S_\#$ (given above) in this case. Then the group $Q^*_{S\cap W_1(R_i)} = SO((\xi_0)^\perp, q_{W_1}) \cap \mathbb{P}_{S\cap W_1(R_i)}^{SO(W_1(R_i))}$ has the form:

$$(GL(\{z_2, \cdots, z_\ell\}) \times SO((z_1^-)^\perp, q_{W_1})) \times U_{\{z_2,\cdots,z_\ell\}}$$

Here the group $U_{\{z_2,\cdots,z_\ell\}}$ is the unipotent radical of $P_{\{z_2,\cdots,z_\ell\}}$, the parabolic subgroup of $SO((\xi_0)^\perp, q_{W_1})$ stabilizing the isotropic subspace $Span\{z_2, \cdots z_\ell\}$; Here $(z_1^-)^\perp$ taken in the space $Sp\{z_1^-, y_t | \cdots\}$; so that $SO((z_1^-)^\perp, q_{W_1})$ is a subgroup of $SO(Sp(z_1^-, y_t \cdots), q_{W_1})$ and in fact $SO((z_1^-)^\perp, q_{W_1}) = SO(W_1(S) \cap W_1(R_i)), q_{W_1})$. By the determination of $W_1(S) = W_1(S) \cap W_1(R_i)$ above we have that

$$SO((z_1^-)^\perp, q_{W_1}) = SO(W_1(S) \cap W_1(R_i), w_{W_1}) = SO(W_1(S), q_{W_1}).$$

We note here that in the case W_1 is the totally split and $j =$ Witt index of W and i odd, then we choose $\{z_1, \ldots, z_{\ell-1}, z_\ell^*\}$ to be the basis of $S \cap W_1(R_i)$ ($\ell = j - i, \dim W_1 - \dim W_1(R_i) = 2j - 2i$ here). In such a case, $W_1(R_i) = \mathrm{Sp}\{z_1, \ldots, z_\ell^*\} \oplus \mathrm{Sp}\{z_1^*, \ldots, z_{\ell-1}^*, z_\ell\}$ and $(\xi_0)^\perp = \mathrm{Sp}\{z_2, \ldots, z_{\ell-1}, z_\ell^*\} \oplus (z_1^-) \oplus \mathrm{Sp}\{z_2^*, \ldots, z_{\ell-1}^*, z_\ell\}$. Here $Q^*_{S\cap W_1(R_i)} = SO((\xi_0)^\perp, q_{W_1}) \cap \mathbb{P}^{SO(W_1(R_i))}_{S\cap W_1(R_i)} = GL(\{z_2, \ldots, z_{\ell-1}, z_\ell^*\} \times U_{\{z_2, \ldots, z_\ell^*\}}$.

Remark 6. The decomposition of the group $Q^*_{S\cap W_1(R_i)} =$

$$GL(\{z_2, \cdots, z_\ell\}) \times SO(W_1(S) \cap W_1(R_i)) \ltimes U_{\{z_2, \cdots, z_\ell\}}$$

is not compatible with the decomposition of $\mathbb{P}^{SO(W_1(R_i))}_{S\cap W_1(R_i)} = GL(\{z_1, \cdots, z_\ell\}) \times SO(W_1(S) \cap W_1(R_i) \ltimes \mathbb{U}^{SO(W_1(R_i))}_{S\cap W_1(R_i)}$. In fact we note that for the first two components of $Q^*_{S\cap w_1(R_i)}$

$$GL(\{z_2, \cdots, z_\ell\}) \times SO(W_1(S) \cap W_1(R_i)) \subseteq GL(\{z_1, \cdots, z_\ell\}) \times SO(W_1(S) \cap W_1(R_i)).$$

However the group $U_{\{z_2, \cdots, z_\ell\}}$ admits a more subtle decomposition. We recall that $\mathbb{U}^{SO(W_1(R_i))}_{S\cap W_1(R_i)}$ has the form

$$N_\ell(((Z_1 | \cdots | Z_\ell), s))$$

where $Z_i \in Span\{y_1, \cdots, y_t\}$ and $s \in Skew_\ell(k)$. We know there exists a mirabolic subgroup $\underset{\sim}{P}_{\{z_1, \cdots, z_\ell\}}$ (fixing *pointwise* the vector z_1^*) in $Sp\{z_1^*, \cdots z_\ell^*\} =$ the dual space of $Sp\{z_1, \cdots, z_\ell\}$ (relative to q_{W_1} pairing)) of $GL(\{z_1, \cdots, z_\ell\})$ having the form $GL(\{z_2, \cdots, z_\ell\}) \ltimes k^{\ell-1}$. Here we identify $k^{\ell-1}$ as the additive row space $= \{(y_1, \cdots, y_{\ell-1}) | y_i \in k\}$ with $(y_1, \ldots, y_{\ell-1})(z_1) = z_1 + \sum_{i \geq 2} y_i z_{i+1}$. Then the group $U_{\{z_2, \cdots, z_\ell\}}$ has the form

$$\{(a, b) \in k^{\ell-1} \times N_\ell | Z_1 = 0, s_{1\nu} = a_{\nu-1} q_{W_1}(\xi_0, \xi_0)^{-1}, \nu \geq 2\}$$

Moreover we then have the simple decomposition of $U_{\{z_2, \cdots, z_\ell\}}$ into a product of two groups. Namely, $U_{\{z_2, \cdots, z_\ell\}} = T_{\ell-1} V_{\ell-1}$, where $T_{\ell-1} = \{(a, b) | b = N_\ell((0, s)$ so that $s_{1\nu} = a_{\nu-1} q_{W_1}(\xi_0, \xi_0)^{-1}$ for $\nu \geq 2$ and $s_{\ell_1, \ell_2} = 0$ for $\ell_1 \geq 2, \ell_2 \geq 2\}$ and $V_{\ell-1} = \{0, b) | b = N_\ell(Z, s)$ with $Z = (0 | Z_2 | \cdots | Z_\ell)$ and $s_{1t} = 0$ for all $t\}$. We note here that $V_{\ell-1}$ lies in $\mathbb{U}^{SO(W_1(R_i))}_{S\cap W_1(R_i)}$.

Again we note here that in the case W_1 is totally split and $j =$ Witt index of W_1 with i odd, then in the above **Remark** we replace z_ℓ everyplace by z_ℓ^*.

Remark 7. We can also express $U_{\{z_2, \cdots, z_\ell\}} \cong N_{\ell-1}((W_1, \cdots, W_{\ell-1})), s))$ with $W_i \in Sp\{z_1^-, y_1, \cdots\}$ and $s \in Skew_{\ell-1}(k)$ we emphasize that

$$N_{\ell-1} \subseteq SO((\xi_0)^\perp, q_{W_1})$$

relative to the splitting $(\xi_0)^\perp = Sp\{z_2, \cdots, z_\ell\} \oplus Sp\{z_1^-, y_1, \cdots\} \oplus Sp\{z_2^*, \cdots, z_\ell^*\}$. Indeed we have that $V_{\ell-1} = \{N_\ell(Z,s)|Z = (0|Z_2|\cdots|Z_\ell)$ with $Z_i \in Sp\{y_1,\cdots\}$ and $S \in Skew_\ell(k)$ with $S_{1t} = 0$ for all $t\} \cong N_{\ell-1}(((W,s))|W = (W_1|\cdots|W_{\ell-1})$ with $W_i \in Sp\{y_1,\cdots\}$ and $s \in Skew_{\ell-1}(k)\}$. The isomorphism is given by the map $N_\ell\left((0|Z_2,\cdots,Z_\ell), \begin{bmatrix} 0 & \cdots & 0 \\ & \ddots & \\ & & s \\ 0 & & \end{bmatrix}\right) \to N_{\ell-1}((Z_2|\cdots|Z_\ell)|s)$. On the other hand $T_{\ell-1} = \{(a,b)|a = (a_1,\cdots,a_{\ell-1})$ and $b = N_\ell((0, \begin{bmatrix} & 0 & a_1^* & a_{\ell-1}^* \\ & -a_1^* & & \\ & & 0 & \\ -a_{\ell-1}^* & & & \end{bmatrix})) \cong$ $N_{\ell-1}(((a_1z_1^-|a_2Z_1^-|a_2z_1^-|\cdots|a_{\ell-1}z_1^-),0)$ with $a_\nu^* = a_\nu q_{W_1}(\xi_0,\xi_0)^{-1}$.

Again we note the caveat that in the case W_1 is totally split and $j =$ Witt index of W_1 with i odd, then in the above **Remark** we replace z_ℓ everyplace by z_ℓ^*.

In fact, we note the same cavaet for the remaining part of this section (i.e., the space A in **Lemma 3.5** etc...).

Lemma 3.5. *Let $\alpha = 0$ and $\dim S - \beta = i$. Then the space*

$$ind_{\mathbb{P}_i^j(0,\beta)}^{\mathbb{P}_i}(w_S \circ \Gamma|_{\tau\otimes\sigma\otimes 1})_{(N^i\mathbb{U}_i,\Lambda_{\xi_0})}$$

fits as the middle term of an exact sequence of $SO((\xi_0)^\perp, q_{W_1})$ modules

$$0 \to A \to C \to B \to 0.$$

where (with unnormalized induction)

$$B = \bigoplus_\eta ind_{GL(\eta^{-1}(S\cap W_1(R_i)) \times SO(\mathcal{M},q_\mathcal{M}) \ltimes \tilde{\mathbb{U}}_{\eta^{-1}(S\cap W_1(R_i))}}^{SO((\xi_0)^\perp q_{W_1})}(|detg_1|^{-i} \otimes (\tau^{w_S})^{(i)}$$
$$(\eta g_1 \eta^{-1}) \otimes (\sigma^{w_S})(\eta g_2 \eta^{-1}) \otimes \mathbb{1}(u))$$

Here $(g_1,g_2)u) \in GL(\eta^{-1}(S\cap W_1(R_i)) \times SO(\mathcal{M},q_\mathcal{M}) \ltimes \tilde{U}_{\eta^{-1}(S\cap W_1(R_i))}$. and η varies over one or two elements (depending on whether $\dim S \cap W_1(R_i) \neq \frac{1}{2}\dim(\xi_0)^\perp$ or $= \frac{1}{2}\dim(\xi_0)^\perp$). Moreover

$$A = ind_{GL(\{z_2,\cdots,z_\ell\}) \times SO(W_1(S)\cap W_1(R_i)) \times T_{\ell-1}\cdot V_{\ell-1}}^{SO((\xi_0)^\perp q_{W_1})}(|detg_1|^{-i}$$
$$\otimes (\tau^{w_S})^{(i)}(g_1 a) \otimes (\sigma^{w_S})(g_2) \otimes 1(b))$$

(Here $(g_1, g_2, (a,b)) \in GL(\{z_2,\cdots,z_\ell\}) \times SO(W_1(S) \cap W_1(R_i)) \ltimes T_{\ell-1} \cdot V_{\ell-1})$).

In the case $\beta = 0$, then $A = 0$ and the space $B = (\tau^{w_S})^{(i)} \otimes (\sigma^{w_S})|_{SO(\mathcal{M},q_\mathcal{M})}$ (in this case $\mathcal{M} = (\xi_0)^\perp$ in $W_1(R_i)$)

Proof. The proof follows exactly the same method used in **Lemma 3** and **Lemma 4**. In this instance $SO((\xi_0)^\perp, q_{W_1})$ admits at most 3 orbits in

$$\mathbb{P}_{S\cap N_1(R_i)}^{SO(W_1(R_i))} \backslash SO(W_1(R_i))$$

We note here that $N^{(0,\beta)}_{(i,j)} \cap SO(W_1(S)) = \{e\}$.

Then the group N_β in this case has one of two forms

$$\eta^{-1} N^{(0,\beta)}_{(i,j)}(\eta) \text{ or } N^{(0,\beta)}_{(i,j)}$$

depending on which $SO((\xi_0)^\perp, w_{W_1})$ orbit is used (η as above).

Then the character ψ_Z on $\begin{bmatrix} (\eta)^{-1} & \\ & e \end{bmatrix} (N^{(0,\beta)}_{(i,j)} \cap GL(S)) \begin{bmatrix} \eta & \\ & e \end{bmatrix}$ becomes the following:

(i) $(\eta)^{-1} N_i((Z',0))\eta \rightsquigarrow \psi(q_{W_1}(Z'_1, \eta\xi_0))$

(ii) $N_i((Z',0)) \rightsquigarrow \psi(q_{W_1}(Z'_1, \xi_0))$.

We note that in this case the group $N^{(0,\beta)}_{(i,j)} \cap GL(S)$ has no skew symmetric part.

Here $Z' = (Z'_1 | \cdots | Z'_i)$ with $Z'_\nu \in S \cap W_1(R_i)$ (for all ν). In the first case ψ_Z becomes trivial. In the second case ψ_Z is no longer trivial. In fact the character $\psi_Z|_{N^{(0,\beta)}_{(i,j)}}$ is fixed by the subgroup $N^i \times GL(\{z_2, \cdots, z_\ell\})$. Thus the character $\psi_i \otimes \psi_Z$ can be defined on the subgroup $N^i \times (N^{(0,\beta)}_{(i,j)} \cap GL(S))$ in such a way that it extends the character ψ_Z on $(N^{(0,\beta)}_{(i,j)} \cap GL(S))$. Moreover we note that $N^{(0,\beta)}_{(i,j)} \cap GL(S) = N^{(0,\beta)}_{(i,j)}$. Then we know that $N^i \times GL(S \cap W_1(R_i)) \times ((N^{(0,\beta)}_{(i,j)})$ is a subgroup of $GL(S)$. In fact if $\Lambda_\ell : GL(S \cap W_1(R_i) \to GL(k^\ell)$ is defined relative to the *basis* $\{z_1, \cdots, z_\ell\}$ used above, it follows that $\Omega_i = N^i \times \Lambda_\ell^{-1}(\begin{bmatrix} 1 & & * \\ & 1 & \\ 0 & & 1 \end{bmatrix}) \times (N^{(0,\beta)}_{(i,j)})$ is a maximal unipotent subgroup of $GL(S)$. Moreover the character $\psi_i \otimes \psi_Z$ (defined on $N^i \times (N^{(0,\beta)}_{(i,j)})$ is *nontrivial* on each

simple root subgroup and trivial on each non simple root subgroup of $\Omega_i \cap (N^i \times (N^{(0,\beta)}_{(i,j)}))$. Thus taking the Jacquet functor of τ^{w_S} relative to $\psi_i \otimes \psi_Z$ on the group $N^i \times w_\ell^{-1}(N^{(0,\beta)}_{(i,j)})w_\ell$ we get the twisted i-th derivative of τ^{w_S}, i.e. $(\tau^{w_S})^{\langle i \rangle}$ (see above)/Q.E.D.

Appendix 1 to §3.

We let $H = S \cdot T$ be a semidirect product where T is the unipotent radical and S is reductive. Let Ψ_T be a character on T so that S stabilizes Ψ_T. Let R be a closed subgroup of H so that $R = R \cap S \cdot R \cap T$. Let σ be a smooth representation of R.

Then

Lemma. *As an S module we have the equivalence:*

$$ind^H_R(\sigma)_{(T,\psi_T)} \cong ind^S_{S\cap R}(|\ | \otimes \sigma_{(T\cap R, \Psi_T)})$$

(with $| \ | = |det_{R\cap R\backslash T}(\ )|$.

Proof. Let P be the projection map of σ to $\sigma_{(T\cap R,\Psi_T)}$. Then for $F \in \mathrm{ind}_R^H(\sigma)$ we consider $P(F(g)) = h_F(g)$. Then we form the integral

$$\sigma_F(s) = \int_{T\cap R\backslash T} h_F(ts)\overline{\Psi_T(t)}dt$$

Then $\sigma_F \in \mathrm{ind}_{S\cap R}^S(| \ | \otimes \sigma_{(T\cap R,\Psi_T)})$. The compact support property follows from the fact that F has compact support mod R. It s also clear that the map $F \to \sigma_F$ factors through $\mathrm{ind}_R^H(\sigma)_{(T,\Psi_T)}$. Thus we must show that this map is an S module isomorphism.

First we note that the module $\mathrm{ind}_{R\cap T}^T(\sigma|_{R\cap T})_{(T,\psi_T)} \cong \mathrm{ind}_{R\cap T}^T(\sigma_{R\cap T,\psi_T}))_{(T,\psi_T)}$. Then if $s \in S$ we also have the

$$\mathrm{ind}_{s^{-1}Rs\cap T}^T(\sigma|_{s^{-1}Rs\cap T})_{(T,\psi_T)} \cong \mathrm{ind}_{s^{-1}Rs\cap T}^T(\sigma_{(s^{-1}Rs\cap T,\psi_T)})_{(T,\psi_T)}\,.$$

Now we assume that $\sigma_F(s) = 0$ for all $s \in S$. In particular this implies that

$$\int_{T\cap s^{-1}Rs\backslash T} P(F(st'))\overline{\psi_T(t')}dt' \equiv 0.$$

Then following the usual stratification of the space $R\backslash H$ into T orbits and the conditions above imply that F has the form

$$\sum F_i * t_i - \Psi_T(t_i)F_i$$

for $F_i \subset \mathrm{ind}_R^H(\sigma)$ and $t_i \in T$. This proves the *injectivity* of the map of $\mathrm{ind}_R^H(\sigma)_{(T,\psi_T)}$ into $\mathrm{ind}_{S\cap R}^S(| \ | \otimes \sigma_{(R\cap T,\psi_T)})$ given by σ above.

Now we analyze more precisely the map $F \to \sigma_F$. We note first that the product $R \cdot T$ is in fact a subgroup of H and that the map

$$r \cdot t \rightsquigarrow \sigma(r)\overline{\Psi}_T(r)$$

defines a representation $\sigma \otimes \Psi_T$ of the group $R \cdot T$. Note that $R \cdot T$ stabilizes (or fixes) Ψ_T. Thus we can form the H module $\mathrm{ind}_{R\cdot T}^H(\sigma \otimes \overline{\Psi}_T)$. In any case there exists a surjective map of $S(H) \otimes \sigma$ into this module given by the following:

$$f \otimes v \rightsquigarrow \int_{R\cdot T} f(r \cdot tx)\sigma(r)(v)\overline{\Psi_T(t)}d\mu(r \cdot t)$$

with $v \in V_\sigma$ = the space of σ and $d\mu$ an invariant measure in $R \cdot T$. In particular when we restrict x to S in the formula above we obtain an element in $\mathrm{ind}_{S\cap(R\cdot T)}^S(\sigma \otimes \overline{\Psi}_T) = \mathrm{ind}_{S\cap R}^S(| \ | \otimes \sigma)$; in fact we obtain all of this induced module. Then we apply P to such a section and we obtain the original formula for σ_F and on the other hand P induces a surjection of $\mathrm{ind}_{S\cap R}^S(| \ | \otimes \sigma)$ to $\mathrm{ind}_{S\cap R}^S(| \ | \otimes \sigma_{(R\cap T,\Psi_T)})$. This proves surjectivity of the map $F \to \sigma_F$ /**Q.E.D.**

Appendix 2 to §3 The local Spherical-Whittaker Uniqueness Principle.

We let $N^i\mathbb{U}_i$ as given in this section. We let $M = SO((\xi_0)^\perp, q_{W_1})$. The character Λ_{ξ_0} is also given as above. We let σ be an admissible irreducible module of M. We form the induced module, $X_\sigma^{\xi_0}$

$$\mathrm{ind}_{(N^i\mathbb{U}_i)\times M}^{\mathbb{G}}(\Lambda_{\xi_0} \otimes \sigma)$$

We let Π be an irreducible admissible $\mathbb{G}$ module. Then the *SO-local Uniqueness Principle* in this case is the following.

$$(\dim Hom(X_\sigma^{\xi_0}, \Pi))(\dim Hom(X_{\sigma^\nu}^{\xi_0}, \Pi^\nu)) \leq 1$$

for it all σ and Π as given above.

Remark. If we replace M by $M' = O((\xi_0)^\perp, q_{W_1})$ and σ by an admissible irreducible representation of M' then the *O-local* Uniquesness Principle in this case is

$$\dim Hom_{\mathbb{G}'}(\mathrm{ind}_{N^i\mathbb{U}_i\times M'}^{\mathbb{G}'}(\Lambda_{\xi_0} \otimes \sigma'), \Pi') \leq 1$$

for all σ and all Π' (irred. admissible of $\mathbb{G}' = O(W, q_W)$ given above).

On the other hand if the pair (σ, Π) is spherical we note here the obvious statements of the generic Uniqueness Principle. In particular we note that by exactly the same argument in the **Proposition** in **Appendix** to §1 we deduce the *SO* local Uniqueness Principle implies

$$\dim Hom_{\mathbb{G}}(X_\sigma^{\xi_0}, \Pi) \leq 1$$

for all spherical Π and σ.

We note that in [R] we prove the O-local Uniqueness Principle and the SO local Uniqueness Principle.

§4. Global Theory

§4.1. Spherical Case.

We let $(W = V \oplus V_1, q_W = q_V \oplus_{V_i})$ be a space provided with a nondegenerate form q_W. Moreover $V \oplus V_1$ is a direct orthogonal sum where q_V and q_{V_1} are also nondegenerate.

Assume that there exists an embedding $i : V \to V_1$ so that q_{V_1} restricted to $i(V)$ is $-q_V$.

We let $O(W, q_W), O(V, q_V)$ and $O(V_1, q_{V_1})$ be the isometry groups of the forms q_W, q_V and q_{V_1} respectively.

Moreover there exists the following embeddings.

$$O(V, q_V) \times O(V_1, q_{V_1}) \hookrightarrow O(W, q_W)$$
$$O(V, q_V) \hookrightarrow O(V_1, q_{V_1})$$

Let X be the variety of isotropic subspaces of W of dimension equal to the Witt index of q_W. Choose $S \in X$ so that $S = \{(v, i(v)) | v \in V\} \oplus (S \cap V_2)$ (where $V_2 = i(V)^{\perp}$ in V_1). Then the stabilizer of S in $O(W, q_W)$ is a parabolic subgroup of the form

$$GL(S) \times O((S \cap V_2)^{\perp}/(S \cap V_2)) \ltimes \mathbb{U}_S$$

Here $(S \cap V_2)^{\perp}$ is taken in the space V_2 and $\mathbb{U}_S$ is the unipotent radical of $\mathbb{P}_S$.

Given an algebraic group R defined over k we let $R(\mathbb{A})$ be the adelic completion.

In particular we form $O(W, q_W)(\mathbb{A}), \mathbb{P}_S(\mathbb{A}), GL(S)(\mathbb{A})$, etc.

We then consider the family of quasi-characters defined on $GL(S)(\mathbb{A})$ via

$$g \rightsquigarrow \|det(g)\|_{\mathbb{A}}^{s}$$

As usual we form the normalized adelic induction space

$$I(s) = Ind_{\mathbb{P}_S(\mathbb{A})}^{O(W,q_W)(\mathbb{A})}((g, \xi, u) \rightsquigarrow \|detg\|^{s})$$

Then starting with an analytic section $f_s \in I(s)$ we form the Eisenstein series

$$E(f_s, x) = \sum_{\mathbb{P}_S(k)\backslash O(W,q_W)(k)} f_s(\gamma x)$$

We then consider $\mathbf{\Pi}$ and σ as cuspidal automorphic irreducible representations of $O(V, q_V)(\mathbb{A})$ and $O(V_1, q_{V_1})(\mathbb{A})$ respectively.

Let $f_{\mathbf{\Pi}} \in \mathbf{\Pi}$ and $f_\sigma \in \sigma$. We form the $O(V, q_V)(\mathbb{A})$ invariant pairing

$$\langle\langle f_{\mathbf{\Pi}} | f_\sigma \rangle\rangle = \int_{O(V,q_V)(k)\backslash O(V,q_V)(\mathbb{A})} f_{\mathbf{\Pi}}(x) f_\sigma(x) dx$$

Definition (I). : We say the $O(V,q_V)(\mathbb{A})$ invariant form $\langle\langle\ |\ \rangle\rangle$ is "factorizable" if the following is valid. That is, given irreducible representations $\mathbf{\Pi}$ and σ and $O(V,q_V)(\mathbb{A})(O(V_1,q_{V_1})(\mathbb{A})$ resp.) embeddings

$$i_{\mathbf{\Pi}}: \mathbf{\Pi} \hookrightarrow L^2_{cusp}(Q(V,q_V)(k)\backslash O(V,q_V)(\mathbb{A}))$$

and $i_\sigma: \sigma \to L^2_{cusp}(O(V_1,q_{V_1})$
$(k)\backslash O(V_1,q_{V_1})(\mathbb{A}))$ then

$$\langle\langle f_{\mathbf{\Pi}} | f_\sigma \rangle\rangle = c(\mathbf{\Pi},\sigma) \quad \prod_v \langle \xi_v | Z_v \rangle_v$$

Here $\langle\ |\ \rangle_v$ is an $O(V,q_V)(k_v)$ invariant pairing between $\mathbf{\Pi}_v$ and σ_v, which is the unique (up to scalar multiple) element of

$$Hom_{O(V,q_V)}(\mathbf{\Pi}_v, \sigma_v).$$

This is the general uniqueness Principle asserted in the **Appendix** *to §1.* In the case $\mathbf{\Pi}_v$ and σ_v are spherical we normalize $\langle\ |\ \rangle_v$ so that $\langle \xi_v | Z_v \rangle_v = 1$ where ξ_v and Z_v are the spherical vectors in $\mathbf{\Pi}_v$ and σ_v. Here $c(\mathbf{\Pi},\sigma)$ is a constant depending only on $\mathbf{\Pi}$ and σ. Moreover $f_{\mathbf{\Pi}} = i_{\mathbf{\Pi}}(\otimes \xi_v)$ and $f_\sigma = i_\sigma(\bigotimes_v Z_v)$.

Then we consider the family of Rankin Selberg integrals of the following form:

$$*(s) \equiv \int\limits_{O(V,q_V)\times O(V_1,q_{V_1})(k)\backslash O(V,q_V)\times O(V_1,q_{V_1})(\mathbb{A})} E(f_s,(x,y)) f_{\mathbf{\Pi}}(x) f_\sigma(y) dx dy.$$

The basic result concerning this family is the following.

Theorem 4.1. *If* $\dim V_1 - \dim V > 1$ *and* V_1/V *admits a split form, then* $*(s) = 0$.

If $\dim V_1 - \dim V = \begin{cases} 0 \\ 1 \end{cases}$ *then it is possible to choose the data* $f_s, f_{\mathbf{\Pi}}$ *and* f_σ *so that*

$$(*)(s) = \frac{L^S(\sigma, st, s+\frac{1}{2})}{d^S_{V_1}(s)} \gamma_\infty(s) \langle\langle f_{\mathbf{\Pi}} | f_\sigma \rangle\rangle$$

where

(i) $L^S(\sigma, st,\ )$ *is the restricted* L *function associated to* σ *and the standard representations of the* L*-group of* $SO(V_1,q_{V_1})$.

(ii) $d^S_{V_1}(s) = \prod_{k=0}^{k=[\frac{\dim V_1 - 2}{2}]} \zeta^S(2s + (\dim V_1 - 1) - 2k)$

(iii) S *is the set of places which include the Archimedean primes and those finite primes where the local component of* $\mathbf{\Pi}$ *and* σ *is not a spherical representation.*

(iv) γ_∞ *is a meromorphic function in* s. *More specifically the data* f_s, f_Π *and* f_σ *can be chosen so that* γ_∞ *is nonvanishing at a given point* $s_0 \in \mathbb{C}$ *(the data depends on* s_0*).*

Proof. The idea of the proof follows the lines of the arguments used in [PS-R]. Namely we consider the set of $O(V, q_V) \times O(V_1, q_{V_1})$ orbits in the space $\mathbb{P}_S(k)\backslash O(W, q_W)(k)$. In particular we note that in the case $\dim V_1 - \dim V > 1$ and V_1/V admits a split form each orbit has the property that the stabilizer of a general point in the orbit is negligible. That is, the stabilizer contains as a *normal subgroup* the unipotent radical of a parabolic subgroup of $O(V, q_V) \times O(V_1, q_{V_1})$. Thus it follows that when we do the Rankin unwinding of $*(s)$ we have that $*(s) = 0$ (for $\dim V_1 > \dim V + 1$ and V_1/V admitting a split form).

In any case the point of the above argument is that there is only one $O(V, q_V) \times O(V_1, q_{V_1})$ orbit that can possibly contribute to the integrals defined by $*(s)$. This is the case precisely when V_1/V is not a split form. Then the $O(V, q_V) \times O(V_1, q_{V_1})$ nonnegligible orbit is characterized as the set $\{S \in S | \dim(S \cap V_1) = 0\}$. This set is a single orbit under $O(V, q_V) \times O(V_1, q_{V_1})$. In fact if we choose as a representative element.

$$\{(v, i(v)) | v \in V\}.$$

Then the stabilizer of this subspace is

$$\{(h, h) | h \in O(V, q_V)\} \cong O(V, q_V)^\Delta \times O(V_2, q_{V_2}) \, .$$

Then unwinding the integral defining $(*)(s)$ we have

$$(*)(s) = \int\limits_{O(V_2,q_{V_2})(\mathbb{A})\backslash O(V_1,q_{V_1})(\mathbb{A})} f_s(1, h_2)$$
$$\left(\int\limits_{(O(V,q_V)(k)\backslash O(V,q_V)(\mathbb{A}))\times(O(V_2,q_{V_2})(k)\backslash O(V_2,q_{V_2})(\mathbb{A}))} f_\sigma(x\gamma h_2) f_\Pi(x) dx d\gamma \right) dh_2 \, .$$

Using the notation above, this integral equals

$$\int\limits_{O(V_2,q_{V_2})(\mathbb{A})\backslash O(V_1,q_{V_1})(\mathbb{A})} f_s(1, h_2) \left(\int\limits_{O(V_2,q_{V_2})(k)\backslash O(V_2,q_{V_2})(\mathbb{A})} \langle\langle f_\sigma * \gamma h_2 \mid f_\Pi \rangle\rangle d\gamma \right) dh_2 \, .$$

In the $\langle\langle \mid \rangle\rangle$ pairing above $*h_1$ denotes left translation applied to f_σ, i.e. $f_\sigma * h(x) = f_\sigma(xh)$.

We note that in the case $\dim V_1 = \dim V$ then the result of the **Theorem** follows from [ ] and [ ].

In the case $\dim V - \dim V_1 = 1$ we require more care.

In this case, $O(V_2, q_{V_2}) \cong Z_2$ (over any field k local or global). Thus, in fact, we deduce that

$$(*)(s) = \int\limits_{Z_2(k)\backslash O(V_1)(\mathbb{A})} f_s(1, h_2)\langle\langle f_\sigma * h_2 \mid f_{\Pi}\rangle\rangle dh_2$$
$$= \frac{1}{2} \int\limits_{O(V_1)(\mathbb{A})} f_s(1, h_2)\langle\langle f_\sigma * h_2 \mid f_{\Pi}\rangle\rangle dh_2 .$$

Here we use the telescoping of the above integral. First we note that the space $W = V \oplus V_1$ can be extended into a bigger space W_1 by adding a one dimensional space (ξ).

$$W_1 = (V \oplus V_1) \oplus (\xi) = (V \oplus (\xi)) \oplus V_1$$

We choose a nondegenerate form q_{W_1} on W_1 so that the embedding $i : V \to V_1$ extends to an embedding of $i_1 : V \oplus (\xi) \hookrightarrow V_1$ (in fact an isomorphism) so that q_{W_1} restricted to $i_1(V \oplus (\xi))$ is $-q_{V_1}$.

If X' = variety of maximal isotropic subspaces of W_1 (relative to q_{W_1}) then the group $O(W, q_W) \times Z_2$ (Z_2 = the orthogonal group of the one dimensional space (ξ)) operates transitively on X'. In particular we have that the space (S' = the isotropic subspace of X' given by $\{(z, i_1(z)) | z \in V \oplus (\xi)\}$)

$$\mathrm{ind}_{P_{S'}}^{O(W_1, q_{W_1})}((g, \xi, u) \to |det g|^s)$$

is isomorphic as a $O(W, q_W) \times Z_2$ module to

$$\mathrm{ind}_{P_S}^{O(W, q_W) \times Z_2}((g, \xi, u) \to |det g|^s)$$

(Here normalized induction in both cases). This implies that an arbitrary section $f_s \in \mathrm{ind}_{P_{S'}}^{O(W_1, q_{W_1})}((g, \xi, u) \to |det g|^s)$

$$f_s(1, h) = Res(f_s|_{O(W, q_W)})(1, h).$$

for $h \in O(V_1, q_{V_1})$. Here Res is the restriction map of f_s to the group $O(W, q_W)$. In any case we note by the results of **Theorem 1** of §1 that for v_0, the unramified vector in σ and $Re(s)$ sufficiently large:

$$\sigma(f_s(1, h))(v_0) = \sigma(Res(f_s|_{O(W, q_W)}(1, h))(v_0) = \left[\frac{L(\sigma, st, s + \frac{1}{2})}{d_{V_1}(s)}\right](v_0)$$

(here f_s is the unique unramified vector in the space $ind_{P_{S'}}^{O(W_1, q_{W_1})}((g, \xi, u) \rightsquigarrow |det g|^s)$ normalized so that $f_s(e) = 1$). We note $Res(f_s|_{O(W, q_W)})$ is the unique unramified vector in the space $\mathrm{ind}_{P_S}^{O(W, q_W)}((g, \xi, u) \to |det g|^s)$.

Thus from **Definition** (I) above we deduce that (for f_s factorizable)

$$(*)(s) = c(\mathbf{\Pi}, \sigma) \prod_v \int_{O(V_1, q_{V_1})(k_v)} f_s(1, (h_1)_v) \langle \xi_v | Z_v * (h_1)_v \rangle dh_{1v}$$

$$= c(\mathbf{\Pi}, \sigma) \left(\frac{L^S(\sigma, st, s + \frac{1}{2})}{d^S_{V_1}(s)} \right)$$

$$\prod_{v \in S} \int_{O(V_1, q_{V_1})(k_v)} f_s(1, (h_1)_v)) \langle \xi_v | Z_v * (h_1)_v \rangle d(h_{1v})$$

We note here the normalization of the form $< \xi_v | Z_v >$ above when the data is unramified (see **Definition** (I)).

Now using **Remark 3** of §1 we can replace the local integral

$$\int_{O(V_1, q_{V_1})(k_v)} f_s(1, h_v) \langle \xi_v | Z_v * h_v \rangle dh_v$$

by

$$\int_{O(V_1, q_{V_1})(k_v)} f_s(1, h_v) m_{\bar{\pi}}(h_v) dh_v$$

(for all the primes v in S) with $m_{\bar{\pi}}$ a matrix coefficient of π. But then following the same arguments used in [K-R] we can choose local data f_s so the the above integral equals

$$\begin{cases} \langle \xi_v | Z_v \rangle & \text{if } v \text{ is finite} \\ \gamma_\infty(s) \langle \xi_\infty | Z_\infty \rangle_\infty & \text{if } v \text{ is Archimedean} \end{cases}$$

Here γ_∞ satisfies the hypotheses stated in the **Theorem**. Thus we deduce that

$$(*)(s) \equiv \frac{L^S(\sigma, st, s + \frac{1}{2})}{d^S_{V_1}(s)} \gamma_\infty(s) \langle \langle f_{\mathbf{\Pi}} | f_\sigma \rangle \rangle$$

for the data chosen above./**Q.E.D.**

In the case $\dim V_1 - \dim V = 1$ the arguments used in **Theorem** 4.1 carry an additional consequence. Namely we have shown locally (for each prime v in K) that

$$\mathrm{ind}_{P_{S'}}^{O(W_1, q_{W_1})}((g, \xi, u) \rightsquigarrow |detg|^s)$$

is isomorphic as a $O(W, q_W) \times Z_2$ module to

$$\mathrm{ind}_{P_S}^{O(W, q_W)}((g, \xi, u) \rightsquigarrow |detg|^s) \otimes \mathrm{ind}_{\{e\}}^{Z_2}(1)$$

Here $\mathrm{ind}_{\{e\}}^{Z_2}(1)$ is the left regular representation Z_2 on Z_2. This isomorphism can be interpreted in global adelic terms . We let Z_2^∞ be the adelization of the group $Z_2(\cong$

orthogonal group of the one dimensional space (ξ) given in the proof of **Theorem** 4.1). Then Z_2^∞ is a compact group.

We have that

$$\mathrm{ind}_{P_{S'}(\mathbb{A})}^{O(W_1,q_{W_1})(\mathbb{A})}((g,\xi,u) \rightsquigarrow \|detg\|_{\mathbb{A}}^s)$$

is isomorphic as a $O(W,q_W)(\mathbb{A}) \times Z_2^\infty$ module to

$$\mathrm{ind}_{P_S(\mathbb{A})}^{O(W,q_E)(\mathbb{A})}((g,\xi,u) \to \|detg\|_{\mathbb{A}}^s) \otimes \mathrm{ind}_{\{e\}}^{Z_2^\infty}(1)$$

We let χ_T be any unitary character of Z_2^∞. We assume that a given section

$$f_s \in \mathrm{ind}_{P_{S'}(\mathbb{A})}^{O(W_1,q_{W_1})}((g,\xi,u) \rightsquigarrow \|detg\|_{\mathbb{A}}^s)$$

transforms on the left according to χ_T.

We let

$$\mathcal{E}(f_s,z) = \sum_{P_{S'}(k)\backslash O(W_1,q_{W_1})(k)} f_s(\gamma z)$$

be the Eisenstein series formed from f_s. Then we deduce that (for $(h_1,h_2) \in O(V,q_V) \times O(V_1,q_{V_1})(\mathbb{A})$)

$$\mathcal{E}(f_s,(h_1,h_2)) = (1+\chi_T(-1))E(Res(f_s|_{O(W,q_W)(\mathbb{A})}),(h_1,h_2))$$

Thus we have the identity:

$$\int_{O(V,q_V)\times O(V_1,q_{V_1}(k)\backslash O(V,q_V)\times O(V_1,q_{V_1})(\mathbb{A})} \mathcal{E}(f_s,(h_1,h_2))f_{\Pi}(h_1)f_\sigma(h_2)dh_1dh_2$$
$$= (1+\chi_T(-1)).$$
$$\int_{O(V,q_V)\times O(V_1,q_{V_1})(k)\backslash O(V,q_V)\times O(V_1,q_{V_1})(\mathbb{A})} E(Res(f_s|_{O(W,q_W)(\mathbb{A})}),(h_1,h_2))$$
$$\cdot f_{\Pi}(h_1)f_\sigma(h_2)dh_1dh_2$$

The left hand side of the identity then has another simple interpretation given below.

In the proof of **Theorem 1** there is a decomposition of the space

$$W_1 = V \oplus (\xi) \oplus V_1$$

We now let $\hat{V} = (\xi) \oplus V_1$ provided with form $q_{\hat{V}_2} = q_{W_1}$ restricted to $\hat{V}$. Thus $\hat{V} \cong V \oplus H_{(1,1)}$ where $H_{(1,1)}$ is a hyperbolic plane. We now consider the embedding of $O(V,q_V) \times O(\hat{V},q_{\hat{V}}) \to O(W_1,q_{W_1})$.

Then given $f_s \in \text{ind}_{P_{S'}(\mathbb{A})}^{O(W_1,q_{W_1})(\mathbb{A})}((g,\xi,u) \rightsquigarrow \|detg\|_{\mathbb{A}}^s)$ above we consider an integral of the form

$$\Lambda(f_s, f_{\mathbf{\Pi}})(h_2) = \int_{O(V,q_V)(\mathbb{A})} f_s(h_1,h_2) f_{\mathbf{\Pi}}(xh_1)dh_1$$

Here $h_2 \in O(\hat{V}, q_{\hat{V}})(\mathbb{A})$ and $f_{\mathbf{\Pi}} \in \mathbf{\Pi}$, an irreducible cuspidal representation of $O(V, q_V)(\mathbb{A})$.

We consider the parabolic subgroup P^ξ of $O(\hat{V}, q_{\hat{V}})$ which stabilizes the isotropic line through the vector $(\xi, i_1(\xi))$ in $\hat{V}$. Then $P^\xi \cong (GL_1 \times O(V, q_V)) \ltimes U$. Moreover using a similar argument as in §1 it is straightforward to verify that

$$\Lambda(f_s, f_{\mathbf{\Pi}}) \in \text{ind}_{P^\xi(\mathbb{A})}^{O(\hat{V},q_{\hat{V}})(\mathbb{A})}(|t|^s \otimes \mathbf{\Pi} \otimes 1)$$

(with normalized induction)

We note that if G_s is any section in the space $\text{ind}_{P^\xi(\mathbb{A})}^{O(\hat{V},q_{\hat{V}})(\mathbb{A})}(|t|^s \otimes \mathbf{\Pi} \otimes 1)$ then we again form the Eisenstein series associated to G_s, i.e.

$$\mathbb{E}(G_s, x) = \sum_{P^\xi(k)\backslash O(\hat{V},q_{\hat{V}})(k)} G_s(\gamma x)$$

Remark 1. We note that the section $\Lambda(f_s, f_{\mathbf{\Pi}})$ is not analytic in s. It is *meromorphic* in s.

Lemma 4.1. *Given $f_s \in ind_{P_{S'}(\mathbb{A})}^{O(W_1,q_{W_1})(\mathbb{A})}((g,\xi,u) \rightsquigarrow \|detg\|^s)$. The integral*

$$\int_{O(V,q_V)(k)\backslash O(V,q_V)(\mathbb{A})} \mathcal{E}(f_s,(h_1,h_2)) f_{\mathbf{\Pi}}(h_1)dh_1 = \mathbb{E}(\Lambda(f_s, f_{\mathbf{\Pi}}), h_2).$$

Proof. The proof is just the usual unwinding of $\mathcal{E}$ and the use of the fact there is only one nonnegligible orbit again./Q.E.D.

Thus we deduce the identity:

$$\int_{O(V_1,q_{V_1})(k)\backslash O(V_1,q_{V_1})(\mathbb{A})} \mathbb{E}(\Lambda(f_s, f_{\mathbf{\Pi}}), h_2) f_\sigma(h_2)dh_2 = (1+\chi_T(-1))$$

$$\int_{O(V,q_V)\times O(V_1,q_{V_1})(k)\backslash O(V,q_V)\times O(V_1,q_{V_1})(\mathbb{A})} \mathcal{E}(Res(f_s|_{O(W,q_W)(\mathbb{A})}),(h_1,h_2))$$

$$f_{\mathbf{\Pi}}(h_1) f_\sigma(h_2)dh_1dh_2$$

(where $f_s \in \mathrm{ind}_{P_{S'}(\mathbb{A})}^{O(W_1,q_{W_1})(\mathbb{A})}((g,\xi,u) \rightsquigarrow \|detg\|_{\mathbb{A}}^{s})$ is χ_T equivariant).

We then consider the family of Rankin Selberg integrals given by the following data.

$$(**)(s) = \int\limits_{O(V_1,q_{V_1})(k)\backslash O(V_1,q_{V_1})(\mathbb{A})} \mathbb{E}(G_s,h_2)f_\sigma(h_2)dh_2$$

where $f_\sigma \in \sigma$ and $G_s \in \mathrm{ind}_{P^\xi(\mathbb{A})}^{O(\hat{V},q_{\hat{V}})(\mathbb{A})}(|t|^s \otimes \mathbf{\Pi} \otimes 1)$. ($G_s$ is an analytic family here).

We then determine $**(s)$ from **Theorem 1**.

Theorem 4.2. *It is possible to choose data G_s and f_σ so that*

$$(**)(s) \equiv \frac{L^S(\sigma, st, s+\frac{1}{2})}{L^S(\mathbf{\Pi}, st, s+1)h^S(s)}\tilde{\gamma}_\infty(s)\langle\langle G_s|_{O(V,q_V)(\mathbb{A})}|f_\sigma\rangle\rangle$$

(1) The function G_s restricted to $O(V,q_V)(\mathbb{A})$ is independent of s.
(2) The set S of primes v here is chosen so that v is either Archimedean or a finite prime where $\mathbf{\Pi}_v$ or σ_v is not spherical (unramified).
(3) $\tilde{\gamma}_\infty()$ is a nonzero meromorphic function in s which is nonvanishing at $s=s_0$
(4) The function $h^S(s) = \begin{cases} \zeta^S(2s+1) & \textit{if } \dim V \textit{ odd} \\ 1 & \textit{otherwise} \end{cases}$

Proof. We know for §1 that each element G_s^v of $\mathrm{ind}_{P^\xi}^{(\hat{V},q_{\hat{V}})}(|t|^s \otimes \mathbf{\Pi}_v \otimes 1)$ can be obtained as a linear combination of an integral of the form

$$\Lambda(f_s,w)(g_1) = \int\limits_{O(V,q_V)} f_s(g,g_1)\mathbf{\Pi}_v(g)wdg$$

where $f_s \in \mathrm{ind}_{P_{S'}}^{O(W_1,q_{W_1})}((g,\xi,u) \rightsquigarrow |detg|^s)$. This is shown in §1 for v finite. If v is Archimedean then each K finite G_s^∞ has the same form (see **Remark 4** of §1).

On the other hand we know from **Theorem 1** of §1 that if f_s is the unique unramified vector in $\mathrm{ind}_{P_{S'}}^{O(W_1,q_{W_1})}((g,\xi,u) \rightsquigarrow |detg|_v^s)$ normalized by $f_s(e)=1$ and w_0 is the $K_{O(V,q_v)}$ fixed vector in $\mathbf{\Pi}_v$, then

$$\Lambda(f_s,w_0)(k) \equiv \frac{L_v(\mathbf{\Pi}_v, st, s+1)}{d_V(s+\frac{1}{2})}w_0.$$

(for $k \in K_{O(\hat{V},q_{\hat{V}})}$).

Thus it follows that we can choose $f_s \in \mathrm{ind}_{P_{S'}(\mathbb{A})}^{O(W_1,q_{W_1})(\mathbb{A})}((g,\xi,u) \rightsquigarrow \|detg\|_{\mathbb{A}}^{s})$ and $f_{\mathbf{\Pi}} \in \mathbf{\Pi}$ so that

$$\Lambda(f_s,f_{\mathbf{\Pi}}) = \rho_\infty(s)\frac{L^S(\mathbf{\Pi}, st, s+1)}{d_V^S(s+1/2)}G_s$$

Here we require that

$$G_s|_{O(V,q_V)(\mathbb{A})} \equiv f_{\mathbf{\Pi}} = i_{\mathbf{\Pi}}(\bigotimes_v \xi_v)$$

(i.e. $f_{\mathbf{\Pi}}$ is factorizable) with $\rho_\infty(s)$ meromorphic in s.

The local problem thus to show is that the data f_s can be chose in a compatible way so that

(1) $\Lambda(f_s, w) \equiv g_s$ where g_s is $Z_2 \cong 0(V_2)$ fixed!

(2) $f_s|_{e \times O(V_1,q_{V_1})}$ = characteristic function of a small compact open subgroup K' of $e_{O(V_1,q_{V_1})}$.

For this we note that the orbit of $O(V, q_V) \times O(\hat{V}, q_{\hat{V}})$ of the subspace S' in X' (the point $P_{S'}$ in $P_{S'} \backslash O(W_2, q_{W_2})$) contains the $O(V_1, q_{V_1}) \times O(V_1, q_{V_1})$ orbit of S'. Here we simply observe that $P_{S'}(O(V_1, q_{V_1}) \times O(V_1, q_{V_1})) = P_{S'}\{e_{O(V_1,q_{V_1})} \times O(V_1, q_{V_1})\} \subseteq P_{S'}\{O(V, q_V) \times O(\hat{V}, q_{\hat{V}})\}$ since $O(V_1, q_{V_1}) \subseteq O(\hat{V}, q_{\hat{V}})$. Then for v finite we can choose a function ϕ on the $1 \times O(V_1, q_{V_1})$ orbit through $P_{S'}$ having the form $1 \otimes \chi_{K'}(\chi_{K'}$ = characteristic function of K'). Here we require that K' is small enough so that $Z_2 \cong O(V_2) \cap K' = \{e\}$. We extend ϕ in the obvious way to a f_s in $\mathrm{ind}_{P_{S'}}^{)(W_1;q_{W_1})}((G, \xi, u) \to |detG|^s))$. Then $f_s|_{e \times O(V_1,q_{V_1})} = \chi_{K'}$ and f_S is $O(V_2) \cong Z_2$ fixed (hence g_s is $O(V_2) \cong Z_2$ fixed).

We can repeat the same argument as above in the Archimedean case to determine ϕ an arbitrary C^∞ function supported close to e in $O(V_1, q_{V_1})$. Again we can extend to an f_s which is Z_2 invariant as above and $f_s|_{exO(V_1,q_{V_2})} = \phi$. Then the associated local Archimedean integral

$$\int_{O(V_1,q_{V_1})} f_s(1, h)\langle \xi | Z * h \rangle dh$$

can be nonvanishing for all $s \in \mathbb{C}$ for an appropriate choice of ϕ. Here the term $\langle \xi | Z * h \rangle_\infty$ is bounded in h since it is a $\langle \ | \ \rangle_\infty$ is a local compact of a global integral which is absolutely convergent! In particular we know that for each $f_s \in I(s)_\infty$ = smooth vectors in $I(s)$ the integral

$$I(f_s) = \int_{O(V_1,q_{V_1})} f_s(1, h)\langle \xi | Z * h \rangle dh$$

can be defined for $Re(s)$ large and has a meromorphic continuation in s; the added condition is that for each $s_0 \in \mathbb{C}$ the functional

$$f_{s_0} \to \text{ max order term in Laurent expansion of } I(f_s) \text{ at } s = s_0$$

determines a continuous linear functional on $I(s_0)_\infty$. Thus it follows we can choose $f_s \in I(s)$ which K_G finite and Z_2 fixed so that the above integral is nonvanishing at $s = s_0$. (See **Remark 3** in §1).

Thus we apply the above construction to $\Lambda(f_s, f_{\mathbf{\Pi}})$ and deduce that

$$\int_{O(V_1,q_{V_1}(k)\backslash O(V_1,q_{V_1}(\mathbb{A})} \mathbb{E}(G_s, h_2)f_\sigma(h_2)dh_2 \equiv$$

$$\left(\frac{1}{\rho_\infty(s)}\right)\left(\frac{d_V^S(s+\frac{1}{2})}{L^S(\mathbf{\Pi}, st, s+1)}\right)(1+\chi_T(-1))$$

$$\int_{O(V,q_V)\times O(V_1,q_{V_1})(k)\backslash O(V,q_V)\times O(V_1,q_{V_1})(\mathbb{A})} \mathcal{E}(Res(f_s|_{O(W,q_W)(\mathbb{A})}), (h_1,h_2)) \cdot f_\pi(h_1)f_\sigma(h_2)dh_1dh_2$$

We assume $\chi_T = 1$.

Then we choose f_s, f_π and f_σ as above so that following **Theorem** 4.1 we have

$$(**)(s) = \gamma_\infty(s)\left(\frac{L^S(\sigma, st, s+\frac{1}{2})}{L^S(\mathbf{\Pi}, st, s+1)}\right)\left(\frac{d_V^S(s+\frac{1}{2})}{d_{V_1}^S(s)}\right)\langle\langle G_s|_{O(V,q_V)(\mathbb{A})}|f_\sigma\rangle\rangle$$

Thus we deduce that

$$(**)(s) = \gamma_\infty(s)\left(\frac{L^S(\sigma, st, s+\frac{1}{2})}{L^S(\mathbf{\Pi}, st, s+1)h^S(s)}\right)\langle\langle G_s|_{O(V,q_V)(\mathbb{A})}|f_\sigma\rangle\rangle.$$

/Q.E.D.

Remark 2. The Eisenstein series G_s used in **Theorem (2)** has the property that the constant term $(G_s)_{U_\xi}$ (along the unipotent radial U_ξ of P_ξ) is given by

$$G_s + M_w(G_s)$$

where $M_w(G_s)$ is the usual intertwining operator of $\text{ind}_{P_\xi(\mathbb{A})}^{O(V_2,q_{V_2})(\mathbb{A})}(|t|^{+s}\otimes\mathbf{\Pi}\otimes 1) \rightsquigarrow \text{ind}_{P_\xi(\mathbb{A})}^{O(V_2,q_{V_2})(\mathbb{A})}(|t|^{-s}\otimes\mathbf{\Pi}\otimes 1)$. The principal part of this intertwining operator is the ratio

$$\left\{\frac{L^S(\mathbf{\Pi}, st, s)h^S(s-1)}{L^S(\mathbf{\Pi}, st, s+1)h^S(s)}\right\}$$

We note that we expect that the function

$$s \rightsquigarrow L^S(\mathbf{\Pi}, st, s+1)h^S(s)\mathcal{E}(G_s, \ )$$

should admit at most a finite number of poles.

§4.2. Spherical-Whittaker Case (I).

The construction of Rankin Selberg integrals given in **Theorem 2** (by $(**)(s)$) admits a generalization to a mixed Spherical and Whittaker model (to be made precise below).

First we present $(**)(s)$ in another light. By unwinding the Eisenstein series $\mathcal{E}(G_s, \ )$ relative to $O(V_1, q_{V_1})(k)$ action we obtain two terms:

$$\mathcal{E}(G_s, x) = \sum_{O(V,q_V)(k)\backslash O(V_1,q_{V_1})(k)} G_x(\gamma x) + \sum_{\tilde{P}\backslash O(V_1,q_{V_1})(k)} G_s(\eta\gamma x)$$

We note that $O(V_1, q_{V_1})$ admits 2 orbits in the space $P_\xi \backslash O(V_2, q_{V_2})$. The geometric construction of these two orbits is given in §2. Here $\tilde{P}$ is a parabolic subgroup of $O(V, q_{V_1})$ (stabilizing an isotropic line) having the form $GL_1 \times O(V^\#, q_{V\#}) \times U'$ (U' the unipotent radical).

Then (assuming $f_\sigma \in \sigma$, a cuspidal representation)

$$\int_{O(V_1,q_{V_1})(k)\backslash O(V_1,q_{V_1}(\mathbb{A})} \mathcal{E}(G_s, h) f_\sigma(h) dh =$$

$$\int_{O(V,q_V)(\mathbb{A})\backslash O(V_1,q_{V_1})(\mathbb{A})} \left(\int_{O(V,q_V)(k)\backslash O(V,q_V)(\mathbb{A})} G_s(h_1 h) f_\sigma(h_1 h) dh_1 \right) dh$$

$$\equiv \int_{O(V,q_V)(\mathbb{A})\backslash O(V_1,q_{V_1})(\mathbb{A})} \langle\langle G_s(\ h) | f_\sigma * h \rangle\rangle dh$$

The second term in the unwinding of $\mathcal{E}(G_s, \ )$ vanishes after integration against f_σ since f_σ is a cusp form.

We note at this point that the Eisenstein series $\mathcal{E}(G_s, \ )$ in fact may be more general. In fact we can assume that $\mathbf{\Pi}$ is an irreducible automorphic representation of $O(V, q_V)(\mathbb{A})$ which embeds

$$i_{\mathbf{\Pi}} : \mathbf{\Pi} \to \mathcal{A}(O(V, q_V)(\mathbb{A}))$$

Then we can extend the statement of Theorem 2 to the more general case when $\mathbf{\Pi}$ is just a "slowly increasing" automorphic representation.

Theorem 4.2$'$.

The data G_s and $f_{\mathbf{\Pi}}$ can be chosen so that

$$(**)(s) = \tilde{\gamma}_\infty(s) \left(\frac{L^S(\sigma, st, s + \frac{1}{2})}{L^S(\pi, st, s+1) h^S(s)} \right) \langle\langle G_s|_{O(V,q_V)(\mathbb{A})} | f_\sigma \rangle\rangle.$$

where $\tilde{\gamma}_\infty(s)$ is a meromorphic function in s which is non vanishing at $s = s_0$.

Proof. We use first the fact again that $\langle\langle \mid \rangle\rangle$ is factorizable (Definition (I) above). Thus the integral above becomes

$$\prod_v \left(\int_{O(V,q_V)_v \backslash O(V_1,q_{V_1})_v} \langle G_s^v(*h_v) | \sigma(h_v)(Z_v)\rangle dh_v \right)$$

where $i_\sigma(\otimes Z_v) = f_\sigma$ and $\langle \mid \rangle$ are the local $O(V, q_V)_v$ invariant pairings between $\mathbf{\Pi}_v$ and σ_v. Using Remark to **Corollary** to **Lemma 2.2** we know that the local calculation for unramified data becomes

$$\begin{aligned}
&\int \langle G_s * L(\ h) | \sigma(h)(Z_v)\rangle dh_v = \\
&\frac{d_V(\ )}{L(\mathbf{\Pi}, st, s+1)} \int \langle \xi_v | \sigma(h)(Z_v)\rangle dh_v \\
&= \frac{d_V(\ )}{L(\mathbf{\Pi}, st, s+1)} \langle \xi_v | Z_v \rangle \left(\int \chi_{K_{O(V,q_V)}}(h) dh \right) \\
&= L(\sigma, st, s + \frac{1}{2}) \int \langle G_s(*\ h) | \sigma(h)(Z_v)\rangle dh
\end{aligned}$$

Then assuming that the volume dh on $O(V, q_V)\backslash O(V_1, q_{V_1})$ is normalized so that the term

$$\int \chi_{K_{O(V,q_V)}}(h) dh = 1$$

Here we have used the standard fact of adjointing the operator L across the form $\langle \mid \rangle$. Thus we have shown that for unramified data G_s, Z_v etc.

$$\begin{aligned}
&\int \langle G_s(\ h) | \sigma(h)(Z_v)\rangle dh_v = \\
&\left(\frac{L_v(\sigma, st, s + \frac{1}{2})}{L_v(\mathbf{\Pi}, st, s+1)} \right) \quad \langle \xi_v | Z_v \rangle
\end{aligned}$$

where ξ_v and Z_v are unramified vectors in $\mathbf{\Pi}_v$ and σ_v.

On the other hand given Z_v it is possible to choose G_s^v so that

$$\text{support } (G_s^v) \subseteq \text{ind}_{O(V,q_V)}^{O(V_1,q_{V_1})}(\mathbf{\Pi}_v)$$

and G_s^v is defined by the formula:

(i) G_s^v vanishes outside $O(V, q_V) \cdot K'$ (K' a small compact open in $O(V_1, q_{V_1})$ to be chosen below))

(ii) $G_s^v(m \cdot k) = c_v \mathbf{\Pi}_v(m)(\xi_v)$ where $\tilde{\xi}_v$ is fixed by $K' \cap O(V, q_V)$

Then for v finite using such a choice of G_s^v we deduce that

$$\int \langle G_s^v(\ h) | \sigma(h)(Z_v)\rangle dh = c_v \langle \tilde{\xi}_v | Z_v \rangle \left(\int \chi_{K'}(h) dh \right)$$

Now we choose c_v so that

$$c_v^{-1} = \int \chi_{K'}(h)dh$$

Then for v finite we deduce that with choice of G_s^v above (with the appropriate K', c_v etc.)

$$\int \langle G_s^v(\ h)|\sigma(h)(Z_v)\rangle dh = \langle \tilde{\xi}_v | Z_v \rangle$$

The proof in the Archimedean case follows exactly the same arguments as in **Theorem 4.2** above.

This completes the proof of the Theorem /**Q.E.D.**

We now extend the particular family Rankin-Selberg integrals to a very general setup.

§4.3. Spherical-Whittaker models (II).

We consider first the family of Spherical-Whittaker models of automorphic forms.

We suppose that the space (W, q_W) admits a decomposition

$$W = R_i \oplus W(R_i) \oplus R_i^*$$

where R_i and R_i^* are isotropic subspaces of dimension equal to i and are nonsingularly paired by q_W. Moreover $W(R_i)$ is the q_W orthogonal complement of $R_i \oplus R_i^*$.

Then we consider the variety X_i of q_W-isotropic subspaces of dimension i of W. The group $O(W, q_W)$ acts transitively on X_i. If $R_i \in X_i$ then the stabilizer $\mathbb{P}_{R_i} = \mathbb{P}_i$ of R_i is a parabolic subgroup of $O(W, q_W)$ having the form

$$(GL(R_i) \times O(W(R_i), q_{W(R_i)})) \ltimes \mathbb{U}_i$$

The unipotent radical $\mathbb{U}_i$ of $\mathbb{P}_i$ has the following form

$$W(R_i) \otimes k^i \oplus \Lambda^2(k^i)$$

The explicit group structure is given in §3. We note that the center $(\mathbb{U}_i) = \Lambda^2(k^i)$ except in the case when $i = 1$ or $R_i = \{0\}$.

In the case when $R_i \neq \{0\}$ then we can define a linear character of $\mathbb{U}_i \to k$

$$N_i[((\xi_1|\xi|\cdots|\xi_i), s)] \overset{\lambda_Z}{\rightsquigarrow} \sum q_W(\xi_i|Z_i)$$

where $Z = (Z_1|\cdots|Z_i)$ is a fixed element of $W(R_i) \otimes k^i$.

Then we construct a unitary character ψ_Z on $\mathbb{U}_i(\mathbb{A})$ via the map

$$N_i((\xi, s)) \overset{\psi_Z}{\rightsquigarrow} \psi(\lambda_Z((\xi, s))).$$

The set of local characters are defined in §3 and are compatible with this global character (in the obvious way).

Definition. We let $Z \in W(R_i) \otimes k^i$ be a rational element. Then the Z-th Fourier coefficient of an automorphic form f in $\mathcal{A}(O(W, q_W)(\mathbb{A}))$ is defined as the Fourier integral:

$$W_Z(f)(g) = \int\limits_{\mathbb{U}_i(k)\backslash\mathbb{U}_i(\mathbb{A})} f(N_i((\xi, s))g)\psi(\lambda_Z(\xi, s))d\xi ds$$

The Levi factor $GL(R_i) \times O(W(R_i), q_{W(R_i)})$ acts on the unipotent radical via the following formula:

$$Ad(g_1, g_2)N_i((\xi, s)) =$$
$$N_i((g_2\xi\Lambda_i(g_1)^{-1}, ({}^T\Lambda_i(g_1)s\Lambda_i(g_1)))$$

Here Λ_i is defined in §3.

We choose $Z = (\xi_0|0|\cdots|0)$ with $q_W(\xi_0) \neq 0$ in order to define the characters above.

Then the stabilizer of λ_Z in $GL(R_i) \times O(W(R_i), q_{W(R_i)})$ equals

$$\{(g_1, g_2)|\lambda_Z(Ad(g_1, g_2)N_i((\xi, s))) = \lambda_Z(N_i((\xi, s)) \text{ for all } (\xi, s)\} =$$
$$\{(g_1, g_2)|\Lambda_i(g_1) = \begin{bmatrix} \nu(g_2) & * & * \\ 0 & & \\ \vdots & & ** \\ 0 & & \end{bmatrix} \text{ with } g_2(\xi_0) = \nu(g_2)\xi_0\}$$

Thus the group $N^i \times O((\xi_0)^\perp, q_{(\xi_0)^\perp})$ stabilizes the character λ_Z where

$$\Lambda_j(N^i) = \left\{ \begin{bmatrix} 1 & & * \\ & \ddots & \\ 0 & & 1 \end{bmatrix} \in GL(i) \right\}.$$

We let ψ_i be the unitary character on the group $N^i(\mathbb{A})$ which is trivial on each nonsimple root subgroup of N^i and satisfies

$$\psi_j : N^i/[N^i, N^i] \cong \underbrace{\mathbb{A} \oplus \cdots \oplus \mathbb{A}}\ \underset{\psi\otimes\cdots\otimes\psi}{\longrightarrow} \mathbb{C}.$$

Then with the specific choice of $Z = (\xi_0|0|\cdots|0)$ it is possible to give a more general Fourier coefficient.

Definition. The $\psi_{i\otimes Z}$ Fourier coefficient of an automorphic form f in $\mathcal{A}(O(W, q_W))(\mathbb{A}))$ is given by the Fourier integral:

$$W_{i\otimes Z}(f)(g) = \left(\int\limits_{N^i(k)\backslash N^i(\mathbb{A})} W_Z(f)(n_i g)\psi_i(n_i)dn_i \right)$$

Now in order to establish a more shorthand notation we let

$$\begin{cases} G & = O(W, q_W) \\ M_i & = O(W(R_i), q_{W(R_i)}) \\ H_i & = O((\xi_0)^{\perp}, q_{(\xi_0)^{\perp}}) \end{cases}$$

We let ϕ be a *rapidly decreasing* automorphic form in the group $H_i(\mathbb{A})$.

Then we form the family of integrals

$$[f, \phi, i \otimes Z] = \int_{H_i(k)\backslash H_i(\mathbb{A})} W_{i\otimes Z}(f)(h)\phi(h)dh$$

Next associated to any parabolic $\mathbb{P}_j(\mathbb{A})$ of $G(\mathbb{A})$ we consider the automorphic representation of $\mathbb{P}_j(\mathbb{A})$ defined in the following way. Now

$$\mathbb{P}_j(\mathbb{A}) \cong GL(R_j)(\mathbb{A}) \times M_j(\mathbb{A}) \ltimes \mathbb{U}_j.$$

Then we consider the representation:

$$\tau \otimes \sigma \otimes 1$$

where τ is an *irreducible cuspidal representation* of $GL_j(\mathbb{A}) \cong GL(R_j)(\mathbb{A})$ and σ is an *irreducible automorphic representation* of $M_j(\mathbb{A})$ so that

$$i_\sigma : \sigma \hookrightarrow \mathcal{A}(M_j(\mathbb{A}))$$

Then we form the induced $G(\mathbb{A})$ module

$$\text{ind}_{\mathbb{P}_j(\mathbb{A})}^{GL(\mathbb{A})}((\tau \otimes \|detg\|_A^s) \otimes \sigma \otimes 1)$$

(normalized induction)

Again we let G_s be an analytic (flat) section associated to this induced module.

Then we form the Eisenstein series associated to G_s.

$$\mathcal{E}_j(G_s, \tau, \sigma)(x) = \sum_{\mathbb{P}_j(k)\backslash G(k)} G_s(\gamma x)$$

To emphasize that $\mathcal{E}_j(\ )$ is built off the j-th parabolic $\mathbb{P}_j$ we use the index j on $\mathcal{E}_j$.

We let $\mathbf{\Pi}$ be a cuspidal automorphic irreducible representation of $H_i(\mathbb{A})$.

Then we consider the family of integrals

$$[\mathcal{E}_j(G_s, \tau, \sigma), f_{\mathbf{\Pi}}, i \otimes Z]$$

The problem is to determine effectively which type of L function can be represented by this family.

We first compute the $i\otimes Z$ Fourier coefficient of the Eisenstein series $\mathcal{E}_j(G_s,\tau,\sigma)$.

First we recall the parametrization of the $\mathbb{P}_i$ orbits in the space $X_j = \mathbb{P}_j\backslash G$ given in §3.

That is, we write

$$\mathcal{E}(G_s,\tau,\sigma) = \sum_S \sum_{(\mathbb{P}_i\cap w_S^{-1}\mathbb{P}_j w_S)\backslash\mathbb{P}_i} G_s(w_S\gamma g)$$

Here S ranges over the representative j-dimensional subspaces of X_j given in §3. In particular S admits a decomposition

$$S = S\cap R_i \oplus S\cap W(R_i) \oplus S\cap R_i^*$$

The invariants (relative to the $\mathbb{P}_i$ action) are $\dim S\cap R_i = \alpha$ and $\dim S\cap (R_i\oplus W(R_i)) = \beta$.

Now we define

$$\mathcal{E}_j^{(\alpha,\beta)}(G_s,\tau,\sigma)(g) = \sum_{(\mathbb{P}_i\cap w_S^{-1}\mathbb{P}_j w_S)\backslash\mathbb{P}_i} G_s(w_S\gamma g)$$

We determine the $i\otimes Z$ Fourier coefficient of the partial series $\mathcal{E}_j^{(\alpha,\beta)}$.

This is done in a series of **Lemmas** that parallel the local theory of §3. In fact most of the arguments run in fact parallel to **Lemma 3** to **Lemma** 5 of §3.

Lemma 4.2. *Let* $\alpha > 0$. *Then*

$$W_{i\otimes Z}(\mathcal{E}_j^{(\alpha,\beta)}) \equiv 0.$$

Proof. We consider the orbits of $N^i\cdot\mathbb{U}_i$ on the space $\mathbb{P}_i^j(\alpha,\beta)\backslash\mathbb{P}_i$ (see **Lemma 3**). In fact the orbits are parametrized by pair (w,g) with $w\in W_{GL_i}$ (the Weyl groups of GL_i) and $g\in H_i$. In fact if $S_{(w,g)}$ is the stabilizer of (w,g) in $N^i\mathbb{U}_i$ then the arguments of **Lemma 3** determine the following points.

(i) The character λ_Z is non-trivial on the subgroup $S_{(w,g)}\cap\mathbb{U}_i$ except if the element w satisfies the property that $x_{1\ell}^* = 0$ where $\ell > i-\alpha$ and

$${}^T[\Lambda_i(w)] = (x_{\nu_1\nu_2}^*)$$

(this is valid for *all places* $v\in k$).

(ii) If w satisfies the property in (i) then $S_{(w,g)}\cap N^i$ contains a root subgroup of a simple positive root of $GL(R_i)$.

Then (i) and (ii) imply the vanishing of $W_{i\otimes Z}$. Indeed the basic integral defining $W_{i\otimes Z}(\mathcal{E}_j^{(\alpha,\beta)})$ can be written as follows:

$$\sum_{(w,g)} \int_{(N^i\cap S_{(w,g)})(k)\backslash N^i(\mathbb{A})} \left(\int_{(\mathbb{U}_i\cap S_{(w,g)})(k)\backslash \mathbb{U}_i(\mathbb{A})} G_s(w_S(w,g)u_i \cdot n_i x)\ \psi_Z(u_i)\psi_i(n_i)du_i \right) dn_i$$

Condition (i) implies that the inner integral vanishes for those (w,g) where w does not satisfy the data in (1). The reason is that the telescoped inner integral contains a term of the form

$$\int_{U_i\cap S_{(w,g)}(k)\backslash U_i\cap S_{(w,g)}(\mathbb{A})} \psi_Z(\tilde{u}_1)d\tilde{u}_1$$

This term vanishes when ψ_Z is nontrivial.

Similarly condition (ii) implies that the outer integral contributes to the vanishing. Indeed by the same argument as above after telescoping the outer integral (the inner telescoped integral equals 1) we obtain a term of the form

$$\int_{N^i\cap S_{(w,g)}(k)\backslash N^i\cap S_{(w,g)}(\mathbb{A})} \psi_i(n_i)dn_1$$

Again the term vanishes when ψ_i is non-trivial. This completes the proof of the **Lemma**. /Q.E.D.

Lemma 4.3. *Let* $\alpha = 0$, $0 < \dim S - \beta < i$, $\beta > 0$, *and* $\dim S \geq i$. *Then*

$$W_{i\otimes Z}(\mathcal{E}_j^{(o,\beta)}) \equiv 0.$$

Proof. The arguments of **Lemma 4** of §3 show that

(i) The character $\psi_i \otimes \psi_Z$ on $N^i\mathbb{U}_i$ is nontrivial on $S_{(w,g)}$ except in the case when $w = w_\ell$ (the "relative" long element of W_{GL_i} defined in §3) and g satisfies the condition that $g(\xi_0) \perp S \cap W(R_i)$.

We let $\eta \in M_i$ so that $\eta^{-1}(S \cap W(R_i)) \subseteq (\xi_0)^\perp$. Thus the set of $g \in M_i$ which satisfy $g(\xi_0) \perp S \cap w(R_i)$ becomes a set of the form

$$\mathbb{P}^{M_i}_{S\cap W(R_i)}\eta H_i$$

Here $\mathbb{P}^{M_i}_{S\cap W(R_i)}$ is the parabolic subgroup of M_i stabilizing the isotropic subspace $S \cap W(R_i)$.

Moreover we know from §3 (Proof of **Lemma 4**) that

$$\mathrm{Stabilizer}_{N^i H_i \mathbb{U}_i}(\mathbb{P}_i^j(0,\beta)(w_\ell,\eta))$$
$$= \left\{ g \in GL(R_i) | \Lambda_i(g) = \begin{bmatrix} \begin{bmatrix} 1 & & * \\ & \ddots & \\ 0 & & 1 \end{bmatrix} & 0 \\ 0 & \begin{bmatrix} 1 & & * \\ & \ddots & \\ 0 & & 1 \end{bmatrix} \end{bmatrix} \right\}$$
$$\times Q_{S \cap W(R_i)} \times (w_\ell \eta)^{-1} N_{(i,j)}^{(0,\beta)} (w_\ell \eta)$$

In the above decomposition of $\Lambda_i(g)$ the first block has size $i - (\dim S - \beta)$ and the second block has size $\dim S - \beta$. Also $Q_{S \cap R}$ is the parabolic subgroup of H_i which stabilizes he isotropic subspace $\eta^{-1}(S \cap W(R_i))$. The group $N_{(i,j)}^{(0,\beta)}$ is defined in §3 (structure of group is given by **Lemma 1** of §3). For notation we let T_i be the component of the stabilizer above in the N^i direction ; also we let $N_\beta = (w_\ell \eta)^{-1} N_{(i,j)}^{(0,\beta)} (w_\ell \eta)$.

Then condition (i) above implies that

$$W_{i \otimes Z}(\mathcal{E}_j^{(0,\beta)})(g) = \sum_{Q_{S \cap R_i} \backslash H_i} \int_{T_i(k) N_\beta(k) \backslash N^i \cdot \mathbb{U}_i(\mathbb{A})} G_s(w_S(w_\ell,\eta) u_i n_i h_i g) \psi_i \otimes \psi_Z(u_i n_i) du_i dn_i \,.$$

First we consider the basic integral

$$\int_{T_i(k) N_\beta(k) \backslash T_i(\mathbb{A}) N_\beta(\mathbb{A})} G_s(w_S(w_\ell,\eta)(un) u_i n_i) h_i g) \psi_i \otimes \psi_Z(un) du dn$$

Then we consider the telescoped inner integral

$$\int_{N_\beta(k) \backslash N_\beta(\mathbb{A})} G_s(w_S(w_\ell,\eta)(un) g) \psi_Z(u) du$$

Then from the structural decomposition of N_β given in the proof of **Lemma 4** of §3

$$N_\beta = (w_\ell \eta)^{-1} A_1 (w_\ell \eta) (w_\ell \eta)^{-1} A_2 (w_\ell \eta) (w_\ell \eta)^{-1} A_3 (w_\ell \eta)$$

where A_1, A_2, and A_3 are the three components in $GL(S), SO(W_1(S))$ and $\mathbb{U}_S$ respectively. Here A_3 is normal in $A_1 A_2 A_3$.

In any case when we compute the above integral the integration over the $(w_\ell \eta)^{-1} A_3 (w_\ell \eta)$ term (since $w_S A_3 w_S^{-1} \subseteq \mathbb{U}_j$) becomes

$$\left(\int_{(w_\ell \eta)^{-1} A_3 (w_\ell \eta)(k) \backslash (w_\ell \eta)^{-1} A_3 (w_\ell \eta)(\mathbb{A})} dz \right) \neq 0.$$

Here we use the fact that ψ_Z is trivial on $(w_\ell\eta)^{-1}A_3(w_\ell\eta)$. Then through telescoping, the integral

$$\int_{(w_\ell\eta)^{-1}A_1(w_\ell\eta)(k)\backslash(w_\ell\eta)(k)\backslash(w_\ell\eta)^{-1}A_1(w_\ell\eta)(\mathbb{A})} G_s(w_S(w_\ell,\eta)u_1g)\psi_Z(u_1)du$$

vanishes in the case where $\dim S-\beta<\dim S$ (since τ is a cusp form on $GL(R_j)(\mathbb{A})$) and ψ_Z is trivial on $(w_\ell\eta)^{-1}A_1(w_\ell\eta)$. This requires that $\beta\neq 0$! In this case the integration over $(w_\ell\eta)^{-1}A_1(w_\ell\eta)(\mathbb{A})$ represents the integration of the τ-cuspidal part of G_s over a unipotent radical of a parabolic subgroup of $GL_j(\mathbb{A})$ (see proof of **Lemma 4** in §3). Again the character ψ_Z is trivial on $(w_\ell\eta)^{-1}A_1(w_\ell\eta)$ /**Q.E.D.**

Thus we are left with case when $\alpha=0$, $\dim S-\beta=i$. (**Lemma 5** of §3). First we note that in this case (following **Lemma 5** of §3).

$$W_{i\otimes Z}(\mathcal{E}_j^{(0,\beta)})(h)=(I)+(II)$$

where

$$(I)=\sum_{Q_{S\cap W(R_i)}\backslash H_i}\ \int_{T_i(k)N_\beta(k)\backslash N^i\mathbb{U}^i(\mathbb{A})} G_s(w_S\eta un\gamma h)\psi_i\otimes\psi_Z(un)dndu$$

and

$$(II)=\sum_{Q^*_{S\cap W(R_i)}\backslash H_i}\ \int_{\tilde{T}_i(k)\tilde{N}_\beta(k)\backslash N^i\mathbb{U}^i(\mathbb{A})} G_s(w_Sun\gamma h)\psi_i\otimes\psi_Z(un)dndu\,.$$

Corollary to Lemma 4.3. *Let* $\alpha=0$ *and* $\dim S-\beta=i$. *Then*

$$(I)=0\,.$$

And if $\alpha=0$, $\dim S-\beta=0$, *and* $\dim S\geq i$, *then* $[\mathcal{E}_j(G_s,\tau,\sigma),\phi,i\otimes Z]=0$ *for all* ϕ *cuspidal on* $H_i(k)\backslash H_i(\mathbb{A})$.

Proof. The first part follows the proof of **Lemma 4.3**. The second part follows (by the usual unwinding) from the fact that $Q_{S\cap W_1(R_i)}$ contains a unipotent radical of H_i and that ϕ is cuspidal. Q.E.D.

The summation defining $W_{i\otimes Z}(\mathcal{E}_j^{(0,\beta)})$ consists of 2 orbits. The first one ((I) above) is similar to a sum as given in **Lemma 2** above. Term (II) is defined by the H_i orbit through $S\cap W(R_i)$. In this case $Q_{S\cap W(R_i)}$ is the stabilizer of $S\cap W(R_i)$ in H_i. The structure of $Q^*_{S\cap W(R_i)}$ is given in §3. We also note here that $T_i=\tilde{T}_i=N^i$ and $N_\beta=(\eta)^{-1}N^{(0,\beta)}_{(i,j)}(\eta)$ and $\tilde{N}_\beta=N^{(0,\beta)}_{(i,j)}$. Using §3 (and the hypothesis that

$\dim S - \beta = i$ and $\alpha = 0$) then $\tilde{N}_\beta = \{N_i(((Z_1|Z_2|\cdots Z_i),0)|Z_\nu \in S \cap W(R_i)\}$. Thus we deduce that

$$\int_{N^i(k)\tilde{N}_\beta(k)\backslash N^i\mathbb{U}_i(\mathbb{A})} G_s(w_S unh)\psi_i \otimes \psi_Z(un)dudn$$

$$\equiv \int_{\tilde{N}_\beta(\mathbb{A})\backslash\mathbb{U}_i(\mathbb{A})} \left(\int_{(N^i\cdot\tilde{N}_\beta)(k)\backslash N^i\cdot\tilde{N}_\beta(\mathbb{A})} G_s(w_S u_1 n_1 u)\psi_i \otimes \psi_Z(u_1 n_1)du_1 dn_1 \right) \psi_Z(u)du$$

We recall from §3 that the group $N^i \cdot \tilde{N}_\beta \subseteq GL(S)$. Recall here that $\ell = j - i$. In the process of choosing the subspaces R_i and R_j etc. we can assume that $R_j = Span\{v_1, \cdots, v_i, z_1, \cdots, z_\ell\}$ and $R_j^* = Span\{v_1^*, \cdots, v_\ell^*, z_1^*, \cdots, z_\ell^*\}$. Then $S =$
$Span\{v_1^*, \cdots v_i^*, z_1, \cdots z_\ell\}$ (except in the case where W_1 is totally split and $j =$ Witt index of W_1 with i odd). The element w_S then can be chosen so that $w_S(v_t) = v_t^*, w_S(v_t^*) = v_t(t = 1, \cdots i), w_S(z_i) = z_i, w_S(z_i^*) = z_i^*$ and $w_S|_{W_1(R_j)}$ is given explicitly by formula in §5 (p.). In any case $w_S(S) = R_j$ (our choice of w_S is such that w_S has determinant 1 and $w_S^2 = 1$).

In the case where W_1 is totally split and $j =$ Witt index of W_1 with i odd, then $S = \text{Span}\{v_1^*, \ldots, v_i^*, z_1, \ldots, z_{\ell-1}, z_\ell^*\}$. Then w_S is also chosen as in §5 where we have the formulae: $w_S(v_t) = v_t^*$, $w_S(v_t^*) = v_t$ $(t \leq i)$ $w_S(z_\nu) = z_\nu$, $w_S(z_\nu^*) = z_\nu^*$ $(\nu < i - j)$ and $w_S(z_{j-i}) = z_{j-i}^*$, $w_S(z_{j-i}^*) = z_{j-i}$.

This implies in particular that the group

$$(w_S)(N^i \cdot \tilde{N}_\beta)(w_S)^{-1} \subseteq GL(R_j).$$

In fact using the explicit isomorphism $\Lambda_j : R_j \to k^j$ constructed by means of the basis $\{v_1, \cdots v_i, z_1, \cdots, z_\ell\}$, then

$$\Lambda_j[w_S N^i w_S] = \left\{ \begin{bmatrix} 1 & 0 & & \\ & \ddots & & 0 \\ * & & 1 & \\ & 0 & & I_{j-1} \end{bmatrix} \Bigg| * \text{ arbitrary} \right\},$$

$$\Lambda_j[w_S \tilde{N}_\beta w_S] = \left\{ \begin{bmatrix} I_i & * \\ 0 & I_{j-i} \end{bmatrix} \Bigg| * \text{ arbitrary} \right\},$$

$$\Lambda_j[w_S N_{\{z_1,\cdots,z_\ell\}} w_S] = \left\{ \begin{bmatrix} I_i & 0 & \\ & 1 & * \\ 0 & & \ddots \\ & 0 & 1 \end{bmatrix} \Bigg| * \text{ arbitrary} \right\}$$

We note here that $w_S N^i \tilde{N}_\beta N_{\{z_1,\cdots,z_\ell\}} w_S = N^j_{+,-}$ given in §5(p.). We note that $N^j_{+,-}$ is the maximal unipotent radical of the Borel subgroup $B_{+,-}$ of $GL(R_j)$ stabilizing the flag $\{v_i, \cdots v_1, z_1, \cdots, z_\ell\} \supseteq \cdots \supseteq \{z_\ell\}$.

We note in the case where W_1 is totally split and $j =$ Witt index of W_1 with i odd, then $N_{\{z_1,\ldots,z_\ell\}}$ is replaced by $N_{\{z_1,z_2,\ldots,z_{\ell-1},z_\ell^*\}}$.

In any case the inner integral of G_s becomes

$$\int\limits_{N^i N_\beta(k)\backslash N^i N_\beta(\mathbb{A})} G_s[w_S u_1 n_1 uh](\psi_i \otimes \psi_Z)(u_1 n_1) du_1 dn_1$$

$$= \int_{N^i \times N_\beta(k)\backslash N^i \times N_\beta(\mathbb{A})} G_s \left[\Lambda_j^{-1} \left(\begin{bmatrix} 1 & & 0 & \\ & \ddots & & y_{ij} \\ x_{ij} & & 1 & \\ & 0 & & I_{j-i} \end{bmatrix} \right) w_S uh \right]$$

$$\psi(x_{21} + \cdots + x_{i,i-1} + y_{1i}) dx_{ij} \otimes dy_{ij}$$

Thus since $G_s \in \mathrm{ind}_{P_j(\mathbb{A})}^{\mathbb{G}(\mathbb{A})}(\tau \otimes \|detg\|_A^S \otimes \sigma \otimes 1)$ it follows that the integral above represents integrating the τ cuspidal component of G_s (defined in terms of the induction data) relative to a degenerate Whittaker Fourier coefficient. In fact relative to the unipotent group $N^j_{+,-}$ we consider the subgroup $X_{r,+,-} =$ the subgroup of $N^j_{+,-}$ which is the semi direct product of the unique unipotent radical containing the root group $X_{\epsilon_1-\epsilon_{i+1}}$, and the group generated by the simple root group $X_{-\alpha_1}, \cdots, X_{-\alpha_{i-1}}$ (if $r = i+1$) and by $X_{-\alpha_1}, \cdots, X_{-\alpha_{i-1}}, X_{\alpha_{i+2}}, \cdots, X_{\alpha_r}$ (if $r > i+1$). In the case $r = i$ w let $X_{r,+,-} =$ the group generated by the simple root subgroups $X_{-\alpha_1}, \cdots, X_{-\alpha_{i-1}}$. Here $N^j_{+,-}$ is the unipotent radical generated by the root subgroups associated to the set $\{-\alpha_1, \cdots, - - \alpha_{i-1}, \epsilon_1 - \epsilon_{i+1}, \alpha_{i+1}, \cdots, \alpha_j\}$ (See §5). In any case we define

$$W^{(r)}(G_s)(x) = \int_{X_{r,+,-}(k)\backslash X_{r,+,-}(\mathbb{A})} G_s[x w_S z] \psi_i(x) dx$$

Then we note that the integral we have constructed above has the form

$$W^{(i+1)}(G_s)[uh].$$

Thus we have that

$$\int\limits_{N^i(k)N_\beta(k)\backslash N^i \mathbb{U}_i(\mathbb{A})} G_s[w_S unh](\psi_i \otimes \psi_2)(un) du dn$$

$$= \int\limits_{N_\beta(\mathbb{A})\backslash \mathbb{U}_i(\mathbb{A})} W^{(i+1)}(G_s)[w_S uh] \psi_Z(u) du$$

Finally we can determine the case when $\alpha = 0$ and $\dim S - \beta = i$.

Lemma 4.4. *Let ϕ be a cusp form on $H_i(k)\backslash H_i(\mathbb{A})$*

(a) First let $\dim S > i$ *and* $\dim S - \beta = j - \beta = i$ *and* $\alpha = 0$. *Then*

$$\int\limits_{H_i(k)\backslash H_i(\mathbb{A})} (I)(h)\phi(h)dh \equiv 0$$

and

$$\int\limits_{H_i(k)\backslash H_i(\mathbb{A})} II(h)\phi(h)dh \equiv$$

$$\int\limits_{Q^*_{S\cap W(R_i)}(\mathbb{A})\backslash H_i(\mathbb{A})} \Bigg(\int\limits_{Q^*_{S\cap W(R_i)}(k)\backslash Q^*_{S\cap W(R_i)}(\mathbb{A})} \phi(h_1 h)$$
$$\left(\int_{\tilde{N}_\beta(\mathbb{A})\backslash U_i(\mathbb{A})} W^{(i+1)}(G_s)(w_S u h_1 h)\psi_Z(u)du \right) dh_1 \Bigg) dh$$

(b) Let $\dim S = i$ *(so that* $\beta = 0$ *and* $\alpha = 0$*) Then*

$$\int\limits_{H_i(k)\backslash H_i(\mathbb{A})} (I)(h)\phi(h)dh \equiv$$

$$\int\limits_{H_i(k)\backslash H_i(\mathbb{A})} \phi(h)\left(\int_{U_i(\mathbb{A})} W^{(i)}(G_s)(w_S u h)\psi_Z(u)du \right) dh$$

Proof. The vanishing of the integral involving (I) in case (a) is similar to the proof in **Corollary** to **Lemma 2**.

The integral involving (II) is deduced from the formula for (II) above. Also we note that the term

$$h \rightsquigarrow \int\limits_{\tilde{N}_\beta(\mathbb{A})\backslash \mathbb{U}_i(\mathbb{A})} W^{(i+1)}(w_S u h)\psi_z(u)du$$

is invariant under $Q^*_{S\cap W(R_i)}(k)$. Thus we deduce the formula in the **Lemma** /**Q.E.D.**

Remark (A). In the case $j = \dim S = i+1$ ($\beta = 1$ above) then $Q^*_{S\cap w(R_i)} = O((z_1^-)^\perp, q_{W_1})$. Then we note that the integral

$$\int\limits_{O((z_1^-)^\perp, q_{W_1})(k)\backslash O((z_1^-)^\perp, q_{W_1})(\mathbb{A})} \phi(h_1 h)$$
$$\left(\int\limits_{\tilde{N}_\beta(\mathbb{A})\backslash \mathbb{U}_i(\mathbb{A})} W^{(i+1)}(G_s)(w_S u h_1 h)\psi_Z(u)du \right) dh_1$$

defines an $O((z_1^-)^\perp, q_{W_1})(\mathbb{A})$ invariant pairing between $L^2_{cusp}(H_i(k)\backslash H_i(\mathbb{A}))$ and the space of functions of the form

$$h_1 \rightsquigarrow \int_{\tilde{N}_\beta(\mathbb{A})\backslash\mathbb{U}_i(\mathbb{A})} W^{i+1}(G_s)(w_S u h_1 x)\psi_Z(u)du$$

The latter space of automorphic functions consists of basically linear combinations of functions of the form

$$m \rightsquigarrow f_\sigma(m) \quad (h_1 \in M_i(\mathbb{A}), m \in M_i(\mathbb{A}))$$

where $f_\sigma \in \sigma^{w_S}$, a space of "slowly increasing" automorphic forms transforming according to σ^{w_S} on $M_i(k)\backslash M_i(\mathbb{A})$. In particular if $\phi \in \mathbf{\Pi}$, a cuspidal irreducible representation of $H_i(\mathbb{A})$, then the integral above defines an $O((z_1^-)^\perp, q_{W_1})(\mathbb{A})$ invariant pairing between $\mathbf{\Pi}$ and σ^{w_S}. We note by the hypothesis stated above (in (1) above) such a pairing is basically "factorizable".

Remark (B). If $\dim S = j = i+\beta$ where $\beta > 1$ then $Q^*_{S\cap W(R_i)} = GL(\{z_2, \cdots z_\ell\})\times O((z_1^-)^\perp, q_{W_1})\times U_{\{z_2,\cdots,z_\ell\}}$. In fact we recall from §3 (**Remark 6**) $O((z_1^-)^\perp, q_{W_1}) = O(W_1(S))$ and the group $U_{\{z_2,\cdots,z_\ell\}}$ admits a decomposition in the group $\mathbb{P}^{O(W(R_i))}_{S\cap W(R_i)}$ in the following way. We use the notation that $\mathbb{U}^{O(W,(R_i))}_{S\cap W(R_i)}$ has the form

$$N_\ell(((Z_1|\cdots|Z_\ell), s))$$

where $Z_i \in Span\{y_t, \cdots\}$ and $s \in Skew_\ell(k)$(§3). Moreover (from §3) let $\underset{\sim}{P}_{\{z_1,\cdots,z_\ell\}}$ be the mirabolic subgroup of $GL(\{z_1, \cdots, z_\ell\})$ which stabilizes z_1^*; then $\underset{\sim}{P}_{\{z_1,\cdots,z_\ell\}} =$

$GL(\{z_2, \cdots, z_\ell\}) \ltimes k^{\ell-1}$. Moreover $U_{\{z_2,\cdots z_\ell\}} \cong$

$$\{(a,b) \in k^{\ell-1} \times N_\ell | Z_1 = 0, s_{1\nu} = a_{\nu-1}q_{W_1}(\xi_0,\xi_0)^{-1}, \nu \geq 2\}$$

$\cong \{(a,b)|b = N_\ell((0,s))$ so that $s_{1\nu} = a_{\nu-1}q_{W_1}(\xi_0,\xi_0)^{-1}$ $(\nu \geq 2)$ and $s_{\nu_1,\nu_2} = 0$ for $\nu_1, \nu_2 \geq 2\} \cdot \{(0,b)|b = N_\ell(Z,s))$ with $Z = (0|z_2|\cdots|z_\ell)$ and $s_{1t} = 0, t \geq 2\} \cdot (\cong T_{\ell-1} \cdot V_{\ell-1}$ in Remark to **Theorem 3**). We note the caveat here that in the case W_1 is totally split and $j =$ Witt index of W_1 with i odd, then we replace z_ℓ by z_ℓ^* in the appropriate places (see §3 **Remark 6, 7**). We also do this in the appropriate places below.

On the other hand we note that the action of an element $(s^*_{1\nu} = s_{1\nu}q_{W_1}(\xi_0, xi_0)^{-1})$

$$((s_{12}, \cdots s_{1\ell}), N_\ell((0, \begin{bmatrix} 0 & s^*_{12}\cdots s^*_{1\ell} \\ -s^*_{12} & \\ & 0 \\ -s^*_{1\ell} & \end{bmatrix})))$$

(in the first group in the decomposition above) on the vector $z_t^*(t \geq 2)$ is given by the formulae $z_t^* \rightsquigarrow z_t^* + s^*_{1t}z_1^- + L_t$ (where $L_t \in Span\{z_2, \cdots, z_\ell\}$.

We also have that the group $U_{\{z_2,\cdots,z_\ell\}}$ (when considered as a subgroup of $SO((\xi_0)^\perp, q_{W_1}))$ admits also a decomposition of the form $N_{\ell-1}(((Y_1|\cdots|Y_{\ell-1}), s))$ where $Y_\nu \in Span\{y_t, \cdots\}$ and $s \in Skew_{\ell-1}(k)$. First note that $N_{\ell-1}$ here is not necessarily compatible with the N_ℓ (in $\mathbb{U}^{O(W(R_i))}_{S\cap W(R_i)}$ above). In any case the above comments show that the Y_ν component of the element

$$((s_{12}, \cdots, s_{1\ell}), N_\ell(0, \begin{bmatrix} 0 & s^*_{12} \cdots s^*_{1\ell} \\ -s_{12} & \\ & 0 \\ -s^*_{1\ell} & \end{bmatrix})))$$

equals $s^*_{1\nu} z_1^-$! This implies that the character of $U_{\{z_2,\cdots,z_\ell\}}$ of the form

$$N_{\ell-1}(((Y_1, \cdots, Y_{\ell-1}), s)) \underset{\psi_{Z_\beta}}{\rightsquigarrow} \psi(q_{W_1}(Y_1, z_1^-))$$

is trivial on the group $V_{\ell-1} =$

$$\{(0, b) | b = N_\ell((Z, s)) \text{ with } Z = (0|z_2|\cdots|z_\ell) \text{ and } s_{1t} = 0, t \geq 2\}$$

and on the group

$$T_{\ell-1} = \{(s_{12}, \cdots, s_{1\ell}), N_\ell((0, \begin{bmatrix} 0 & s^*_{12} \cdots s^*_{1\ell} \\ -s^*_{12} & \\ & 0 \\ -s^*_{1\ell} & \end{bmatrix}))\}$$

the character equals

$$(s_{12}, \cdots, s_{1\ell}) \rightsquigarrow \psi(-s_{12})$$

Moreover we note the group $\underset{\sim}{P}_{\{z_1,\cdots,z_\ell\}}$ stabilizes the character on the group $N^i \times \tilde{N}_\beta$ (used to define $W^{(i+1)}$ above). This implies that

$$W^{(i+1)}(G_s)(h) = \sum_{N_{\{z_1,\cdots,z_\ell\}} \backslash \underset{\sim}{P}_{\{z_1,\cdots,z_\ell\}}} W^{(j)}(G_s)(\gamma h)$$

(here $N_{\{z_1,\dots,z_\ell\}}$ is a maximal unipotent radical of $\underset{\sim}{P}_{\{z_1,\cdots,z_\ell\}}$). Since $\underset{\sim}{P}_{\{z_1,\cdots,z_\ell\}} = GL(\{z_2, \cdots, z_\ell\}) \ltimes k^{\ell-1}$ we may replace the formula for $W^{(i+1)}$ by summation over

$$\begin{aligned} &N_{\{z_2,\cdots,z_\ell\}} \cdot T_{\ell-1} \backslash GL(\{z_2, \cdots, z_\ell\}) \cdot T_{\ell-1} \\ &\quad = (N_{\{z_2,\cdots,z_\ell\}} U_{\{z_2,\cdots,z_\ell\}} \backslash GL(\{z_2, \dots, z_\ell\}) U_{\{z_2,\cdots,z_\ell\}}) \end{aligned}$$

(we use here that G_s is invariant on the left by $\mathbb{U}^{O(W(R_i))}_{S\cap W(R_i)}(\mathbb{A})$ and $N_{\{z_1,\cdots,z_\ell\}} \cap GL(\{z_2, \cdots, z_\ell)) = N_{\{z_2,\cdots,z_\ell\}}(\mathbb{A}))$.

Then we define

$$Z_\nu(G_s)(h) = \int_{\tilde{N}_\beta(\mathbb{A}) \backslash \mathbb{U}_i(\mathbb{A})} W^{(\nu)}(G_s)(w_S u \cdot h) \psi_Z(u) du.$$

(recall $W^{(\nu)}$ defined above prior to **Lemma 3**)

$$Z_j(G_s)(h) = \sum_{N_{\{z_2,\cdots,z_\ell\}}\cdot T_{\ell--1}\backslash GL(\{z_2,\cdots,z_\ell\})\cdot T_{\ell-1}} Z_{i+1}(G_s)(\gamma h)$$

Then the integral

$$\int_{Q^*_{S\cap W(R_i)}(k)\backslash Q^*_{S\cap W(R_i)}(\mathbb{A})} \phi(h_1h)Z_{i+1}(G_s)(h_1h)dh \equiv$$

$$\int_{(N_{\{z_2,\cdots z_\ell\}}\cdot O((z_1^-)^\perp,q_W)\cdot U_{\{z_2,\cdots,z_\ell\}}(k)\backslash(GL(\{z_2,\cdots,z_\ell\})O((z_1^-)^\perp,q_{W_1})U_{\{z_2,\cdots,z_\ell\}}(\mathbb{A})}$$
$$\phi(gg_1u_1h)Z_j(G_s)(gg_1u_1h)dgdg_1du_1$$

On the other hand we note that the telescoped inner integral

$$\int_{N_{\{z_2,\cdots,z_\ell\}}\cdot U_{\{z_2,\cdots,z_\ell\}}(k)\backslash N_{\{z_2,\cdots,z_\ell\}}\cdot U_{\{z_2,\cdots,z_\ell\}}(\mathbb{A})} \phi(nug)Z_j(G_s)(nug)dndu \equiv$$
$$W_{(\ell-1)\otimes Z_\beta}(\phi)(g)Z_j(G_s)(g)$$

Here Z_β determines (as in §3) the character on $U_{\{z_2,\cdots,z_\ell\}}$ given by $\psi\circ\lambda_{Z_\beta}$ given above. Moreover $W_{(\ell-1)\otimes Z_\beta}$ is the character given above. (The $(\ell-1)$ component is taken relative to the generic Whittaker character on the group $N_{\{z_2,\cdots,z_\ell\}}(\mathbb{A})$).

Then using **Lemma 4.1, Lemma 4.2, Lemma 4.3** and **Remarks** (A) and (B) above we establish the final formula for the calculation of the term

$$[\mathcal{E}_j(G_s),\phi,i\otimes Z]$$

Theorem 4.3. *Let ϕ belong to $L^2_{cusp}(H_i(k)|H_i(\mathbb{A}))$. Then $[\mathcal{E}_j(G_s),\phi,i\otimes Z]$ equals*

(a) $$\int_{H_i(k)\backslash H_i(\mathbb{A})} \phi(h)Z_i(G_s)(h)dh \text{ if } i=j$$

(b) $$\int_{O((z_1^-)^\perp,q_{W_1})(\mathbb{A})\backslash H_i(\mathbb{A})} \left(\int_{O((z_1^-)^\perp,q_W)(k)\backslash O(z_1^-)^\perp,q_W)(\mathbb{A})} \phi(h_1h)Z_j(G_s)(h_1h)dh_1\right)dh$$

if $i+1=j$.

$$(c)\quad \int_{(N_{\{z_2,\cdots z_\ell\}}\cdot U_{\{z_2,\cdots z_\ell\}}\cdot O((z_1^-)^\perp,q_W))(\mathbb{A})\backslash H_i(\mathbb{A})} \left(\int_{O((z_1^-)^\perp,q_W)(k)\backslash O((z_1^-)^\perp,q_W)(\mathbb{A})} W_{(\ell-1)\otimes Z_\beta}(\phi)((h_1h)Z_j(G_s)(h_1h)dh_1 \right) dh$$

if $j > i+1$.

Proof. Cases (a) and (b) follow from the comments above. Finally Case (c) follows when we substitute the formula in Remark (C) into the integrals in **Lemma 3.** /**Q.E.D.**

Remark (C). We know by the comments in Remark (A) that the integrals above in (b) is Eulerian. This follows from the "factorizability" property given in Definition (I) above (at beginning of this section). On the other hand to see that the inner integral of (c) is Eulerian we require a stronger notion of factorizability.

Definition (II). Let $M_i = O(W(R_i))$, $H_i = O((\xi_0)^\perp, q_{W_1})$ and $L_i = O((z_1^-)^\perp, q_W)$. We consider the parabolic subgroup of H_i given by $P_{\{z_2,\cdots,z_\ell\}}$. We consider the unipotent group of $P_{\{z_2,\cdots,z_\ell\}}$ given by $N_{\{z_2,\cdots,z_\ell\}} \times U_{\{z_2,\cdots,z_\ell\}}$. Moreover we let $\psi_{\ell-1}\otimes\Psi_{Z_\beta}$ be a character on this group constructed as in the beginning of §3. Then let σ be any irreducible admissible representation of L_i. We consider the non compactly induced module: $X(\sigma,\psi_{\ell-1}\otimes\psi_{Z_\beta}) =$

$$\mathrm{Ind}_{L_i\times N_{\{z_2,\cdots,z_\ell\}}\times U_{\{z_2,\cdots x_\ell\}}}^{H_i}(\sigma\otimes\psi_{\ell-1}\otimes\psi_{Z_\beta}).$$

We require that

$$\dim Hom_{H_i}(\mathbf{\Pi}, X(\sigma,\psi_{\ell-1}\otimes\psi_{Z_\beta}) \leq 1$$

for all $\mathbf{\Pi}$ irreducible admissible modules for L_i. We refer the reader here to the comments in the **Appendix 2** to §3. If this condition is valid for all primes v, then we say that the global representations $\mathbf{\Pi}$ and σ admit the "factorizable" property.

Remark 3. If Definition (II) holds, then the integral given in (c) of **Theorem 3** is an Euler product!

The problem for the rest of this paper is the explicit determination of the integrals in the above **Theorem**.

We now define the various L functions that will be studied in this section.

Namely the given groups are GL_t and H where H is a special orthogonal group. We consider those primes v where H_v is a quasi split group. In particular this means that H_v has one of three forms

(1) $H_v = SO(V, q_V)$ where (V, q_V) is an even dimensional totally split form.

(2) $H_v = SO(V, q_V)$ where (V, q_V) is an odd dimensional form with the Witt index $(V, q_V) = \frac{1}{2}(\dim V - 1)$ and discriminant (V) a unit.

(3) $H_V = SO(V, q_V)$ where (V, q_V) is an even dimensional form where $V = H_{2r-2} \oplus \lambda K$ where H_{2r-2} is a direct sum of $(r-1)$ hyperbolic planes and λK corresponds to a 2 dimensional anisotropic form, i.e. $\lambda N_{K/k}$ where K/k is an unramified extension of degree A and $\lambda N_{K/k}$ represents the λ-multiple of the norm form $N_{K/k}$ on $K(\lambda \in k^x)$.

In each case the L group of H_v becomes (with W_k the usual Weil group).

(1) $SO(2n, \mathbb{C}) \times W_k$ (direct product) with $SO(2n, \mathbb{C})$ = the unique complex special orthogonal group.

(2) $Sp_n(\mathbb{C}) \times W_k$ (direct product) with $Sp_n(\mathbb{C})$ = the subgroup of GL_{2n} fixing a nondegenerate symplectic form.

(3) $SO(2n, \mathbb{C}) \ltimes W_k$ (semi-direct product). Here the Weil group W_k acts on $SO(2n, \mathbb{C})$ via the outer automorphism which comes from conjugation by an element γ_0 in $O(2n, \mathbb{C})$. Then W_k acts through the following diagrams:

(i) : $1 \to W_K \to W_k \to Gal(K/k) \to 1$

(ii): $W_k \to W_K \backslash W_k \overset{\simeq}{\to} \langle \sigma, 1 \rangle \overset{\sigma}{\to}$ conjugation by γ_0 (Fr = some choice of the Frobenius element in W_k is sent to σ).

Then we define the tensor product L function associated to $GL_t \times H$.

In cases (1) and (2) above an unramified representation π of H_v corresponds to a diagonal matrix $D_H(\pi)$ in $SO(2n, \mathbb{C})$ or $Sp_n(\mathbb{C})$. In case (3) an unramified representation π of H_v corresponds to a specific element $(D_H^{(\pi)}, Fr)$ where $D_H^{(\pi)}$ is a diagonal matrix in $SO(2n, \mathbb{C})$ having the form

$$\begin{bmatrix} D_{n-1} & & & 0 \\ & 1 & & \\ & & 1 & \\ 0 & & & D_{n-1}^{-1} \end{bmatrix}$$

where D_{n-1} is diagonal ($n-1 \times n-1$ matrix); we note here that we can choose

$$\gamma_0 = \begin{bmatrix} I_{n-1} & & \\ & \begin{matrix} 0 & I \\ I & 0 \end{matrix} & \\ & & I_{n-1} \end{bmatrix}$$

We note that the L group of GL_t is a direct product $GL_t(\mathbb{C}) \times W_k$. We let $D(\tau)$ be the diagonal matrix in $GL_t(\mathbb{C})$ which corresponds to the unramified representation τ of $GL_t(\mathbb{C})$.

Then the "tensor product" L function of $GL_t \times H_v$ associated to the pair $\tau \otimes \pi$ becomes the following:

(1) $det(I_{2nt} - D(\tau) \otimes D_H(\pi) q^{-s})^{-1}$

(2) $det(I_{2nt} - D(\tau) \otimes D_H(\pi) q^{-s})^{-1}$

(3) $det(I_{2nt} - D(\tau) \otimes \tilde{D}_H(\pi)q^{-s})^{-1}$ where

$$\tilde{D}_H(\pi) = \begin{bmatrix} D_{n-1} & & & 0 \\ & 1 & & \\ & & 1 & \\ 0 & & & D_{n-1} \end{bmatrix} \begin{bmatrix} I_{n-1} & & & \\ & 0 & I & \\ & I & 0 & \\ & & & I_{n-1} \end{bmatrix}$$

In each of the cases above we denote $L_v(\tau \otimes \pi, st \otimes st, s)$ as the specific L function. In more concrete terms if

$$D(\tau) = \begin{pmatrix} q^{\chi_1} & & \\ & \ddots & \\ & & q^{\chi_t} \end{pmatrix}, D_H = \begin{bmatrix} q^{\gamma_1} & & & & & \\ & \ddots & & & & \\ & & q^{\gamma_n} & & & \\ & & & q^{-\gamma_n} & & \\ & & & & \ddots & \\ & & & & & q^{-\gamma_1} \end{bmatrix}$$

then in case (1) and (2) above

$$det(I_{2nt} - D(\tau) \otimes D_H(\pi)q^{-s})^{-1}$$
$$\left[\prod_{\substack{1 \leq i \leq t \\ 1 \leq j \leq n}} (1 - q^{-\chi_i - \gamma_j - s})(1 - q^{-\chi_i + \gamma_j - s}) \right]^{-1}$$

In case (3) if $D_{n-1} = \begin{bmatrix} q^{\gamma_1} & & 0 \\ & \ddots & \\ 0 & & q^{\gamma_{n-1}} \end{bmatrix}$ then

$$det(I_{2nt} - D(\tau) \otimes \tilde{D}_H(\pi)q^{-s})^{-1} =$$
$$\left[\prod_{\substack{1 \leq i \leq t \\ 1 \leq j \leq n-1}} (1 - q^{-\chi_i - \gamma_j - s})(1 - q^{-\chi_i + \gamma_j - s}) \right]^{-1}$$
$$\left[\prod_{1 \leq i \leq t} (1 - q^{-2\chi_i - 2s}) \right]^{-1}$$

ormula in cases (b), (c) and (d) in **Theorem 4.3.**

Namely in cases (b), (c) and (d) we have the following formulae. Recall here the form $\langle\langle \mid \rangle\rangle$ defined at the beginning of this section.

Corollary 1 to Theorem 4.3.

In case (b) the integral equals $(i = j)$

$$\langle\langle \phi | Z_i(G_s) \rangle\rangle_{H_i(\mathbb{A})}$$

In case (c) the integral equals $(i + 1 = j)$

$$\int_{O((z_1^-)^{\perp}, qw_1)(\mathbb{A}) \backslash H_i(\mathbb{A})} \langle\langle \phi * h | Z_j(G_s)(h) \rangle\rangle_{O((z_1^-), qw_1)(\mathbb{A})} dh$$

In case (d) the integral equals ($j > i+1$).

$$\int_{N_{\{z_2,\cdots z_\ell\}}\cdot U_{\{z_2,\cdots,z_\ell\}}O((z_1^-)^\perp,q_{W_1})(\mathbb{A})\backslash H_i(\mathbb{A})} \langle\langle W_{(\ell-1)\otimes Z_\beta}(\phi)*h|Z_j(G_s)(h)\rangle\rangle_{O((z_1^-)^\perp,q_{W_1})(\mathbb{A})}$$

In particular the above Corollary is used to deduce (from the factorizability principle of definitions (I) and (II) above) that each of the cases (b), (c) and (d) is represented by an Euler product.

To make the pairing more natural in case (c) (in order to determine the factorizability) we can rewrite the above integral in **Remark (A)** as

$$\int_{N_\beta(\mathbb{A})\backslash U_i(\mathbb{A})} \left(\int_{(O(z_1)^\perp,q_{W_1})(k)\backslash O((z_1)^\perp,q_{W_1})(\mathbb{A}))} \phi(h_1h)W^{i+1}(G_s)(h_1^{w_S}w_Suh)dh_1 \right) \psi_2(u)du\,.$$

The fact that the integration can be switched follows from the facts that (i) ϕ is cuspidal and (ii) the function

$$h_1 \rightsquigarrow \int_{N_\beta(\mathbb{A})\backslash U_i(\mathbb{A})} |W^{i+1}(G_s)[w_Suh_1h]|du$$

is slowly increasing in h_1. Fact (ii) is the global analogue of the local arguments used in **Appendix 2** to §**5**. In any case the inner integral

$$\phi\otimes G_s \rightsquigarrow \int_{O((z_1^-),q_{W_1})(k)\backslash O((z_1^-),q_{W_1})(\mathbb{A})} \phi(h_1)W^{i+1}(G_s)(h_1^{w_S}w_Su)dh_1$$

represents an $O((z_1)^\perp, q_{W_1})$ invariant pairing between Π and σ^{w_S}. Thus it is possible to factorize this integral in the following way. We let $\langle\ |\ \rangle_v$ represent the local pairing between $\sigma_v^{w_S}$ and Π_v; we "normalize" $\langle\ |\ \rangle_v$ so that if v_0 and w_0 represent unramified vectors in $\sigma_v^{w_S}$ and Π_v, then $\langle v_0 \mid w_0\rangle = 1$.

Then we let $G_s \in I_j(\tau,\sigma,s)$ be identified to the function $\bigotimes_v (G_s)_v$ in such a way if α_0^ν and v_0^ν represent the unramified vectors in τ_v^ν and σ_v^ν where $\langle\alpha_0,\alpha_0^\nu\rangle = \langle v_0, v_0^\nu\rangle = 1$, then $\langle (G_s)_v(e) \mid \alpha_0^\nu\otimes v_0^\nu\rangle = 1$ (for the primes v where the "data" is all unramified).

Let $\phi\in\Pi$ be identified to $\bigotimes_v \phi_v \in \otimes\Pi_v$ with $\phi_v = (w_0)_v$ (almost everywhere).

Let $(L_j)_v$ be a local Whittaker embedding of τ_v. We require the normalization of $(L_j)_v$ for the unramified data so that $\langle (L_j)_v((G_s)_v)(e) \mid v_0^\nu\rangle = 1$.

Thus by factorizability the integral

$$\int\limits_{O((z_1^{\perp}),q_{W_1})(k)\backslash O((z_1)^{\perp},q_{W_1})(\mathbb{A})} \phi(h_1h)W^{i+1}(h_1^{w_S}w_Suh)dh_1$$
$$= \prod_v \langle \phi_v * h_v \mid (L_j)_v[(G_s)_v][w_Su_vh_v]\rangle_v .$$

Thus the integral

$$\int\limits_{N_\beta(\mathbb{A})\backslash U_i(\mathbb{A})} \left(\int\limits_{O((z_1)^{\perp},q_{W_1})(k)\backslash O((z_1^{-}),q_{W_1})(\mathbb{A})} \phi(h_1h)W^{i+1}(G_s)(h_1^{w_S}w_Suh)dh_1 \right) \psi_Z(u)du$$
$$\equiv \prod_v \int\limits_{(N_\beta)_v\backslash(U_i)_v} \langle \phi_v * h_v \mid (L_j)_v[(G_s)_v][w_Su_vh_v]\rangle_v \psi_{Z_v}(u_v)du_v .$$

On the other hand we consider the space $\mathrm{Ind}_{(H_i)_v}^{\mathbb{G}_v}(\sigma^{w_S})$. Then we note it is possible to define the local integral

$$V_j(F)(x) = \int\limits_{N_\beta\backslash U_i} L_j(G_s)[w_Sux]\Lambda_{\xi_0}(u)du$$

as the element in $\mathrm{Ind}_{(H_i)_v}^{\mathbb{G}_v}(\sigma^{w_S})$ which satisfies the property that for *all* $\xi \in (\sigma^{w_S})^{\nu}$

$$(a)\colon \{\xi \mid V_j(F)(x)\} = \int\limits_{N_\beta\backslash U_i} \{\xi \mid L_j(G_s)[w_Sux]\}\Lambda_{\xi_0}(u)du .$$

Here $\{\ \mid\ \}$ is the nondegenerate pairing between σ and σ^{v}. We observe that all we need for this is that

$$\int\limits_{N_\beta\backslash U_i} |\{\xi \mid L_j(G_s)[w_Sux]\}|du < \infty$$

for all $\xi \in (\sigma^{w_S})^{\nu}$ (valid for $Re(s) \gg 0$). Then we see that there exists $V_j(F)(x)$ in σ^{w_S} $((\sigma^{w_S})^{\nu\nu} \cong \sigma^{w_S})$ so that the above identity (a) is valid. Then it is formal to verify that $V_j(F)(x)$ as we vary x in $\mathbb{G}_v$ determines uniquely an element in $\mathrm{Ind}_{(H_1)_v}^{\mathbb{G}_v}(\sigma_v^{w_S})$.

On the other hand, we note that the pairing $\langle\ \mid\ \rangle_v$ between Π_v and $\sigma_v^{w_S}$ determines for a given element $Z \in \Pi_v$ a unique element Z^0 in $(\check{\sigma}_v)^{w_S}$ so that

$$\langle Z \mid w\rangle_v = \{Z^0 \mid w\}$$

(for *all* $w \in \sigma^{w_S}$). Hence we have that

$$\begin{aligned}\langle Z \mid V_j(F)(x)\rangle_v &= \{Z^0 \mid V_j(F)(x)\} \\ &= \int\limits_{N_\beta \backslash U_i} \{Z^0 \mid L_j(G_s)[w_S ux]\} \Lambda_{\xi_0}(u)du \\ &= \int\limits_{N_\beta \backslash U_i} \langle Z \mid L_j(G_s)[w_S ux]\rangle_v \Lambda_{\xi_0}(u)du\,.\end{aligned}$$

Thus we deduce that the integral

$$\begin{aligned}&\int\limits_{N_\beta(\mathbb{A})\backslash U_i(\mathbb{A})} \left(\int\limits_{O((z_1)^\perp, q_{W_1})(k)\backslash O((z_1^\perp), q_{W_1})(\mathbb{A})} \phi(h_1 h) W^{i+1}(G_s)[h_1^{w_S} w_S uh] dh_1 \right) \\ &\qquad \Lambda_{\xi_0}(u)du \\ &\equiv \prod_v \langle \phi_v * h_v \mid V_j[(G_s)_v](h_v)\rangle_v\,.\end{aligned}$$

Thus we deduce that the integral in case (c) above (**Corollary** to **Theorem 4.3**) equals

$$\prod_v \left(\int\limits_{(M_i)_v \backslash (H_i)_v} \langle \phi * h_v \mid V_j[(G_s)_v](h_v)\rangle_v dh_v \right)\,.$$

We can repeat the same considerations for case (d) in **Corollary 1** to **Theorem 4.3**.

Indeed we note that there is a $\mathbb{G}(\mathbb{A})$ equivariant map from $I(\tau \otimes ||^s \otimes \sigma) = \otimes I_{v,j}(\tau_v, \sigma_v, s)$ to the space

$$\mathrm{Ind}_{H_i(\mathbb{A})(N^i \mathbb{U}_i)(\mathbb{A})}^{G(\mathbb{A})} (\tilde{R}_{\ell-1}(\sigma) \otimes \Lambda_{\xi_0})$$

where

$$\tilde{\mathbb{R}}_{\ell-1}(\sigma) = (\mathrm{Ind})_{L_i(\mathbb{A})N_{\{z_1 \cdots, z_\ell\}} U_{\{z_1, \cdots, z_\ell\}}}^{H_i(\mathbb{A})} (\sigma^{w_S} \otimes \psi_{\ell-1} \otimes \psi_{Z_\beta})$$

($R_{\ell-1}(\ )$ defined locally in §5 with $\tilde{\mathbb{R}}_{\ell-1}(\sigma) = \bigotimes_v \tilde{R}_{\ell-1}(\sigma_v)$) given by the map $G_s \rightsquigarrow Z_j(G_s)$. We note that the space $\tilde{R}_{\ell-1}(\sigma_v)$ is defined relative to the group $O((\xi_0)^\perp, q_{W_1}) = H_i$ and $O(W_1(S) \cap W_1(R_i)) = L_i$. That is,

$$\tilde{R}_{\ell-1}(\sigma_v) = \mathrm{ind}_{L_i N_{\{z_2, \cdots, z_\ell\}} U_{\{z_2, \cdots, z_\ell\}}}^{H_i} (\psi_{\ell-1} \otimes \psi_{Z_\beta} \otimes \sigma_v^{W_S})$$

We note the difference with the space $R_{\ell-1}(\sigma_v)$ which is defined in §5 where in place of H_i and L_i we have $SO((\xi_0)^\perp, q_{W_1})$ and $SO(W_1(S) \cap W_1(R_i))$. At the level of local components this map

$$I_{v,j}(\tau_v, s) \text{ to } \mathrm{ind}_{(H_i)_v (N^i \mathbb{U}_i)_v}^{\mathbb{G}_v} (\tilde{R}_{\ell-1}(\sigma_v) \otimes \Lambda_{\xi_0, v})$$

is given by sending $F \subset I_{v,j}(\tau_v, \sigma_v, s)$ to the integral

$$V_j(F)(x) = \int_{(N_\beta \backslash \mathbb{U}_i)} L_j(F)[w_S u x] \Lambda_{\xi_0}(u) du$$

V_j is the precisely the map used in §5.

Then the integral

$$\int_{(L_i(k)\backslash L_i(\mathbb{A}))} W_{\ell-1\otimes Z_\beta}(\phi)(h_1 h) Z_j(G_s)(h_1 h) dh_1$$

can be factorized as an Euler product

$$\prod_v \langle \phi_v * h_v | V_j(G_s)(h_v) \rangle_v$$

Here $\langle \mid \rangle_v$ represents the *unique element* (up to scalar factor) in

$$Hom_{(H_i)_v}(\Pi_v, \mathrm{ind}_{L_i N_{\{z_2,\cdots,z_\ell\}} U_{\{z_2,\cdots,z_\ell\}}}^{H_i} (\check{\sigma}_v^{w_S} \otimes \overline{\psi}_{\ell-1} \otimes \overline{\psi}_{Z_\beta}).$$

(See **Definition** (II) above). Hence we can compute the integral in case (d) as follows.

$$\prod_v \int_{L_i(\mathbb{A}) N_{\{Z_2,\cdots,Z_\ell\}} U_{\{Z_2,\cdots,Z_\ell\}}(\mathbb{A}) \backslash H_i(\mathbb{A})} \langle \phi_v * h_v | V_j(F_s)(h_v) \rangle_v dh_v.$$

Thus the basic problem becomes to compute each of the local factors in cases (b), (c) and (d) above. At least we should compute the local integrals when the data is always unramified.

In fact we prove in §5 to §8 that if v is a prime satisfying the following:

(1) v is finite and v does not have residual characteristic 2

(2) $(H_i)_v$ and $(L_i)_v$ are both quasi-split

(3) The local components Π_v, σ_v and τ_v of Π, σ and τ are spherical representations (admitting a K fixed vector). Then we prove below the following **Theorem**.

(4) The additive character ψ_v at v has order 0.

Theorem 4.4. *Let v be a prime given as above (satisfying (1), (2), (3), and (4) above.) Let $j = i+1$. Then the local integral*

$$\int_{(L_i \backslash H_i)} \langle \phi_v * h_v | V_j(F)(h_v) \rangle_\sigma dh_v =$$

$$\left[\frac{L_v(\tau_{i+1} \otimes \sigma(\gamma), st, s + \frac{1}{2})}{L_v(\tau_{i+1} \otimes \sigma', st, s+1) L_v(\tau_{i+1} \, {}^{\Lambda^2}_{sym^2}, 2s+1)} \right] \langle \phi_v | F(e) \rangle_\sigma$$

The factor $\langle \phi_v | F_s(e) \rangle_\sigma$ *is independent of* s. *The local* L *factors used above are defined earlier in this section.*

The term $L_v(\tau_{i+1}, {}^{\Lambda^2}_{sym^2}, \ )$ *represents either the exterior square* $L_v(\tau_{i+1}, \Lambda^2, \ )$ *or symmetric square* $L_v(\tau_{i+1}, sym^2, \ )$. *In particular it is exterior square when* $\dim O(W)$ *is odd; otherwise it is symmetric square.*

Proof. In order to compute the integral above we must reduce the integral to the one given in **Theorem 5.2 of** §**5**. We must show that the integral above has exactly the form $\langle\langle V_j(F_1'(\ ))|T(w_0)\rangle\rangle$. In particular this latter is an integral over $L_i^0\backslash H_i^0$ where $L_i^0 = SO((z_1)^\perp, q_{W_1})$ and $H_i^0 = SO((\xi_0)^\perp, q_{W_1}\rangle$. In fact the integral (in **Theorem** 4.3)

$$\int\limits_{L_i\backslash H_i} \langle \phi_v * h_v | V_j(F)(h_v)\rangle_{\sigma_v} dh_v =$$

$$\int\limits_{L_i^0\backslash H_i^0} \langle \phi_v * h_v | V_j(F)(h_v)\rangle_{\sigma_v} dh_v^0$$

The normalization of measures is given by $dh_i^0 = \frac{1}{2} dh_i$ and $d\ell_2^0 = \frac{1}{2} d\ell_i$. Thus the only point that must be checked is whether the L_i invariant bilinear form defined by the integral

$$\int\limits_{L_i^0\backslash H_i^0} \langle \phi_v * h_v | V_j(F)(h_v)\rangle_{\sigma_v} dh_v^0$$

satisfies the properties of this Theorem.

In particular we assume $\phi_v \in \Pi_v$ = the unramified constituent of $\mathrm{ind}_{B_{O((\xi_0)^\perp, q_{W_1}}}^{O((\xi_0)^\perp, q_{W_1})}(\beta_1' \otimes \cdots \otimes \beta_r')$ and F lies in the space

$$\mathrm{ind}_{GL(R_j)\times O(W_1(R_j)\ltimes \mathbb{U}_j}^{O(W)} (\tau_j \otimes |\ |^s \otimes \sigma_j \otimes \mathbb{1})\,.$$

We know that upon restriction to $SO((\xi_0)^\perp, q_{W_1})$ and $SO(W)$ the two modules above (Π_v and $\mathrm{ind}_{GL(R_j)\times O(W_1(R_j))\times \mathbb{U}_j}^{O(W)} (\tau_j \otimes |\ |^s \otimes \sigma_j \otimes 1)$ with s large beomce either irreducible or a direct sum of 2 irreducible modules each of which is spherical. Moreover, the $K_{O(\ )}$ spherical vector as in the latter case a direct sum of the spherical vectors in each piece.

Thus replacing (if necessary) φ_v and F_v by its components in the $SO(\ )$ modules we deduce from **Theorem 5.2 of** §**5** that **Theorem 4.4** is valid for the case where ϕ_v and F_v lie in irreducible components. But then we note that L factors arising in the identity in **Theorem 4.4** remains the same for the components which occur in the restriction of σ_v and σ_v' to the special orthogonal groups in question. This follows from the fact that the possible extra component is obtained by twisting a fixed component by an element in $O - SO$. Then by just adding the components together, we get **Theorem 4.4** completely. □

One subtle point we note here is that for the global calculation in the next **Theorem** we must now normalize $\langle\ |\ \rangle_{\sigma_v}$ so that $\langle\varphi_v \mid \beta_v\rangle_{\sigma'_v} = 1$ for φ_v and β_v normalized unramified vectors in σ_v and σ'_v.

Theorem 4.5. *Let $j = i+1$. Let S be the set of primes which is complementary to the set defined by (1), (2) and (3) above. There is the identity:*

$$\int\limits_{L_i(\mathbb{A})\backslash H_i(\mathbb{A})} \langle\langle \phi * h | Z_j(G_s)(h)\rangle\rangle dh =$$
$$\left[\frac{L^S(\tau\otimes\pi, s+\frac{1}{2})}{L^S(\tau\otimes\sigma, s+1)L^S(\tau, {}^{\Lambda^2}_{sym^2}, 2s+1)}\right]$$
$$\left(\int\limits_{L_i(\mathbb{A}_S)\backslash H_i(\mathbb{A}_S)} \langle\langle \phi * h | Z_j(G_s)(h)\rangle\rangle dh\right)$$

Remark 4. We note that the expected identity (which is the generalization of **Theorem 4.5**) is the following:

There is a choice of data ϕ and G_s so that

$$\int\limits_{L_i(\mathbb{A})\backslash H_i(\mathbb{A})} \langle\langle \phi * h | Z_j(G_s)(h)\rangle\rangle dh =$$
$$\left[\frac{L^S(\tau\otimes\pi, s+\frac{1}{2})}{L^S(\tau\otimes\sigma, s+1)L^S(\tau, {}^{\Lambda^2}_{sym^2}, 2s+1)}\right] z_\infty(s)\langle\langle\phi | G_s\rangle\rangle$$

Here $z_\infty(s)$ represents the contribution of the Archimedean components and the factor

$$\langle\langle\phi | G_s\rangle\rangle$$

is independent of s.

Remark 5. We note that the factor in the denominator

$$L^S(\tau\otimes\sigma, s+1)L^S(\tau, {}^{\Lambda^2}_{sym^2}, 2s+1)$$

can be interpreted as the possible normalizing factor of the Eisenstein series $\mathcal{E}(G_s, \tau, \sigma)$. In particular assuming that σ is a cuspidal representation of $O(M_j(\mathbb{A})))$ then the constant term of $\mathcal{E}(G_s, \tau, \sigma)$ along the unipotent radical $\mathbb{U}_j$ is given by the sum of 2 terms:

$$G_s + \frac{L^S(\tau\otimes\sigma, s)L^S(\tau, {}^{\Lambda^2}_{sym^2}, 2s)}{L^S(\tau\otimes\sigma, s+1)L^S(\tau, {}^{\Lambda^2}_{sym^2}, 2s+1)}$$
$$\left[\prod_{v\in S} M^v_{w_j}((G_s)_v)\right] \otimes \left(\bigotimes_{v\notin S} G_{-s}\right)$$

Here $M^v_{w_j}$ is the associated local intertwining operator associated to the Weyl element w_j. We expect that the term

$$L^S(\tau\otimes\sigma, s+1)L^S(\tau, {}^{\Lambda^2}_{sym^2}, 2s+1)G_s+$$
$$L^S(\tau\otimes\sigma, s)L^S(\tau, {}^{\Lambda^2}_{sym^2}, 2s)\Big(\prod_{v\in S} M^v_{w_j}((G_s)_v)\Big)\otimes\Big(\bigotimes_{v\notin S} G_{-s}\Big)$$

has a meromorphic continuation in s with at most a finite number of poles. In particular this implies that $L^S(\tau\otimes\sigma, s+1)L^S(\tau, {}^{\Lambda^2}_{sym^2}, 2s+1)\mathcal{E}(G_s,\tau,\sigma)$ admits a meromorphic continuation in s with at most a finite number of poles. We note that if σ is *not cuspidal* then the constant term of $\mathcal{E}(G_s,\tau,\sigma)$ contains more than two terms. However from the general we still expect that the function:

$$s \rightsquigarrow L^S(\tau\otimes\sigma, s+1)L^S(\tau, {}^{\Lambda^2}_{sym^2}, 2s+1)\mathcal{E}(G_s,\tau,\sigma)$$

to still admit a meromorphic continuation with at most a finite number of poles.

Remark 6. The importance of the factor $\langle\langle\phi|G_s\rangle\rangle$is that it insures that by some choice of $\phi\in\Pi$ and $f_\sigma\in\sigma$ (which defines G_s) the pairing $\langle\langle\phi|G_s\rangle\rangle$ is a nonzero multiple of $\langle\langle\phi|f_\sigma\rangle\rangle$ which can be chosen to be *nonzero*!!

Appendix to §4.3.

There is a degenerate case here which we include for the sake of completeness.

Namely we consider again the example in §2. Here $H = O(V, q_V)$ and $M = O(V_1, q_{V_1})$. In this case $W = V_1$. Moreover we consider for any parabolic $\mathbb{P}_\ell$ of M the family of Eisenstein series of the form $\mathcal{E}_\ell(G_s)$ constructed as above.

Let $\mathbf{\Pi}$ be a cuspidal irreducible representation of $H(\mathbb{A})$. Then we consider the family of integrals of the form

$$\int_{H(k)\backslash H(\mathbb{A})} f_{\mathbf{\Pi}}(h)\mathcal{E}_\ell(G_s)(h)dh$$

Thus generalizes the construction in §2 above where we basically consider the case $\ell = 1$.

Then this family of integrals can be computed as above. Indeed given ℓ there is a decomposition of V_1 and V as follows :

$$V_1 = Span\{z_1,\cdots,z_\ell\}\oplus Y\oplus Span\{z_1^*,\cdots z_\ell^*\}$$

and

$$V = Span\{z_2,\cdots,z_\ell\}\oplus((z_1^-)\oplus Y)\oplus Span\{z_2^*,\cdots,z_\ell^*\}$$

with $q_W(z_i,z_j^*)=\delta_{ij}, q_W(z_i,z_j)=q_W(z_i^*,z_j^*)=0$ $Y=(Span\{z_1,\cdots z_\ell\}$ $\oplus\, Span\{z_1^*,\cdots,z_\ell^*\})^\perp, \xi_0 = z_1^+$ and $z_1^-\perp\xi_0$.

Then using the same notation as above we note that

$$G_s(h) = \sum_{N_{\{z_1,\cdots,z_\ell\}}\backslash P_{\{z_1,\cdots,z_\ell\}}} W^{(\ell)}(\gamma h)$$

But as in Remark (C) the sets

$$\begin{aligned} N_{\{z_1,\cdots,z_\ell\}}\backslash P_{\{z_1,\cdots,z_\ell\}} &= N_{\{z_2,\cdots,z_\ell\}}T_{\ell-1}\backslash GL(\{z_2,\cdots,z_\ell\})T_{\ell-1} \\ &= N_{\{z_2,\cdots,z_\ell\}}U_{\{z_2,\cdots,z_\ell\}}\backslash GL(\{z_2,\cdots,z_\ell\})U_{\{z_2,\cdots,z_\ell\}} \end{aligned}$$

Then doing the usual unwinding of $\mathcal{E}_\ell$ in the integral above we know that H admits 2 orbits. Again the small orbit contributes zero since f_Π is a cusp form. The large orbit contributes in the same way as Remark (C). Since the appropriate stabilizer ($Q^*_{S\cap W(R_i)}$ here) is given by $GL(\{z_2,\cdots,z_\ell\})\cdot O((z_1^-)^\perp,q_W)\cdot U_{\{z_2,\cdots,z_\ell\}}$. In any case we deduce the following calculation analogous to **Theorem 3** above.

Proposition. *Let $\ell \geq 2$. Let ϕ be a cusp form on $H(k)\backslash H(\mathbb{A})$. Then the integral*

$$\int_{H(k)\backslash H(\mathbb{A})} \phi(h)\mathcal{E}_\ell(G_s)(h)dh \equiv$$

$$\int_{(N_{\{x_2,\cdots,z_\ell\}}U_{\{z_2,\cdots,z_\ell\}}O((z_1^-)^\perp,q_W))(\mathbb{A})\backslash H_i(\mathbb{A})} \Big(\int_{O((z_1^-)^\perp,q_W)(k)\backslash O(z_1^-)^\perp,q_W)(\mathbb{A})}$$
$$W_{(\ell-1)\otimes Z_\beta}(\phi)(h_1h)W^{(\ell)}(G_s)(h_1h)dh_1\Big)dh$$

Remark 1. In the case V is a totally split form and $\ell = \frac{\dim V}{2}$, then the above **Proposition** is basically Method (B) given in [G-PS-R].

Remark 2. The integral above is "factorizable" if we assume **Definition (II)** above.

§5. Support Ideals (II)

We recall the representation

$$\mathrm{ind}_{\mathbb{P}_j}^{\mathbb{G}}(\tau\cdot|\det x|^{\frac{1}{2}(\dim W_1-j-1)}\otimes\sigma\otimes 1)$$

given in §3. In particular we replace τ by $\tau\otimes|detx|^s$. We call such a representation

$$I_j(\tau,\sigma,s).$$

We consider the $SO((\xi)^\perp,q_{W_1})$ module

$$(I_j(\tau,\sigma,s))_{(N^i\times\mathbb{U}_i,\Lambda_\xi)}$$

constructed in §3. In this section we assume that τ and σ are unramified representations in the sense that τ and σ are irreducible and admit nonzero fixed vectors under the maximal compact subgroups

$$K_{GL(R_j)} \text{ and } K_{SO(W_1(R_j))} \text{ of } GL(R_j) \text{ and } SO(W_1(R_j))$$

Note here we are now assuming that $\mathbb{G}, SO(W_1(R_j))$, and $SO((\xi)^\perp,q_{W_1})$ are all quasi-split groups.

In particular we analyze the structure of

$$I_j^*(\tau,\sigma,s)=[(I_j(\tau,\sigma,s))_{N^i\times\mathbb{U}_i,\Lambda_\xi)}]^{K_{SO(\xi)^\perp,q_{W_1})}}$$

as a module for the Hecke algebra

$$\mathbb{A}_{SO((\xi)^\perp,q_{W_1})}=\mathbb{C}[X,X^{-1}]\otimes\mathcal{H}$$

with $\mathcal{H}=\mathcal{H}(SO((\xi)^\perp,q_{W_1})//K_{SO((\xi)^\perp,q_{W_1})})$. We consider the generalization of the support ideal in $\mathbb{A}_{SO((\xi)^\perp,q_{W_1})}$ given in §2. First we note that $\mathbb{A}_{SO((\xi)^\perp,q_{W_1})}$ acts on $I_j^*(\tau,\sigma,s)$ as follows: $\varphi_s*(X\otimes\varphi)=q^{-s}(\varphi_s*\varphi)$ with $*\varphi$ denoting the usual induced action of φ on the Jacquet module $I_j^*(\tau,\sigma,s)$, $(\varphi_s\in I_j^*(\tau,\sigma s))$.

Namely we define the support ideal as follows:

$$\mathcal{I}_{supp}(\tau,\sigma)=\{F\in\mathbb{A}_{SO((\xi)^\perp,q_{W_1})}|$$
$$I_j^*(\tau,\sigma,s)*F\subseteq\left\{\begin{matrix}(A) & i<j\\ (B) & i=j\end{matrix}\right\}$$

where A and B are defined in **Lemma 3. 5** (with the twisting by $|\ |^s$). In particular in the case where $i=j=\dim S$ then

$$A=0 \text{ and } B=\begin{cases}0 & \text{if } (\tau^{w_S})^{(i)}=0\\ \sigma|_{SO((\xi)^\perp,q_{W_1})} & \text{otherwise}\end{cases}$$

We note that if $(\tau^{w_S})^{(i)} \neq 0$ (with $I = j$) then τ^{w_S} admits a Whittaker model and hence by the irreduciblity of τ we have $(\tau^{w_S})^{(i)} = \mathbb{C}$, a one dimensional space.

For the rest of this section we assume that $j = \dim S \geq i$.

We recall from **Lemmas 3.3, 3.4** and **3.5** that the boundary components of the Jacquet functor of $I_j(\tau, \sigma, s)$ relative to $(N^i\mathbb{U}_i, \Lambda_{\xi_0})$ are given as $SO((\xi)^\perp, q_w)$ modules as (unnormalized induction)

$$\begin{aligned}
&\mathrm{ind}_{GL(\eta^{-1}(S\cap W_1(R_i)))\times O(\mathcal{M},q_{\mathcal{M}})\times \tilde{U}_{\eta^{-1}(S\cap w_1(R_i))}}^{SO((\xi)^\perp,q_{W_1})}\\
&(|detg|^{s-i+\frac{1}{2}(\dim W_1-j-1)}(\tau^{w_S})^{(\dim S-\beta)}(\eta g_1\eta^{-1})\otimes\\
&(\sigma^{w_S})^{\langle i-(\dim S-\beta)\rangle}(\eta g_2\eta^{-1})\otimes \mathbb{1}(u))
\end{aligned}$$

Here (g_1, g_2, u) lies in the inducing group $GL(\eta^{-1}(S \cap W_1(R_i)) \times O(\mathcal{M}, q_{\mathcal{M}}) \times \tilde{U}_{\eta^{-1}(S\cap W_1(R_i))}$. Here we vary all S so that $\alpha = 0$ and $\dim S - \beta \leq i$. We call each such module $\mathcal{I}^{(S)}_{j,(\alpha,\beta)}(s)$. Then we have that $\mathcal{I}_{supp}(\tau, \sigma) = \{F \in \mathbb{A}_{SO((\xi_0)^\perp, q_{W_1})} \mid \mathcal{I}^{(S)}_{j,(\alpha,\beta)} * F = 0$ for all S so that $\alpha \geq 0$ and $\dim S - \beta \leq i\}$.

Thus $\mathcal{I}_{supp}(\tau, \sigma)$ = the ideal annihilating all the boundary components of the Jacquet module

$$(\mathcal{I}_j(\tau, \sigma, s))_{(N^i\mathbb{U}_i, \Lambda_{\xi_0})}.$$

It is possible to compute this ideal (as in §2) explicitly. First it is possible to compute the derivative of τ^{w_S} of $GL(S)$ for a specific family of $GL(S)$ modules (using **Remark** 5 of §3). Namely we let

$$\tau^{w_S} = \mathrm{ind}_{B_S}^{GL(S)}(\chi_1 \otimes \cdots \otimes \chi_j) = \text{“}\chi_1 \otimes \chi_2 \otimes \cdots \otimes \chi_j''$$

By construction of the flag defining B_{R_j} in §8 and **Remark 5** in §3 we have that (as a $GL_{j-\nu}$ module) after semisimplification

$$(\tau^{w_S})^{(v)} = \oplus(\chi_{z_1} \otimes \chi_{z_2} \otimes \cdots \otimes \chi_{z_{j-\nu}}) \otimes |detg|^{\nu/2}$$

where the sum is taken over all tuples $(z_1, \cdots, z_{j-\nu})$ where $1 < z_1 < \cdots < z_{j-\nu} \leq j$. Thus for τ^{w_S} we deduce that the semisimplication of $\mathcal{I}_{j,(\alpha,\beta)}(s)$ equals as a $SO((\xi_0)^\perp, q_{W_1})$ module the direct sum of spaces (with unnormalized induction)

$$\begin{aligned}
&\mathrm{ind}_{B_{\eta^{-1}(S\cap W_1(R_i)}\times O(\mathcal{M},q_{\mathcal{M}})\ltimes\tilde{U}_{\eta^{-1}(S\cap W_1(R_i))}}^{SO((\xi_0)^\perp,q_{W_1})}\\
&\left(|\ |^{s+\lambda}\delta_B^{1/2}\otimes(\chi_{z_1}\otimes\cdots\otimes\chi_{z_{j-\nu}})\otimes(\sigma^{w_S})^{\langle i-(j-\beta)\rangle}\otimes \mathbb{1}\right)
\end{aligned}$$

(δ_B = usual Jacobian of the Borel group $B_{\eta^{-1}(S\cap W_1(R_i))}$ and $\lambda = \frac{1}{2}(\dim W_1 - j - 1) - i + (j - \beta)/2$). In particular we note that each such module can be expresssed as follows. Indeed we consider the parabolic $P_1(\xi_0) = GL_1 \times M(\xi_0) \ltimes U_1(\xi_0)(\cong$

stabilizing an isotropic line). Then the $SO((\xi_0)^{\perp}, q_{W_1})$ modules above have the form (unnormalized induction

$$\mathrm{ind}_{P_1(\xi_0)}^{SO((\xi_0)^{\perp}, q_{W_1})}\left(\chi(t_1)\cdot|t_1|^{(s+\frac{1}{2})+(\frac{\dim W_1(R_i)-3}{2})}\otimes\tilde{\sigma}\otimes\mathbb{1}\right)$$

Here $\tilde{\sigma}$ is a smooth representation of $M(\xi_0)$ (depending possibly on s). Also χ is a character of the form χ_i. Then by using the same arguments as in §2 we consider the specific element in $\mathbb{A}_{SO((\xi_0)^{\perp}, q_{W_1})}$ which annihilates the $K_{SO((\xi_0)^{\perp}, q_{W_1})}$ vectors in the above space given by

$$L(\chi, X, Z) = \prod_{L=1}^{i=r}(q^{-\chi-\frac{1}{2}}XZ_i - 1)(q^{-\chi-\frac{1}{2}}XZ_i^{-1} - 1)$$

($r = \mathrm{rank}\ SO((\xi_0)^{\perp}, q_{W_1})$. Here $q^{-\chi} = \chi(\pi_\nu)$.

Here the Z_i are generators of the Hecke algebra

$$\mathcal{H}(SO((\xi_0)^{\perp}, q_{W_1})//K_{SO((\xi_0)^{\perp}, q_{W_1})}) \cong \mathbb{C}[Z_1, Z_1^{-1}, \cdots, Z_r, Z_r^{-1}]^{W_{SO((\xi_0)^{\perp}, q_{W_1})}}$$

(via Satake isomorphism) (See §**1** and §**2**).

Lemma 5.1. *Let τ be as above. Then the ideal $\mathcal{I}_{supp}(\tau, \sigma)$ contains the element*

$$\prod_{\ell=1}^{\ell=j} L(\chi_\ell, X, Z)$$

Proof. We note that the above element annihilates all the possible constituents that occur in the semi-simplification of the module mentioned above. □

Remark 1. We note that in the case where the χ_i are not distinct, then we can replace the element in the above **Lemma** by $\prod L(\chi_i, X, Z)$ where the product is taken over all those χ_i so that the numbers $\{q^{-\chi_i}\}$ are distinct. We note here that the element $\prod_{\ell=1}^{\ell=j} L(\chi_\ell, X, Z)$ is completely determined by the annihilation of the space $I_{j,(0,\beta)}(s)$ where $\dim S - \beta$ is smallest and greater than zero! The space $I_{j,(0,\beta)}(s)$ then corresponds to the smallest possible cell $I_j^{(0,\beta)}$ (see below) which can carry a nontrivial $(N^i\mathbb{U}_i, \Lambda_{\xi_0})$ Jacquet functor.

Remark 2. We emphasize here that the factor

$$\prod_{\ell=1}^{\ell=j} L(\chi_\ell, q^{-s}, \beta) = \prod_{\substack{\ell=1,\dots,j \\ t=1,\dots,r}} (1 - q^{-\chi_\ell+\beta_t-\frac{1}{2}-s})(1 - q^{-\chi_\ell--\beta_t--\frac{1}{2}-s})$$

(recall that $\prod = \mathrm{ind}_{B_{SO((\xi_0)^\perp, q_{W_1})}}^{SO((\xi_0)^\perp, q_{W_1})}(\beta)$ so that $\beta = (\beta_1, \ldots, \beta_r)$ and $\beta_t(\pi) = q^{-\beta_t}$). In any case the above polynomial is connected to L functions in the following way. Indeed we let $D(\tau_v) = D_j(q^{-\chi_1}, \ldots, q^{-\chi_j})$ and $D(\Pi_v)$ be the matrix in ${}^L SO((\xi_0)^\perp, q_{W_1})$ given by

$$D(\Pi_v) = \begin{cases} D_{2r}(q^{\beta_1}, \ldots, q^{\beta_r}, q^{-\beta_r}, \ldots, q^{-\beta_1}) & \text{if the space } ((\xi_0)^\perp, q_{W_1}) \text{ is an even split form or an odd dimensional space} \\ \begin{bmatrix} q^{\beta_1} & & & & & & \\ & \ddots & & & & 0 & \\ & & q^{\beta_{r-1}} & & & & \\ & & & \begin{matrix} 0 & 1 \\ 1 & 0 \end{matrix} & & & \\ & & & & q^{-\beta_{r-1}} & & \\ & 0 & & & & \ddots & \\ & & & & & & q^{-\beta_1} \end{bmatrix} & \text{if the space } ((\xi_0)^\perp, q_{W_1}) \text{ is a even dimensional nonsplit space.} \end{cases}$$

Then the standard L function $L(\tau \otimes \Pi, \lambda)$ associated to the tensor product of τ and Π is given by

$$\det[I - D(\tau_v) \otimes D(\Pi_v) q^{-\lambda}]^{-1}.$$

Thus the relation is that

$$\prod_{\ell=1}^{\ell=j} L(\chi_\ell, q^{-s}, \beta) = \begin{cases} L_v(\tau \otimes \Pi, s + \frac{1}{2})^{-1} & \text{if the space } ((\xi_0)^\perp, q_{W_1}) \text{ is even split or odd} \\ L_v(\tau \otimes \Pi, s + \frac{1}{2})^{-1} \prod_{i=1}^{i=j} \zeta(2s + \chi_i + 1) & \text{if the space } ((\xi_0)^\perp, q_{W_1}) \text{ is even of the form } (r-1, r-1) \oplus W_0 \\ & W_0 \text{ a 2 dimensional space with the norm form on the field } K. \end{cases}$$

We recall (§3) the decomposition of the group $\mathbb{G}$ into Bruhat cells.

$$\mathbb{G} = \bigcup_{w_S} \mathbb{P}_j w_S \mathbb{P}_i$$

We note that the open cell corresponds to the subspace S (in $\mathbb{P}_j \backslash \mathbb{G} = X_j$) which satisfies $\alpha = 0$ and $\dim S - \beta = j - \beta = i$. Then we let

$$I_j^{(\alpha,\beta)}(\tau, \sigma, s) = \mathrm{ind}_{\mathbb{P}_i^j(\alpha,\beta)}^{\mathbb{P}_i}(w_S \circ \Gamma|_{\tau \otimes |\ |^s \otimes \sigma \otimes 1})$$

and

$$I_j^r(\tau,\sigma,s) = \{f \in I_j(\tau,\sigma,s) | supp(f) \subseteq \bigcup_S P_j w_S \mathbb{P}_i \text{ where } \dim(\mathbb{P}_j w_S \mathbb{P}_i) \geq r\}$$

Then there is an exact sequence of $\mathbb{P}_i$ modules:

$$0 \to I_j^{(r+1)}(\tau,\sigma,s) \to I_j^{(r)}(\tau,\sigma,s) \to \bigoplus_{\{(\alpha,\beta)|\dim(\mathbb{P}_j w_S \mathbb{P}_i)=r\}} I_j^{(\alpha,\beta)}(\tau,\sigma,s) \to 0$$

Suppose we are given a function $F \in I_j^{(r)}(\tau,\sigma,s)$ so that for each (α,β) with $\dim(\mathbb{P}_j w_S \mathbb{P}_i) = r$ the image $\rho_{(\alpha,\beta)}(F)$ in $I_j^{(\alpha,\beta)}(\tau,\sigma,s)$ lies in the kernel of the Jacquet module relative to $(N^i\mathbb{U}_i, \Lambda_{\xi_0})$. That is, $\rho_{(\alpha,\beta)}(F) = \sum F_i * \nu_i - \Lambda_{\xi_0}(\nu_i)F_i$ for $\nu_i \in N^i\mathbb{U}_i$ and $F_i \in I_j^{(\alpha,\beta)}(\tau,\sigma,s)$. Then it follows that there exists a sufficiently large compact open subgroup $N_0^i\mathbb{U}_{i,0}$ of $N^i\mathbb{U}_i$ so that

$$\rho_{\Lambda_{\xi_0}}(F)(g) = \frac{1}{meas(N_0^i\mathbb{U}_{i,0})}\left(\int_{N_0^i\mathbb{U}_{i,0}} F(gu)\Lambda_{\xi_0}(u)du\right)$$

vanishes under the $\rho_{(\alpha,\beta)}$ map. Then repeating this construction for each r and with the assumption that $\rho_{(\alpha,\beta)}(F)$ lies in the kernel of the Jacquet module for $(N^i\mathbb{U}_i, \Lambda_{\xi_0})$ for all (α,β) (except when S determines the open orbit in $\mathbb{G}$), we deduce that

$$\rho_{\Lambda_{\xi_0}}(F) \in I_j^{(0,\beta)}(\tau,\sigma,s)$$

(where $(0,\beta)$ determines the *open cell*).

On the other hand we look at the orbits of the group

$$G_{\xi_0} = (N^i\mathbb{U}_i) \cdot SO((\xi_0)^{\perp}, q_{W_1})$$

in the space $\mathbb{P}_i^j(0,\beta)\backslash\mathbb{P}_i((0,\beta)$ which determines the open cell. We know from §3 there exist finitely many orbits of G_{ξ_0} in this space. In fact the orbits are parameterized by the orbits of $SO((\xi_0)^{\perp}, q_{W_1})$ in the space $\mathbb{P}_{S\cap W_1(R_i)}^{SO(W_1(R_i))}\backslash SO(W_1(R_i))$. Then the following ideas used in §3 (See data before **Remark 6**) we have that there is a unique G_{ξ_0} open orbit (largest dimensional orbit) in $\mathbb{P}_i^j(0,\beta)\backslash\mathbb{P}^i$. In fact when $(\beta \neq 0)$ the open orbit condition is that $S \cap W_1(R_i) \cap (\xi_0)^{\perp}$ is an isotropic subspace of codimension 1 in $S \cap W_1(R_i)$. In the case where $S \cap W_1(R_i) = \{0\}$ then $\mathbb{P}_{S\cap W_1(R_i)}^{SO(W_1(R_i))} = SO((\xi_0)^{\perp}, q_{W_1})$. In any case we note here that $e =$ the identity element is a representative of the open G_{ξ_0} orbit in $\mathbb{P}_i^j(0,\beta)\backslash\mathbb{P}_i$ (i.e. $\mathbb{P}_i^j(0,\beta)$. G_{ξ_0} is open).

By using exactly the same arguments as in the above paragraph we can deduce the following statement. If $F \in I_j^{(0,\beta)}(\tau,\sigma,s)$ satisfies the property that the projection of F onto the induced modules (associated to all the G_{ξ_0} orbits in $\mathbb{P}_i^j(0,\beta)\backslash\mathbb{P}^i$ except the open orbit) vanishes relative to the Jacquet functor $(N^i\mathbb{U}_i, \Lambda_{\xi_0})$, then

$$\rho_{\Lambda_{\xi_0}}(F) \in$$
$$\operatorname{ind}_{\mathbb{P}_i^j(0,\beta)\cap G_{\xi_0}}^{G_{\xi_0}}(w_S \circ \Gamma|_{\tau|\ |^s\otimes\sigma\otimes 1})$$

We note here how the choice of $N_0^i\mathbb{U}_{i,0}$ varies with the variable s. In fact, we note by a similar argument used in [CS] (using a Baire category argument) that for a fixed analytic section F as above, there exists a large enough $N_0^i\mathbb{U}_{i,0}$ so that $\rho_{\Lambda_{\xi_0}}(F) \in \mathrm{ind}_{\mathbb{P}_i^j(0,\beta)\cap G_{\xi_0}}^{G_{\xi_0}}(w_S \circ \Gamma\mid_{\tau|\,|^s\otimes\sigma\otimes 1})$ which is *valid for all* $s \in \mathbb{C}$!

In order to construct the space A (given in **Lemma 3.5**, see also the argument used in **Lemma 3.4**) we recall that

$$\mathbb{P}_i^j(0,\beta) \cap G_{\xi_0} = N^i \cdot Q^*_{S\cap W_1(R_i)} \cdot N_\beta$$

where $Q^*_{S\cap W_1(R_i)}$ is defined in §3 (See proof of **Lemma 3.4** and **Remark 6**)

$$N_\beta = N_{(i,j)}^{(0,\beta)} = (N_{(i,j)}^{(0,\beta)} \cap GL(S)).$$

In any case when we compute the Jacquet functor of $w_S \circ \Gamma \circ |_{\tau\otimes|\,|^S\otimes\sigma\otimes 1}$ restricted to $(\hat{N}_i N_\beta, \Lambda_{\xi_0})$ as a $GL(\{z_2,\cdots z_\ell\}) \cdot SO(W_1(S)\cap W_1(R_i)) \ltimes T_{\ell-1}V_{\ell-1}$ module we have

$$(\tau^{w_S})^{\langle i\rangle}(g_1 a) \otimes (\sigma^{w_S})(g_2) \otimes 1(bu)$$

for $(g_1 g_2(a,b)u') \in GL(\{z_2,\cdots z_\ell\})\cdot SO(W_1(S)\cap W_1(R_i)) \ltimes T_{\ell-1}V_{\ell-1}$ (See **Lemma 3.5**). Then we let J be the quotient map from

$$w_S \circ \Gamma \circ |_{\tau\otimes|\,|^s\otimes\sigma\otimes 1} \to (w_S \circ \Gamma|_{\tau\otimes|\,|^s\otimes\sigma\otimes 1})_{(N^i\cdot N_\beta,\Lambda_{\xi_0})}$$

we deduce that the space

$$\begin{aligned}&\mathrm{ind}_{(\mathbb{P}_i^j(0,\beta)\cap G_{\xi_0}}^{G_{\xi_0}}(w_S \circ \Gamma|_{\tau\otimes|\ |^s\otimes\sigma\otimes 1})_{(N^i\mathbb{U}_i,\Lambda_{\xi_0})}\\ &\cong \mathrm{ind}_{GL(\{z_2,\cdots,z_\ell\})\times SO(W_1(S)\cap W_1(R_i)\ltimes T_{\ell-1}V_{\ell-1}}^{SO((\xi_0)^\perp,q_{W_1})}(\text{image } (J)) \cong A\end{aligned}$$

(given in **Lemma 3.5**). We emphasize that in this case $(j-\beta=i)$ that $(W_1(S)\cap W_1(R_i) = W_1(S)$ and that $w_S^{-1}W_1(S)w_S = W_1(R_j)$.

Thus we have now shown the following

Proposition 5.1. *Let* $\phi \in \mathcal{I}_{supp}(\tau,\sigma)$. *Let* $f \in I_j(\tau,\sigma,s)^{K_{SO((\xi_0)^\perp,q_{W_1})}}$. *Then the function* $g \rightsquigarrow J\left[\rho_{\Lambda_0}(f*\phi)(g)\right]$ *lies in the space*

$$\begin{aligned}&ind_{GL(\{z_2,\cdots z_\ell\})\times SO(W_1(R_i)\cap W_1(S))\ltimes T_{\ell-1}V_{\ell-1}}^{SO((\xi_0)^\perp,q_{W_1})}((g_1,g_2,(a,b)u) \rightsquigarrow\\ &((\tau\otimes|\ |^s)^{w_S})^{\langle i\rangle}(g,a)\otimes(\sigma^{w_S})(g_2)\otimes 1(bu))^{K_{SO((\xi_0)^\perp,q_{W_1})}}\end{aligned}$$

Proof. We know $\rho_{\Lambda_0}(f*\phi) \in$

$$\mathrm{ind}_{\mathbb{P}_i^j(0\beta)\cap G_{\xi_0}}^{G_{\xi_0}}(w_S \circ \Gamma|_{\tau\otimes|\,|^s\otimes\sigma\otimes 1})$$

□

Remark 3. In the case $\beta = 0(\dim S = i)$ then we have that J is the quotient map from $w_S \circ \Gamma \circ |_{\tau\otimes||^s\otimes\sigma\otimes 1}$ to $(w_S \circ \Gamma|_{\tau\otimes||^s\otimes\sigma\otimes 1})_{(N^iN_\beta,\Lambda_{\xi_0})}$; this latter space equals $(\tau^{w_S} \otimes |\ |^s)^{(i)} \otimes \sigma^{w_S}$ as a $SO(W_1(R_i) \cap W_1(S))$ module. Hence

$$J[\rho_{\Lambda_0}(f * \phi)(g)] \in$$
$$(\tau \otimes |\ |^s)^{(i)} \otimes [\mathrm{ind}_{SO(W_1(R_i)\cap W_1(S))}^{SO((\xi_0)^\perp,q_{W_1})}(\sigma^{w_S})]^{K_{SO((\xi_0)^\perp,q_{W_1})}}$$

We let Π be an unramified representation of the group $SO((\xi_0)^\perp, q_{W_1})$

We consider the representation of $G_{\xi_0} = N^i\mathbb{U}_i \cdot SO((\xi_0)^\perp, q_{W_1})$ given by $\Lambda_{\xi_0} \otimes \Pi$. Then we form the non compactly induced module

$$\mathrm{Ind}_{G_{\xi_0}}^{\mathbb{G}}(\Pi \otimes \Lambda_{\xi_0})$$

Then by **Definition (II)** in §4 we know that if the data τ, s and σ is sufficiently generic so that $I_j(\tau, \sigma, s)$ is irreducible (as a $\mathbb{G}$ module), then

$$Hom_{\mathbb{G}}(I_j(\tau, \sigma, s), \mathrm{Ind}_{G_{\xi_0}}^{\mathbb{G}}(\Pi \otimes \Lambda_{\xi_0}))$$

is at most one dimensional. It is possible to construct a nonzero element in the above open space for sufficiently general parameters τ, σ and s.

We note here that the space

$$W_1 = R_i \oplus W_1(R_i) \oplus R_i^*.$$

Here $R_i = \mathrm{Sp}\{v_1, \ldots, v_i\}$ and $R_i^* = \mathrm{Sp}\{v_1^*, \ldots, v_i^*\}$. Then we have that $W_1(R_i) = S \cap W_1(R_i) \oplus S_\# \oplus (S \cap W_1(R_i))^*$ where $S \cap W_1(R_i) = \mathrm{Sp}\{z_1, \ldots, z_{j-i}\}$, $(S \cap W_1(R_i))^* = \mathrm{Sp}\{z_1^*, \ldots, z_{j-i}^*\}$ and $S_\# = W_1(S) = W_i(S) \cap W_1(R_i) = \mathrm{Sp}\{y_1, \ldots\}$. We note that $\xi_0 \in \mathrm{Sp}\{z_1, z_1^*\}$, i.e., $\xi_0 = z_1^+$ and $z_1^- \in (\xi_0)^\perp$ (in $W_1(R_i)$). Indeed $(z_1^-)^\perp = \mathrm{Sp}\{z_2, \ldots, z_{j-i}\} \oplus S_\# \oplus \mathrm{Sp}\{z_2^*, \ldots, z_{j-i}\}$. Then we define w_S explicitly here. In fact, if $2j < \dim W_1$ ($W_1(S) \neq \{0\}$) then we let w_S be given by the following formula: $w_S(v_t) = v_t^*$, $w_S(v_t^*) = v_t$, $1 \leq t \leq i$, $w_S(z_t) = z_t$, $w_S(z_t^*) = z_t^*$ $(1 \leq t \leq j - i)$ and w_S restricted to $\mathrm{Sp}\{v_1, \ldots, v_i, z_1, \ldots z_{j-i}, v_1^*, \ldots, v_i^*, z_1^*, \ldots, z_{j-i}^*\}^\perp$ to have the appropriate determinant so that $\det(w_S) = 1$. In fact, we choose $w_S\ |_{S_\#}$ more concretely below.

Indeed we define $w_S\ |_{S_\#}$ as follows: (See **Appendix** 3 to this section.)

(1) $S_\#$ is totally split, i.e. $S_\# = \mathrm{Sp}\{z_{j-i+1}, \ldots, z_\nu, z_\nu^*, \ldots, z_{j-i+1}^*\}$. Then we let $w_S = I$ if i is even and $w_S(z_\nu) = z_\nu^*$ and $w_S(z_\nu^*) = z_\nu$ and I on $\{z_\nu, z_\nu^*\}^\perp$ (in $S_\#$) if i is odd.

(2) $S_\#$ has an anisotropic part S_{anis} which is either one or two dimensional. In any case, $w_S\ |_{S_\#}$ is defined in **Appendix** 3 in 4 cases.

We note that in the specific case $2j = \dim W_1$, then w_S has the form $w_S(v_t) = v_t^*$, $w_S(v_t^*) = v_t$ $(t \leq i)$ and $w_S(z_\ell) = z_\ell$, $w_S(z_\ell^*) = z_\ell$ for $\ell < \nu$ and $w_S = I$ in $\{z_\nu, z_\nu^*\}$ if i even and $w_S(z_\nu) = z_\nu^*$, $w_S(z_\nu^*) = z_v$ if i is odd.

We also emphasize by such a choice of w_S that $w_S^2 = e$ and that w_S does not depend on j. That is w_S is the same for all $j > i$!

We note the above discussion is valid except for the case where $j = \dim W_1$ and i is odd. Then the space $S \cap W_1(R_i) = Sp\{z_1, \dots, z_{j-i}^*\}$. and in such an instance we must make the appropriate adjustments (i.e., in $(S \cap W_1(R_i))^*, \dots$). However, we note with this adjustment w_S still remains the same.

In fact, we recall that $R_j = \mathrm{Sp}\{v_1, \dots, v_i, z_1, \dots, z_{j-i}\}$ (where $S = \mathrm{Sp}\{v_1^*, \dots, v_i^*, z_1, \dots, z_{j-i}\}$ with $W_1(R_i) \cap S = \mathrm{Sp}\{z_1, \dots, z_{j-i}\}$ (in the case where $j = \dim W_1$ and i odd, then v_j in the formula above is replaced by v_j^*)). We let $v_{i+1} = z_1, \dots, v_j = z_{j-i}$. Then B_{R_j} = Borel subgroup of $GL(R_j)$ stabilizing the flag $\{v_1, \dots, v_j\} \supseteq \dots \supseteq \{v_j\}$. Let N^j be the associated maximal unipotent group. On the other hand, we consider the flag

$$\{v_i, \dots, v_1, v_{i+1}, \dots, v_j\} \supseteq \dots \supseteq \{v_j\}.$$

We let $B_{+,-}$ be the Borel group of $GL(R_j)$ stabilizing this flag; let $N^j_{+,-}$ be the corresponding maximal unipotent group. Then, in particular, if $\{\alpha_1, \dots, \alpha_j\}$ is the usual set of positive simple roots associated to N^j it follows that the associated set of simple roots for $N^j_{+,-}$ is $\{-\alpha_1, \dots, -\alpha_{i-1}, \varepsilon_1 - \varepsilon_{i+1}, \alpha_{i+1}, \dots, \alpha_j\}$ (here $\alpha_t = \varepsilon_t - \varepsilon_{t+1}$). We let $\mathbf{\Psi}_{i,j}$ be the usual Whittaker character on $N^j_{+,-}$. That is, if we let $\Lambda_j \colon GL(R_j) \to GL(k^j)$ be the isomorphism adapted to the basis $\{v_1, \dots, v_j\}$ then

$$N^j_{+,-} = \Lambda_j^{-1}\left(\left\{\left[\begin{matrix} 1 & & 0 & & \\ & \ddots & & & * \\ * & & 1 & & \\ & & & 1 & * \\ & 0 & & & \vdots \\ & & & 0 & 1 \end{matrix}\right] \middle| \; * \text{ arbitrary}\right\}\right).$$

Moreover

$$\mathbf{\Psi}_{i,j}\left(\Lambda_j^{-1}\left(\left[\begin{matrix} 1 & & & & & & \\ x_{21} & & & & & & \\ \vdots & \ddots & & & & y_{ij} & \\ \vdots & \dots & x_{i-1,i} & 1 & & & \\ & & & & 1 & z_{11} & \\ & 0 & & & & \ddots & \ddots \\ & & & & 0 & & 1 \quad z_{j-i-1,j-i} \end{matrix}\right]\right)\right)$$

$$= \psi(x_{21} + \dots + x_{i-1,i})\psi(y_{1i+1}\gamma)\psi(z_{11} + \dots + z_{j-i-1,j-i})$$

(γ *chosen below*).

Then we know that there exists a nonzero $L_j \in \mathrm{Hom}_{GL(R_j)}(\tau, \mathrm{Ind}_{N^j_{+,-}}^{GL(R_j)}(\mathbf{\Psi}_{i,j}))$. In **Appendix I** we define L_j explicitly. Thus for $F \in I_j(\tau, \sigma, s)$ we have $L_j(F)(\,) \in \mathrm{Ind}^{\mathbb{G}}_{N^j_{+,-} \times SO(W_1(R_j)) \times \mathbb{U}_j}(\overline{\mathbf{\Psi}}_{i,j} \otimes \sigma \otimes \mathbb{1})$.

We recall that $\mathbb{U}_i \cap GL(S) = N_\beta = \{N_i((M_1,\ldots,M_i),0) \mid M_\mu \in S \cap W_1(R_i)\}$. Then we note, if

$$M_\nu = \sum_{j-i \geq t \geq 1} \lambda_{\nu t} v_{i+t}$$

$$\Lambda_j(w_S N_i((M_1,\ldots,M_i),0)w_S) = \begin{bmatrix} I_i & \lambda_{\nu t} \\ 0 & I_{j-i} \end{bmatrix},$$

(in the case where $j = \dim W_1$ and i odd, then v_j in the formula above is replced by v_j^*).

Thus we have that $(u_1 \in N_\beta)$

$$L_j(F)[w_S u_1 x] = \overline{\Psi}_{i,j}(w_S u_1 w_S) L_j(F)[w_S x] = \overline{\psi}(\lambda_{11}\gamma) L_j(F)[w_S x].$$

On the other hand, we have that the character Λ_{ξ_0} restricted to N_β has the following form. If $u_1 = N_i((M_1,\ldots,M_i),0))$ then $\Lambda_{\xi_0}(u_1) = \psi(q_{W_1}(M_1,\xi_0))$. If $\xi_0 = z_1 + \gamma z_1^*$, then $\Lambda_{\xi_0}(u_1) = \psi(\lambda_{11}\gamma)$.

In particular, this means we can define formally the integral

$$V_j(F)(x) = \int_{N_\beta \backslash \mathbb{U}_i} L_j(F)[w_S u x] \Lambda_{\xi_0}(u) du.$$

Our goal here is to determine first the formal invariance properties of the integral defining V_j. We determine the convergence issue in **Appendix 2**.

For this discussion we recall some notation. First $\ell = j - i$ *here*. We let $P_{\{z_2,\ldots,z_\ell\}}$ be the parabolic subgroup of $SO((\xi_0)^\perp, q_{W_1})$ which stabilizes $\mathrm{Sp}\{z_2,\ldots,z_\ell\}$ (again in the case $j = \dim W_1$ and i odd, we replace z_{j-i} by z_{j-i}^* here in the ensuing discussion). We also let $\underset{\sim}{P}_{\{z_1,\ldots,z_\ell\}}$ be the mirabolic subgroup of $GL(\{z_1,\ldots,z_\ell\})$ (the subgroup of $GL(\{z_1,\ldots z_\ell\})$ fixing z_1^*). We let $U_{\{z_2,\ldots,z_\ell\}}$ be the unipotent radical of $P_{\{z_2,\ldots,z_\ell\}}$. Then $Q^*_{S\cap W_1(R_i)}$ = the stabilizer in $SO((\xi_0)^\perp, q_{W_1})$ of the space $\mathrm{Sp}\{z_2,\ldots,z_\ell\} = GL(\{z_2,\ldots,z_\ell\}) \times SO(W_1(S) \cap W_1(R_i)) \rtimes U_{\{z_2,\ldots,z_\ell\}}$. Moreover, $\underset{\sim}{P}_{\{z_1,\ldots,z_\ell\}} \cong GL(\{z_2,\ldots,z_\ell\}) \rtimes k^{\ell-1}$ and we let $N_{\{z_1,\ldots,z_\ell\}}$ ($N_{\{z_2,\ldots,z_\ell\}}$ resp.) be the maximal unipotent subgroup of $B_{\{z_1,\ldots,z_\ell\}}$ ($B_{\{z_2,\ldots,z_\ell\}}$ resp.) where $B_{\{z_1,\ldots,z_\ell\}}$ stabilizes (in $GL(\{z_1,\ldots,z_\ell\})$ the flag $\{z_1,\ldots,z_\ell\} \supseteq \cdots \supseteq \{z_\ell\}$ ($B_{\{z_2,\ldots,z_\ell\}}$ is the Borel subgroup of $GL(\{z_2,\ldots,z_\ell\})$ which stabilizes the flag $\{z_2,\ldots,z)_\ell\} \supseteq \cdots \supseteq \{z_\ell\}$). At this point we note that

$$\Lambda_j(w_S N_{\{z_1,\ldots,z_\ell\}} w_S) = \left\{ \left. \begin{bmatrix} I_i & 0 \\ 0 & \begin{matrix} 1 & & * \\ & \ddots & \\ 0 & & 1 \end{matrix} \end{bmatrix} \right| * \text{ arbitrary} \right\}$$

and

$$\Lambda_j(w_S N_{\{z_2,\ldots,z_\ell\}} w_S) = \left\{ \left[\begin{array}{cc} I_i & 0 \\ 0 & \begin{matrix} 1 & 0 & \cdots & 0 \\ 0 & 1 & & * \\ \vdots & & \ddots & \\ 0 & 0 & & 1 \end{matrix} \end{array} \right] \middle| \; * \text{ arbitrary} \right\}$$

First we determine how the function $V_j(F)$ transforms when $g = r_1 \cdot u_1 \in N_{\{z_2,\ldots,z_\ell\}} \cdot U_{\{z_2,\ldots,z_\ell\}}$. Indeed we have that

$$V_j(F)[r_1 u_1 g] = \psi_{\ell-1}(w_S r_1 w_S)\psi_{Z_\beta}(w_S u_1 w_S) V_j(F)[g]\,.$$

Here $\psi_{\ell-1}$ is the generic Whittaker character on the group $\Lambda_i(w_S N_{\{z_2,\ldots,z_\ell\}} w_S)$ given by the restriction of $\overline{\mathbf{\Psi}}_{i,j}$ to $w_S N_{\{z_2,\ldots,z_\ell\}} w_S$. Moreover, the character ψ_{Z_β} is given by the following formula. We know from §3 that $U_{\{z_2,\ldots,z_\ell\}} = T_{\ell-1} V_{\ell-1}$. Then ψ_{Z_β} sends $V_{\ell-1}$ to 1 and

$$T_{\ell-1} = \left\{ (s_{12},\ldots,s_{1\ell}) \cdot N_\ell \left(0, \begin{bmatrix} 0 & s_{12}^* & \cdots & s_{1\ell}^* \\ -s_{12}^* & & & \\ & & 0 & \\ -- s_{1\ell}^* & & & \end{bmatrix} \right) \right\}$$

(with $s_{1t}^* = s_{1t} q_W(\xi_0,\xi_0)^{-1}$) goes to the character $\psi(-s_{12})$. In particular we note from

Remark 7 of §3 that via the identification of $T_{\ell-1}$ with

$$N_{\ell-1}((s_{12}^* z_1^-,\ldots,s_{1\ell}^* z_1^-),0)$$

$\psi(-s_{12})$ equals $\psi(q_{W_1}(s_{12}^* z_1^-, z_1^-))$. This implies that ψ_{Z_β} (defined on $U_{\{z_2,\ldots,z_\ell\}}$) has exactly the same form as ψ_Z (defined on $\mathbb{U}_i$ in §3) where $Z_\beta = (z_1^-, 0, \cdots, 0)$. We note here that

$$\Lambda_j(w_S(a_1,\ldots,a_{\ell-1})w_S) = \left[\begin{array}{cc} I_i & 0 \\ 0 & \begin{matrix} 1 & +a_1 & \cdots & \cdots & +a_{\ell-1} \\ & 1 & & & \\ & & \ddots & & \\ & & & \ddots & \\ & & & & 1 \end{matrix} \end{array} \right].$$

Also, if $\gamma_1 \in SO(W_1(S) \cap W_1(R_i))$, then

$$V_j(F)(\gamma_1 g) = \sigma^{w_S}(\gamma_1) V_j(F)(g)\,.$$

Thus we have shown that $V_j(F)$ belongs to the $SO(\xi_0)^\perp, q_{W_1})$ module (unnormalized induction)

$$R_{\ell-1}(\sigma) = \mathrm{Ind}_{N_{\{z_2,\dots,z_\ell\}}\times U_{\{z_2,\dots,z_\ell\}}\times SO(W_1(S)\cap W_1(R))}^{SO((\xi_0)^\perp, q_{W_1})}(\psi_{\ell-1}\otimes\psi_{Z_\beta}\otimes\sigma^{w_S}).$$

In fact, from the considerations above the character ψ_{Z_β} on $U_{\{z_2,\dots,z_\ell\}}$ is fixed by $SO(W_1(S)\cap W_1(R_i))$. Hence the above induced module corresponds to a "Whittaker-Spherical" model as discussed in §3 and §4. In particular, since σ is an irreducible module for $SO(W_1(R_j))$, then σ^{W_S} is an irreducible module for $SO(W_1(S)) = SO(W_1(S)\cap W_1(R_i))$.

We let $R_{\ell-1}(\sigma)^*$ be defined exactly as $R_{\ell-1}(\sigma)$ except where $\psi_i\otimes\psi_{Z_\beta}$ is replaced by $\overline{\psi}_i\otimes\overline{\psi}_{Z_\beta}$ and σ replaced by σ^ν, the centragredient of σ.

Moreover we know that

$$\mathrm{Hom}_{SO((\xi_0)^\perp, q_{W_1})}(\Pi, R_{\ell-1}(\sigma)^*)$$

has dimension at most one (again using **Definition (II)** in §4). This implies that we can construct a pairing $\langle\langle\ |\ \rangle\rangle$ between the space given by

$$\{\mathrm{Span} V_j(F) \mid F\in I_j(\tau,\sigma,s)\}$$

and the space

$$\{\mathrm{Span} T(w) \mid w\in\Pi\}$$

($T\in \mathrm{Hom}_{SO((\xi_0)^\perp, q_{W_1})}(\Pi, R_{\ell-1}(\sigma)^*)$. We let $\{\ |\ \}_\sigma$ be an invariant pairing of σ with σ^ν.

Specifically we note here that the $SO(W_1(S)\cap W_1(R_i))$ invariant $\{\ |\ \}_\sigma$ constructs a $SO((\xi_0)^\perp, q_{W_1})$ invariant pairing $\langle\ |\ \rangle_\sigma$ between Π and σ given by the following relation: $\{\xi|T(w)(e)\}_\sigma = \langle\xi|w\rangle_\sigma$ for all $\xi\in\sigma$ and $w\in\Pi$. Then define

$$\langle\langle V_j(F)\mid T(w)\rangle\rangle = \int_{(N_{\{z_2,\dots,z_\ell\}}\cdot U_{\{z_2,\dots,z_\ell\}}\cdot SO(W_1(S)\cap W_1(R_i)))\backslash SO((\xi_0)^\perp, q_{W_1})} \{V_j(F)(h)\mid T(w)(h)\}_\sigma dh$$

where dh is *invariant* measure on the space $N_{\{z_2,\dots z_\ell\}}\cdot U_{\{z_2,\dots,z_\ell\}}SO(W_1(S)\cap W_1(R_i))\backslash\ SO((\xi_0)^\perp, q_{W_1})$. We note that if the above integral exists, then $\langle\langle\ |\ \rangle\rangle$ defines an element in the space

$$\mathrm{Hom}_{G_{\xi_0}}(I_j(\tau,\sigma,s), \Lambda_{\xi_0}\otimes\Pi).$$

This in turn by (Frobenius reciprocity) defines an element in $\mathrm{Hom}_{\mathbb{G}}(I_j(\tau,\sigma,s), \mathrm{Ind}_{G_{\xi_0}}^{\mathbb{G}}(\Lambda_{\xi_0}\otimes\Pi))$. In fact we have the identity

$$\begin{aligned}&\langle\langle V_j(F)|T(w)\rangle\rangle =\\ &\int \{V_j(F)(h)|T(w)(h)\}_\sigma dh =\\ &\int \langle V_j(F)(h)|\pi(h)(w)\rangle_\sigma dh\end{aligned}$$

The domain of integration in both cases is the same (which we omit). Thus we note the inductive nature of the pairing $\langle\langle\ |\ \rangle\rangle$ with the pairing $\langle\ |\ \rangle_\sigma$.

We prove (**Theorem A** of **Appendix (I)** of this section) that the integral defining the above pairing is absolutely convergent for suitable data in (χ, γ, s). In fact, in**Theorem A** we prove also that the above integral can be made equal to a distribution of the form $T^{\xi}_{\chi+s,\gamma}(\varphi_1 \otimes \varphi_2)$ defined in §8 (for the case $j = i+1$).

For notation we recall that $\tau =$ "$\chi_1 \otimes \cdots \otimes \chi_j$" and $\Pi = \mathrm{ind}_{B_{SO((\xi_0)^\perp, q_{W_1}}}^{SO((\xi_0)^\perp, q_{W_1})}(\beta_1 \otimes \cdot \otimes \beta_r)$. Let w_0 be the unique unramified vector in Π.

Theorem 5.1. *Let $j > i$. Let $\phi_0 \in \mathcal{I}_{\mathrm{supp}}(\tau, \sigma, s)$ be the element given in* **Lemma 5.1**.

Let $F \in I_j(\tau, \sigma, s)^{K_{SO((\xi_0)^\perp, q_{W_1})}}$. Then

$$\langle\langle V_j(F * \phi_0) \mid T(w_0)\rangle\rangle \equiv \left(\prod_{\ell=1}^{\ell=j} L(\chi_\ell, q^{-s}, \beta)\right) \langle\langle V_j(F) \mid T(w_0)\rangle\rangle\,.$$

In the case $j = i+1$ the term $\langle\langle V_j(F * \phi_0) \mid T(w_0)\rangle\rangle$ is a polynomial in q^s and q^{-s}.

Proof. We note that the convolution operator determined by ϕ can be adjointed across the $\langle\langle\ |\ \rangle\rangle$ form to determine the first statement in the **Theorem**.

On the other hand, we also note

$$\langle\langle V_j(\rho_{\Lambda_0}(F * \phi_0)) \mid T(w_0)\rangle\rangle = \langle\langle V_j(F * \phi_0) \mid T(w_0)\rangle\rangle\,.$$

Thus we can work with $\rho_{\Lambda_0}(F * \phi_0)$ to analyze the holomorphicity of the term

$$\langle\langle V_j(F * \phi_0) \mid T(w_0)\rangle\rangle\,.$$

In fact if we assume that $V_j(\rho_\Lambda(F * \phi_0))(h)$ is given by the integral

$$\int_{\widehat{N}_\beta \backslash \mathbb{U}_i} L_j(\rho_{\Lambda_0}(F * \phi_0))[w_S u g]\Lambda_{\xi_0}(u)du\,.$$

Then we must determine the set

$$\{g \in SO((\xi_0)^\perp, q_{W_1}) | V_j(\rho_{\Lambda_0}(F * \phi_0))(g) \neq 0\}\,.$$

In fact, the above set is contained in the set $\{g \in SO((\xi_0)^\perp, q_{W_1} \mid$ there exists $u_g \in \mathbb{U}_i$ so that $\rho_{\Lambda_0}(F * \phi_0)[w_S w_\ell u_g g] \not\equiv 0\}$.

However, we know that $\rho_{\Lambda_0}(F * \phi_0)$ is contained in the space (from the proof of **Proposition 5.1**)

$$\mathrm{ind}_{N^i N_\beta \cdot Q^*_{S \cap W_1(R_i)}}^{G_{\xi_0}}(w_S \circ \Gamma\ |_{\tau \otimes |\ |^s \otimes \sigma \otimes 1})\,.$$

In the case $j = i+1$, then $Q^*_{S\cap W_1(R_i)} = SO(W_1(S)\cap W_1(R_i))$. This implies that the closure of the set support $(\rho_{\Lambda_0}(F*\phi_0))$ (in the space $SO(W_1(S)\cap W_1(R_i))\backslash SO((\xi_0)^\perp, q_{W_1}))$ has compact support $\mod SO(W_1(S)\cap W_1(R_i))$. Thus

$$\text{support}\,(V_j(\rho_{\Lambda_0}(F*\phi_0))(\ ))$$

has compact support $\mod SO(W_1(S)\cap W_1(R_i))$.

Specifically we note here that (for any s), if F_1 lies in $\text{ind}^{G_{\xi_0}}_{N_iN_\beta Q^*_{S\cap W_1(R_i)}}(w_S\circ \Gamma\mid_{\tau\otimes|\ |^s\otimes\sigma\otimes 1})$, then $V_j(F_1)$ belongs to the space

$$\text{ind}^{SO((\xi_0)^1, q_{w_1})}_{SO(W_1(S)\cap W_1(R_i))}(\sigma)$$

This means concretely that the integral defining V_j is absolutely convergent for all s! Thus let the section F_1 above depend analytically on s (that is, F_1 can be expressed as a finite linear combination of standard sections of $I_j(\tau,\sigma,s)$ with analytic coefficients). Moreover, suppose F_1 lies in $\text{ind}^{G_{\xi_0}}_{N^iN_\beta Q^*_{S\cap W_1(R_i)}}(w_S\circ \Gamma\mid_{\tau\otimes|\ |^s\otimes\sigma\otimes 1})$, i.e. where for each s the support of F_1 remains in a set of the form $N^iN_\beta SO(W_1(S)\cap W_1(R_i))\Omega$, with Ω compact in $\mathbb{U}_iN^i_jSO((\xi_0)^\perp, q_{W_1})$. Then we let A_r be an increasing set of compact open subsets in $\mathbb{U}_i\cdot SO((\xi_0)^\perp, q_{W_1})$ so that

$$\bigcup_{r\geq 0} N_\beta\cdot SO(W_1(S)\cap W_1(R_i))A_r = \mathbb{U}_i\cdot SO((\xi_0)^\perp, q_{W_1})\,.$$

We consider the closed subset of $\mathbb{C}$ given by

$$\rho_r = \{s\in\mathbb{C}\mid \text{support}\,(F_1)\subseteq P_jw_SA_r\}\,.$$

Thus $\bigcup_{r\geq 0}\rho_r = \mathbb{C}$ and by Baire Category Principle there is an r_0 so that ρ_{r_0} contains a nonempty open set. Thus $(F_1)_s(g) = 0$ for $g\notin P_jw_SA_{r_0}$ and s in some open set. Since $(F_1)_s$ is analytic in s, we deduce $(F_1)_s(g)\equiv 0$ for $g\notin P_jw_SA_{r_0}$ and *all* s!

Now A_{r_0} is contained in a compact subset $U_{r_0}\times K_{r_0}\subseteq \mathbb{U}_i\times SO((\xi_0)^\perp, q_{W_1}$ where U_{r_0} and K_{r_0} are compact open in $\mathbb{U}_i$ and $SO((\xi_0)^\perp, q_{W_1})$. Thus it follows that if $L_j(F_1(w_S\mathcal{Z}))\neq 0$ for $\mathcal{Z}\in\mathbb{U}_i\cdot SO((\xi_0)^\perp, q_{W_1}))$ then $F_1(w_S\cdot\mathcal{Z})\neq 0$. In particular it follows that the integral

$$\begin{aligned}V_j(F_1)(g) &= \int_{N_\beta\backslash\mathbb{U}_i} L_j(F_1(w_Su\cdot g))\Lambda_{\xi_0}(u)d\mu\\ &= \int_{N_\beta\backslash N_\beta A_{r_0}} L_j(F_1(w_Sug))\Lambda_{\xi_0}(u)d\mu\end{aligned}$$

is absolutely convergent for *all* s and

$$\text{supp}V_j(F_1)\subseteq SO(W_1(S)\cap W_1(R_i))K_{r_0}\ \text{(for all } s)\,!$$

Thus given F_1 we let $(U_i)_0$ and $(K_i)_0$ be compact open subgroups of $\mathbb{U}_i$ and $SO((\xi_0)^\perp, q_{W_1})$ which fix the function F_1. We also assume $(K_i)_0$ fixes $T(w_0)(\ )$ on the left. Then we can choose K_{r_0} to be a finite union of the form $K_{r_0} = \cup\xi(K_i)_0$. In particular, let $(\mathbb{U}_i)_{00}$ be a sufficiently small compact open subgroup so that $(\xi(K_i)_0)^{-1}(\mathbb{U}_i)_{00}(\xi(K_i)_0)$ lies in $(\mathbb{U}_i)_0$ for all ξ as above. Moreover, let U_{r_0} then be a finite disjoint union $\cup t(\mathbb{U}_i)_{00}$. Then

$$V_j(F_1)(\xi) = \sum_t c_t L_j(F_1(w_S t\xi))\Lambda_{\xi_0}(t)$$

with c_t some constant depending only on volume considerations. This shows that $V_j(F_1)(\xi)$ is analytic in the s variable. In fact, each $L_j(F_1(w_S t\xi))$ is analytic in the s variable. Thus since F_1 can be expressed in terms of standard sections, it follows that for each g, $L_j(F_1(g))$ lies in the finite dimensional vector space of σ which is fixed by a small compact open subgroup $\tilde{K}$ of $SO(W_1(R_j)) = SO(W_1(S) \cap W_1(R_i))$. This proves that $L_j(F_1(g))$ can be written as a finite sum

$$\sum \lambda_v^g(s) a_\nu .$$

where $\{a_\nu\}$ is a basis of $\sigma^{\tilde{K}}$ and $\lambda_\nu(s)$ a polynomial in q^s and q^{-s}.

Then we consider the basic integral

$$\int_{SO(W_1(S)\cap W_1(R_i))\backslash SO((\xi_0)^\perp, q_{W_1})} \{V_j(F_1)(h) \mid T(w_0)(h)\} dh .$$

If F_1 admits the properties as above, we deduce that the integral equals

$$\begin{aligned} &\sum_{t,\xi} c_t \{L_j(F_1(w_S t\xi) \mid T(w_0)(\xi)\}\Lambda_{\xi_0}(t) \\ &= \sum_{t,\xi,\nu} c_t \{a_\nu \mid T(w_0)(\xi)\}\lambda_\nu^{w_S t\xi}(s)\Lambda_{\xi_0}(t) . \end{aligned}$$

Thus we have shown how the integral above can be expressed as a linear combination of terms with coefficient polynomials in q^s and q^{-s}.

We now apply the above considerations to the function

$$F_1 = \rho_{\Lambda_0}(F * \phi_0) .$$

We note that the choice of ρ_{Λ_0} is canonical (does not depend on s). Also we note that

$$V_j(\rho_{\Lambda_0}(F * \phi_0))(g) = V_j(F * \phi_0)(g)$$

for $g \in SO((\xi_0)^\perp, q_{W_1})$. This implies that $V_j(\rho_\Lambda(F * \phi_0))$ is analytic in s and that the integral

$$\int_{SO(W_1(S)\cap W_1(R_i))\backslash SO((\xi_0)^\perp, q_{W_1})} \{V_j(F * \phi_0)(h) \mid T(w_0)(h)\} dh$$

can be expressed as a polynomial in q^s and q^{-s}. $\square$

Remark 4. For the general case of $j > i$ in the proof above, we note that we can only deduce that

$$\text{support}\,(V_j(\rho_{\Lambda_0}(F * \phi_0))(\;))$$

is compact mod $GL(\{z_2, \ldots, z_\ell\})U_{\{z_2,\ldots,z_\ell\}} \cdot SO(W_1(S) \cap W_1(R_i))$ (and not necessarily mod $N_{\{z_2,\ldots,z_\ell\}} \cdot U_{\{z_2,\ldots,z_\ell\}} \cdot SO(W_1(S) \cap W_1(R_i))$ which is what is required to finish the proof). The subtle obstruction can be explained in the following way. We note that the map $F \to V_j(F)(\;)$ from $I_j(\tau, \sigma, s)$ to the non-compactly induced module

$$\mathrm{Ind}_{N_{\{z_2,\ldots,z_\ell\}}U_{\{z_2,\ldots,z_\ell\}}SO(W_1(S)\cap W_1(R_i))}^{SO((\xi)_0)^\perp, q_{W_1})}(\psi_{\ell-1} \otimes \psi_{Z_\beta} \otimes \sigma^{ws})$$

induces a $SO((\xi_0)^\perp, q_{W_1})$ intertwining map Ω from $I_j(\tau, \sigma, s)_{N^i\mathbb{U}_i, \Lambda_{\xi_0})}$ to the space above. However, it is not clear that this map is *surjective*. In fact, we know that the subspace

$$\mathrm{ind}_{GL(\{z_2,\ldots,z_\ell\})\times SO(W_1(S)\cap W_1(R_i))\rtimes T_{\ell-1}V_{\ell--1}}^{SO((\xi_0)^\perp, q_{W_1})}$$

$$((g_1, g_2, (a, b)u) \rightsquigarrow (\tau^{ws})^{\langle i\rangle}(g_1, a) \otimes (\sigma^{ws})(g_2) \otimes 1(bu))$$

is a $SO((\xi_0)^\perp, q_{W_1})$ submodule of $I_j(\tau, \sigma, s)_{(N^i\mathbb{U}_i, \Lambda_{\xi_0})}$ and it is not clear the image of this subspace under Ω is necessarily contained in the compactly induced module

$$\mathrm{ind}_{N_{\{z_2,\ldots,z_\ell\}}U_{\{z_2,\ldots,z_\ell\}}SO(W_1(S)\cap W_1(R_i))}^{SO((\xi_0)^\perp, q_{W_1})}(\psi_{\ell-1} \otimes \psi_{Z_\beta} \otimes \sigma^{ws}).$$

In particular this comment replies to the function $V_j(\rho_{\Lambda_0}(F * \phi_0))(\;)$.

Remark 5. It follows from **Theorem 5.1** (for the case $j = i + 1$) and **Remark 2** that the family of inner products $\langle\langle V_j(F) \mid T(w_0)\rangle\rangle$ can be expressed as the product of

$$\left(\prod_{\ell=1}^{\ell=j} L(\chi_\ell, q^{-s}, \beta)\right)^{-1}$$

times the factor $\langle\langle \Lambda_j(F * \phi_0) \mid T(W_0)\rangle\rangle$, which is a holomorphic function in s. The universal denominator term

$$\left(\prod_{\ell=1}^{\ell=j} L(\chi_\ell, q^{-s}, \beta)\right)$$

equals the product of 2 terms:

$$P(s) \cdot L\left(\tau \otimes \Pi, s + \frac{1}{2}\right)$$

where $P(s)$ is a certain rational function in q^{-s} and q^{χ_i} (given above) and $L(\tau\otimes\Pi, \lambda)$ denotes the standard L function of the tensor product of the representations τ and Π, i.e.,

$$\det(I - D(\tau_v) \otimes D(\Pi_v)q^{-\lambda})^{-1}.$$

Here $D(\tau_v)$ and $D(\Pi_v)$ represent the conjugacy classes of elements in $GL_j(\mathbb{C})$ and $SO((\xi_0)^{\perp}, q_{W_1})$ (the dual L groups of GL_j and $SO((\xi_0)^{\perp}, q_{W_1})$ repsecitvely) associated to τ and Π via the Satake isomorphism.

Thus we have presented in **Theorem 5.1** (for the case $j = i+1$) how the support idea developed in §2 makes it possible to determine the poles of the term

$$\langle\langle V_j(F) \mid T(w_0)\rangle\rangle$$

where F satisfies the hypothesis of **Theorem 5.1**. We note that we are specifically interested in the case where $I_j(\tau, \sigma, s)$ have a nonzero $K_{\mathbb{G}}$ fixed vector (spherical case). In that case F is the $K_{\mathbb{G}}$ fixed vector. However, we would expect more in the case. We recall that w_S conjugates $K_{SO((\xi_0)^{\perp}, q_{W_1})}$ and $K_{SO(W_1(S)\cap W_1(R_i))}$ to itself. Then we define the function

$$F^*(g) = \begin{cases} \tau(\gamma_1)(v_0)|\det \gamma_1|^{s+\lambda_j}\sigma(\gamma_2)(v^0) & \text{if } g = \gamma_1\gamma_2 u w_S u' g' \\ & \text{with } \gamma_1 \in GL_j,\ \gamma_2 \in SO(W_1(R_j)) \\ & u \in \mathbb{U}_j,\ u' \in \mathbb{U}_i(\mathcal{O}),\ g' \in K_{SO(\xi_0)^{\perp}, q_{W_1})} \\ 0 & \text{otherwise} \end{cases}$$

Here v_0 and v^0 are the unramified spherical vectors in τ and σ respectively and $\lambda_j = \frac{1}{2}(\dim W_1 - j - 1)$. Also $\mathbb{U}_i(\mathcal{O}) = K_{\mathbb{G}} \cap \mathbb{U}_i$ here.

Then $F^* \in I_j(\tau, \sigma, s)$ and, in fact,

$$F^* \in \operatorname{ind}_{N^i N_\beta Q_{S\cap W_1(R_i)}}^{G_{\xi_0}} (w_S \circ \Gamma \mid_{\tau| \ |^s \otimes\sigma\otimes 1}) .$$

Moreover, by direct calculation

$$\operatorname{supp}(V_j(F^*)) \subseteq SO(W_1(R_j))K_{SO(\xi_0)^{\perp}, q_{W_1})}$$

with $V_j(F^*)(k) = V_j(F^*)(e)$ where $k \in K_{SO((\xi_0)^{\perp}, q_{W_1})}$. More specifically,

$$\sigma^{w_S}(k')(V_j(F^*)(e)) = V_j(F^*)(e)$$

with $k' \in K_{SO(W_1(R_j))}$.

Now the basic question that is similar to **Proposition 2.1** is, does the following *identity hold*:

$$(*)\colon V_j(F * \phi_0)(g) = \lambda_j(\tau, \sigma, s)V_j(F^*)(g)$$

with

$$\lambda_j(\tau, \sigma, s) = \frac{1}{L(\tau \otimes \sigma, s+1)L(\tau, {}^{\text{sym}^2}_{\Lambda^2}, 2s+1)}$$

where we choose Λ^2 if $\mathbb{G}$ corresponds to an even dimensional form and to an odd dimensional form (in case $SO((\xi_0)^{\perp}, q_{W_1})$ corresponds to an even dimensional form with nonzero 2 dimensional anisotropic part). Otherwise we choose sym^2.

The point here is that by use of the above identity, we deduce that

$$\langle\langle V_j(F) \mid T(w_0)\rangle\rangle = \frac{L(\tau \otimes \Pi, s + \frac{1}{2})}{L(\tau \otimes \sigma, s+1) L(\tau, {}^{\mathrm{sym}^2}_{\Lambda^2}, 2s+1)} \langle\langle V_j(F^*) \mid T(w_0)\rangle\rangle \,.$$

Next we know by direct calculation that $V_j(F^*)(e) = L_j(F^*)(w_S)) = L_j(v_0)v^0$.

We now normalize L_j so that $L_j(v_0) = 1$. Thus we have that $\langle\langle V_j(F^*) \mid T(w_0)\rangle\rangle = \langle v^0 \mid w_0\rangle_\sigma$.

Thus we can state the Main Theorem of the section.

Theorem 5.2. *Let F be the unramified vector in $I_j(\tau, \sigma, s)$ normalized so that $F(e) = 1$.*

Then for $g \in SO((\xi_0)^\perp, q_{W_1})$

$$V_j(F * \phi_0)(g) = \lambda_j(\tau, \sigma, s) V_j(F^*)(g)$$

where ϕ_0, λ_j are as given above.

Moreover, we have the identity:

$$\langle\langle V_j(F) \mid T(w_0)\rangle\rangle = \frac{L(\tau \otimes \Pi, s + \frac{1}{2})}{L(\tau \otimes \sigma, s+1) L(\tau, {}^{\mathrm{sym}^2}_{\Lambda^2}, 2s+1)} \langle v^0 \mid w_0\rangle_\sigma \,.$$

Proof. The strategy in establishing the above identity is to analyze more concretely the function

$$F_1 = F_1 * \phi_0 - \lambda_j(\tau, \sigma, s) F^* \,.$$

First we modify F_1 by F_1' defined by

$$F_1' = \rho_{\Lambda_0}(F * \phi_0) - \lambda_j(\tau, \sigma, s) F^* \,.$$

What we know is that the function F_1' is analytic in s and $F_1' \in \mathrm{ind}_{N^i N_\beta SO(W_1(R_j))}^{G_{\xi_0}}(w_S \circ \Gamma)_{\tau \otimes |\ |^s \otimes \sigma \otimes 1})$. Moreover we know that $\mathrm{supp}(F_1')$ does not depend on s (i.e., $\mathrm{supp}(F_1') \subseteq N^i N_\beta SO(W_1(R_j)) K_\Lambda$, where K_Λ is some compact subset of G_{ξ_0}).

The point now is to establish that

$$V_j(F_1)(g) = V_j(F_1')(g) = 0 \qquad \text{(for all } s) \,.$$

First we note here that if Λ_0 is large enough, then $V_j(F_1) = V_j(F_1')$.

In particular, if $\tau = \tau(\chi)$, $\sigma = \sigma(\chi')$ and $\Pi = \Pi(\gamma)$, then we must analyze $\langle\langle V_j(F_1) \mid T(w_0)\rangle\rangle$ for *all* choices of triples (χ, χ', γ). The choice of the map T discussed above is *thus the essential* starting point here. The point here is that we finesse the proof in the following way. Basically we show for (χ, χ', γ) in a *suitable dense open set* the validity of the formula

$$\langle\langle V_1(F_1) \mid T(w_0)\rangle\rangle_{(\chi,\chi',\gamma)} \equiv 0\,.$$

Here $T = T_{(\chi,\chi',\gamma)}$ controls the form $\langle\langle\ \mid\ \rangle\rangle_{(\chi,\chi',\gamma)}$ (we have set $s = 0$ here with no loss of generality).

The fact that $V_j(F_1)(x) \equiv 0$ $(x \in SO((\xi_0)^\perp, q_{W_1})$ will then follow from the *Density Principle* (given in **Appendix 4 of** §**5**). Indeed what we show there is the following. Let H_1 be an analytic section in $I_j(\tau_j(\chi) \otimes \sigma(\chi'))$ which is $K_{SO((\xi_0)^\perp, q_{W_1})}$ invariant. Again this means that H_1 is a linear combination of standard sections of $I_j(\tau_j(\chi) \otimes \sigma(\chi'))$ with analytic coefficients (i.e., polynomials in the q^{χ_i} or $q^{-\chi_i}$). Then we assume that $V_j(H_1)$ is analytic also in (χ, χ') and that support $(V_j(H_1)) \subseteq \bigcup_\xi SO(W_1, (R_j))\xi K_{SO((\xi_0)^\perp, q_{W_1})}$ (a finite union) which is independent of (χ, χ'). Then we assert that if (χ, χ', γ) lies in a suitable open subset and

$$\langle\langle V_j(H_1) \mid T_{(\chi,\chi',\gamma)}(w_0)\rangle\rangle_{(\chi,\chi',\gamma)} \equiv 0$$

then $V_j(H_1)(x) = 0$ (for all $x \in SO((\xi_0)^\perp, q_{W_1})$).

Then in **Appendix 4 of** §**5** we show that $F_1' = F * \phi_0 - F^*$ can be chosen as such an H_1 in order to apply the *Density Principle*. Thus we have established $(*)$ and hence the second identity follows directly. □

Appendix I.

We make a choice of a Borel subgroup $B_{\mathbb{G}}$. Namely we look at the flag

$$\{z_\nu, \ldots, z_1, v_1, \ldots, v_i\} \supseteq \cdots \supseteq \{v_i\}.$$

This specific basis $\{z_\nu, \ldots, z_1, v_1, \ldots, v_i\}$ has the added property that $\{v_1, \ldots, v_i\}$ is a basis of R_i and $\{v_1, \ldots, v_i, z_1, \ldots, z_\ell\}$ (with $\ell = j - i$) is a basis of R_j. Note that $\{z_1, \ldots, z_\ell\}$ is a basis of $S \cap W_1(R_i)$ which is given in §4, §7 and §8. Moreover, $\{z_\nu, \ldots, z_{(j-i)+1}\}$ is chosen so that the associated flag $\{z_\nu, \ldots, z_{(j-i)+1}\} \supseteq \cdots \supseteq \{z_{(j-i)+1}\}$ is adapted to the Borel $B_2 \subseteq SO(\{y_1, \ldots\}) = SO(W_1(S) \cap W_1(R_i)) = SO(W_1(R_j))$ given below. Moreover, we note here $\xi_0 = z_1 + \gamma z_1^*$ (so that $q_{W_1}(\xi_0, \xi_0) \neq 0$. Then $B_{\mathbb{G}}$ is the stabilizer of this flag. We let $v_{i+1} = z_1, \ldots, v_j = z_{j-i}$. Then we consider another flag given by $\{z_\nu, \ldots, z_{(j-i)+1}, v_1, \ldots, v_j\} \supseteq \cdots \supseteq \{v_j\}$. Then we let $B'_{\mathbb{G}}$ be the Borel subgroup of $\mathbb{G}$ stabilizing this flag.

Again we note here that the torus T_j of B_{R_j} is given by $T_j = T_j(t_j, \ldots, t_1)$ where $T_j(t_j, \ldots, t_1)(v_s) = t_{j-s+1} v_s$ and $T_j(t_j, \ldots t_1)(v_s^*) = t_{j-s+1}^{-1} v_s^*$. Moreover the basic character $\chi + s$ is given by the map

$$T_j(t_j, \ldots, t_1) \rightsquigarrow |t_1 \ldots t_j|^s \prod \chi_i(t_i)$$

Also we note that the torus $T_\mu = T_\mu(s_1 \ldots s_\mu)$ of $B_{W_1(R_j)}$ is given by the data: $T_\mu(s_1 \ldots s_\mu)(z_t) = s_{t-(j-1)} z_t$ and $T_\mu(s_1 \ldots s_\mu)(z_t^*) = s_{t-(j-i)}^{-1} z_t^*$ (here $t = \nu, \ldots, (j-i)+1$.)

Moreover the basic character χ' on T_μ is given by

$$T_\mu(s_1 \ldots s_\mu) \to \prod_t \chi'_{j+t}(s_t)$$

We discuss here the issue of normalization of measures used in ensuring discussion.

The group $\mathbb{G}$, H, M, etc., are given. The maximal compacts $K_{\mathbb{G}}$, K_H and K_M are defined as stabilizers of appropriate lattices. In fact, we give below the choice of lattices given in the **Remark** preceding **Theorem (A)**.

We note that the various Borels $B'_{\mathbb{G}}$, $B_{\mathbb{G}}$, $'B_{\mathbb{G}}$, $B_{W_1(R_i)}$, etc. That are used are all adapted to the same K_X below. For instance $B'_{\mathbb{G}}$, $B_{\mathbb{G}}$ and '$B_{\mathbb{G}}$ are all adapted to the same $K_{\mathbb{G}}$.

Moreover, when we choose a measure db or du on B_X or U_X with U_X a unipotent group so that $K_X \cap B_X$ and $K_X \cap U_X$ have measure 1.

On the groups M and H' ($H' = SO(W_1(R_j))$ and $M = SO((\xi_0)^\perp, q_{W_1})$) we are given *specific Haar measures* dh' and dm which come from the global constructions given in §4 (see **Theorem 4.3**).

We note that the specific decomposition $dk = d_\ell(b_X) \otimes dk_X$ has the property that $\mathrm{vol}_X(K_X)$ is not necessarily equal to one.

Then we consider the module $I_j(\tau, \sigma, s)$. Here we take B_{R_j} = Borel subgroup of $GL(R_j)$ stabilizing the flag $\{v_1, \dots, v_j\} \supseteq \cdots \supseteq \{v_j\}$. We let $B_{W_1(R_j)}$ = Borel subgroup of $SO(W_1(R_j))$ stabilizing the flag $\{z_\nu, \dots, z_{(j-i)+1}\} \supseteq \cdots \supseteq \{z_{(j-i)+1)}\}$. Then in the construction of τ and σ above we first let $\tau \otimes |\ |^s = \mathrm{ind}_{B_{R_j}}^{GL(R_j)}(\chi + \boldsymbol{s})$ Then we let $\sigma = \mathrm{ind}_{B_{W_1(R_j)}}^{SO(W_1(R_j))}(\chi' \mid_{T_{W_1(R_j)}})$. Thus the module $I_j(\tau, \sigma, s) \cong \mathrm{ind}_{B'_{\mathbb{G}}}^{\mathbb{G}}((\chi + \boldsymbol{s}) \mid_{T_j} \otimes \chi' \mid_{T_{W_1(R_j)}})$. We emphasize here again the choice of $B'_{\mathbb{G}}$ in this construction. Also we let $F'_1(\chi + \boldsymbol{s} \otimes \chi', \varphi)$ be the contruction of the element in $\mathrm{ind}_{B'_{\mathbb{G}}}^{\mathbb{G}}(\chi + \boldsymbol{s} \mid_{T_j} \otimes \chi' \mid_{T_{R_j}})$ coming from $\varphi \in S(\mathbb{G})$ (see §7 for the definition of F'_1).

Let L_j be the embedding of $\mathrm{ind}_{B_{R_j}}^{GL(R_j)}(\chi + \boldsymbol{s})$ into the Whittaker space for $GL(R_j)$ adapted to the unipotent group $N^j_{+,-}$. In fact, let $w_{(i,j)}$ be the Weyl group element in $GL(R_j)$ so that $B_{R_j} w_{(i,j)} N^j_{+,-}$ is open dense in $GL(R_j)$. We let $\Psi_{i,j}$ be the generic character on $N^j_{+,-}$ defined above in this section. Then L_j can be defined via an integral:

$$L_j(F'_1(\chi + \boldsymbol{s} \otimes \chi', \varphi))(x) = \int_{N^j_{+,-}} F'_1(\chi + \boldsymbol{s} \otimes \chi', \varphi)[w_{(i,j)} n x] \Psi_{i,j}(n) dn \,.$$

Here we assume that $\mathrm{Re}(\chi + \boldsymbol{s}, \check{\alpha}_i) >> 0$ (α_i simple roots in $GL(R_j)$). Then the above integral is absolutely convergent and in fact for arbitrary data $F'_1(\chi + \boldsymbol{s} \otimes \chi', \varphi)$ admits a holomorphic continuation in (χ, s, χ'). Here again $F'_1(\chi + \boldsymbol{s} \otimes \chi', \varphi)$ lies in the induced module $I_j(\tau, \sigma, s)$.

Then we consider the basic integral:

$$V_j(F'_1(\chi + \boldsymbol{s} \otimes \chi', \varphi))(z) = \int\limits_{N_\beta \backslash \mathbb{U}_i} L_j(F'_1(\chi + \boldsymbol{s} \otimes \chi', \varphi))[w_S u_i z] \Lambda_{\xi_0}(u_i) du_i \,.$$

Here w_S is defined in this section. It is straightforward to verify that we put together the integrals for L_j and V_j to obtain an integration over $w_S N^j_{+,-} w_S \mathbb{U}_i$. We let

$$\begin{aligned} \Sigma_{i,j} &= (w_S N^j_{+,-} w_S) \mathbb{U}_i = (w_S N^i_- w_S)(w_S N^{j-i}_+ w_S) \cdot \mathbb{U}_i \\ &= w_S N^i_- w_S N^{j-i} \mathbb{U}_i = N^i N^{j-i} \mathbb{U}_i \,. \end{aligned}$$

Each of the first two groups normalizes $\mathbb{U}_i$ and $w_S N^i_- w_S$ commutes pointwise with $w_S N^{j-i}_+ w_S$. This $w_S N^j_{+,-} w_S \mathbb{U}_i$ is a subgroup. Then we consider the function on

$$(w_S N^i_- w_S)(w_S N^{j-i}_+ w_S) U_i \underset{\psi_{i,j,\xi_0}}{\rightsquigarrow} \psi_i(w_S x_i w_S) \otimes \psi_{j-i}(w_S x_{j-i} w_S) \otimes \Lambda_{\xi_0}(u)$$

where ψ_i and ψ_{j-i} are generic characters on the groups $w_S N^i_- w_S$ and $w_S N^{j-i}_+ w_S$. Thus Ψ_{i,j,ξ_0} is simply the function on $N^i \times N^{j-i} \times \mathbb{U}_i$ given by the product $\psi_i \otimes \psi_{j-i} \otimes \Lambda_{\xi_0}$. Thus the integral of V_j becomes

$$\int_{\Sigma_{i,j}} F'_1(\chi + \boldsymbol{s} \otimes \chi', \varphi)[w_{(i,j)} w_S u x] \psi_{i,j,\xi_0}(u) du \,.$$

Remark. In the case $j = i+1$ and $j = i$, $N^{j-i} = \{e\}$, then $\Sigma_{ij} = w_S N^i_- w_S \mathbb{U}_i = N^i \mathbb{U}_i$. In fact, if $j \geq i+1$, then $\Sigma_{ij} \supseteq w_S N^i_- w_S \cdot \mathbb{U}_i = N^i \mathbb{U}_i$. Specifically we note that ψ_{i,j,ξ_0} is a character on $N^i \mathbb{U}_i$ but not on Σ_{ij} (if $j > i+1$).

We note that the question of convergence of the above integral is connected to the convergence of the integral:

$$\int_{\Sigma_{ij}} F_1'(|\chi| + |\boldsymbol{s}| \otimes |\chi'|, |\varphi|)[w_{(i,j)} w_S ux] du \, .$$

However, we require a more delicate version (with appropriate support conditions).

We recall that $V_j(F(\cdots))$ lies in the space

$$\mathrm{Ind}^{SO((\xi_0)^{\perp}, q_{W_1})}_{N_{\{z_2,\ldots,z_\ell\}} U_{\{z_2,\ldots,z_\ell\}} SO(W_1(S) \cap W_1(R_i))} (\psi_{\ell--1} \otimes \psi_{Z_\beta} \otimes \sigma^{w_S}) \, .$$

In fact, we determine support properties of the function $V_j(F(\tau, \sigma, s)$ when restricted to the torus $T_{\{z_2,\ldots,z_\ell\}}$ (given in the Borel subgroup $B_{\{z_2,\ldots,z_\ell\}}$ which stabilizes in $GL(\{z_2,\ldots,z_\ell\})$ the flag $\{z_2,\ldots,z_\ell\} \supseteq \cdots \supseteq \{z_\ell\}$). Here $\ell = j - i$ (again in the case where $j = \dim W_1$ and i odd, we replace z_{j-i} by z^*_{j-i} in the ensuring discussion). Indeed using the definition of V_j, given above we deduce that for $g \in SO((\xi_0)^{\perp}, q_{W_1})$ having the form:

$$g = D(t_2, \ldots, t_\ell) g_1$$

with $D(t_2,\ldots,t_\ell) \in T_{\{z_2,\ldots,z_\ell\}}$ and $g_1 \in SO(\{z_1^-, y_1, \ldots\})$, then $V_j(F(\tau,\sigma,s))$ $(D(t_2,\ldots,t_\ell) \cdot g_1) \neq 0$ implies

$$|t_\ell| \leq C|t_{\ell-1}|, \ldots, |t_3| \leq C|t_2| \, , \; |t_2| \leq C \, .$$

Here C depends only on $F'(\tau,\sigma,s)$ and not on g_1. Thus we have an estimate on the function $V_j(F'(\tau,\sigma,s))$ on the domain $N_{\{z_2,\ldots,z_\ell\}} U_{\{z_2,\ldots,z_\ell\}} \backslash SO((\xi_0)^{\perp}, q_{W_1})$. Since the function $F'(\tau,\sigma,s)$ is $K_{\mathbb{G}}$ finite, it suffices to estimate $V_j(F(\; , \; , \;))$ on the domain $T_{\{z_2,\ldots,z_\ell\}} \cdot SO(\{z_1^-, y_1, \ldots\})$. Indeed by the above considerations

(1) $V_j(F'(\tau,\sigma,s))(D(t_2,\ldots,t_\ell)g_1) \neq 0$ implies there exists $C > 0$ so that

$$|t_\ell| \leq C|t_{\ell-1}|, \ldots, |t_3| \leq C|t_2| \, , \quad |t_2| \leq C$$

(2) On the domain in (1) above

$$|V_j(F_1'(\tau,\sigma,s))(D(t_2,\ldots,t_\ell)g_1)| \leq |t_2 \ldots t_\ell|^{Re(s)} \left(\prod_2^\ell \rho_i(t_i) \right)$$
$$\int_{\Sigma_{ij}} F_1'((|\chi| + |\boldsymbol{s}|) \otimes |\chi|', |\varphi|)[w_{(i,j)} w_S u g_1] du \, .$$

Here ρ_i are positive quasicharacters on k^x.

We next consider the map $T \in \mathrm{Hom}_{SO((\xi_0)^\perp, q_{W_1})}(\Pi, R_{\ell--1}(\sigma)^*)$ where $\prod = \mathrm{ind}_{B'}^{SO((\xi_0)^\perp, q_{W_1})}(\gamma)$ where B' is to be chosen below and γ the appropriate quasicharacter. First we consider the parabolic $P_{\{z_2,\dots,z_\ell\}}$ in $SO((\xi_0)^\perp, q_{W_1})$ given by the decomposition

$$P_{\{z_2,\dots,z_\ell\}} = GL(\{z_2,\dots,z_\ell\})SO(\{z_1^-, y_1,\dots\}) \rtimes U_{\{z_2,\dots,z_\ell\}}\,.$$

Then we consider Borel subgroups B_1 of $SO(\{z_1^-, y_1,\dots\})$ and B_2 of $SO(\{y_1,\dots\})$ so that a representative of $B_1 \times B_2$in $SO(\{z_1^-, y_1,\dots\})$ is η_{B_1,B_2} (that is $B_1\eta_{B_1,B_2}B_2$ is open in $SO(\{z_1^-, y_1,\dots\})$). In particular we can choose B_1 and B_2 so that B_1 and B_2 satisfy the hypotheses of **Remark 1** of §7. Then we choose the set $\{z_\nu,\dots,z_{j-i+1}\}$ so that $B_2 = B_{W_1(R_j)}$ is defined by the flag $\{z_\nu,\dots,z_{j-i+1}\} \supseteq \cdots \supseteq \{z_{j-i+1}\}$. Thus $B_{\mathbb{G}} \supseteq B_2$. Moreover, the Borel group B' is chosen so that $B' \supseteq B_1$. (In case $\ell = 1$, i.e., $P_{\{z_2,\dots,z_\ell\}} = e$, then $B' = B_1$.) In particular, we choose B' to be the stabilizer of the flag given by $\{\lambda_1,\dots,\lambda_t, z_2,\dots,z_\ell\} \supseteq \cdots \supseteq \{z_\ell\}$ (here the flag $\{\lambda_1,\dots,\lambda_t\} \supseteq \cdots \supseteq \{\lambda_t\}$ defines the Borel subgroup B_1). Then we denote $\eta_{B_1,B_2} = w_0^H \eta_H$ in this case (see **Remark 1** of §7). Moreover, we choose $w_{\ell-1}^\#$ in $SO(\xi_0^\perp, q_{W_1})$ which is the analogue of w_i constructed in the beginning of §8. Then following the definition of $T_{\chi\otimes\gamma}^{\mathbb{G}}(\ )$ given in §8 we let

$$T(F(\gamma,\varphi_2))[x] = \int\limits_{B_2\times N_{\{z_2,\dots,z_\ell\}}\times U_{\{z_2,\dots,z_\ell\}}} F(\gamma,\varphi_2)[w_0^H\eta_H w_{\ell-1}^\# nub_2x] \cdot (\chi^{**})^{-1}\delta_B^{1/2}(b_2)(\psi_{\ell-1}\otimes\psi_{Z_\beta})(nu)d_\ell(b_2)dndu\,.$$

We choose the character χ^{**} above so that $\chi^{**} = ((\chi')^{w_s})^{-1}$ and

$$T(F(\gamma,\varphi_2))(b_2'x) = [[(\chi')^{w_S}]^{-1}\delta_{B_2}^{1/2}](b_2)T(F(\gamma,\varphi_2))(x)\,.$$

We recall here that $\sigma = \mathrm{ind}_{B_{W_1(R_j)}}^{SO(W_1(R_j))}(\chi' \mid_{T_{W_1(R_j)}})$. Thus the transformation property of $T(F(\gamma,\varphi_2))$ is such that in the $x \in SO(W_1(R_i))$ it lies in the space $(\sigma^v)^{w_S}$! We emphasize here that $(\chi')^{w_S}$ is independent of the character $\chi + \boldsymbol{s}$.

We note here in the specific case where $j = i+1$ and $j = \mathrm{rank}(\mathbb{G})$, then the group $B_2 = SO(W_1(S)\cap W_1(R_i))$ is compact. In fact, the map T is given by

$$T(F(\gamma,\varphi_2))(x) = \int\limits_{(SO(W_1(S)\cap W_1(R_j))\times N_{\{z_2,\dots,z_\ell\}}\times U_{\{z_2,\dots,z_\ell\}})} F(\gamma,\varphi_2) [w_{\ell-1}^\# numx]\psi_{\ell-1}\otimes\psi_{Z_\beta}(nu)dndudm\,.$$

In such an instance $\sigma =$ the one dimensional trivial representation.

Thus we have that $T(F(\gamma,\varphi_2))(\ )$ lies in the space

$$\mathrm{Ind}_{N_{\{z_2,\dots,z_\ell\}}U_{\{z_2,\dots,z_\ell\}}SO(W_1(S)\cap W_1(R_i))}^{SO((\xi_0)^\perp,q_{W_1})}(\overline{\psi}_{\ell--1}\otimes\overline{\psi}_{Z_\beta}\otimes(\sigma^\nu)^{w_S}).$$

We let $K_{SO(W_1(S)\cap W_1(R_i))}$ be the standard maximal compact subgroup of $SO(W_1(S)\cap W_1(R_i))$. Then the pairing between σ^{w_S} and $(\check{\sigma})^{w_S}$ can be achieved as an integral pairing given as follows:

$$\langle f_\chi^1 \mid f_{\chi^v}^2\rangle_\sigma = \int_{K_{SO(W_1(S)\cap W_1(R_i))}} f_\chi^1(k)f_{\chi^v}^2(k)dk.$$

Thus if we put $f_\chi^1 = V_j(F(\tau,\sigma,s))$ and $f_{\chi^v}^2 = T(F(\gamma,\varphi_2))$ (again using notation from §7) we have

$$\begin{aligned}\langle f_\chi^1 \mid f_{\chi^v}^2\rangle_\sigma = &\int_{K_{SO(W_1(S)\cap W_1(R_i))}} \left(\int_{\Sigma_{ij}} F'{}_1(\chi+\boldsymbol{s}\otimes\chi',\varphi_1)[w_{(i,j)}w_S uk]\psi_{i,j,\xi_0}(u)du\right)\\ &\left(\int_{B_2\times N_{\{z_2,\dots,z_\ell\}}\times U_{\{z_2,\dots,z_\ell\}}} F(\gamma,\varphi_2)[w_0^H\eta_H w_{\ell-1}^{\#}nub_2k)\right.\\ &\left.\cdot(\chi^{**})^{-1}\delta_B^{1/2}(b_2)\Lambda_\xi(nu)d_\ell(b_2)dndu\right)dk.\end{aligned}$$

Now we know that the Haar measure $d_{SO(W_1(S)\cap W_1(R_i))}(x) = d_\ell(b_2)dk$. Hence the above integral equals

$$\begin{aligned}&\int_{SO(W_1(S)\cap W_1(R_i))} \left(\int_{\Sigma_{i,j}} F_1'(\xi+\boldsymbol{s}\otimes\chi',\varphi_1)[w_{(i,j)}w_S um]\psi_{i,j,\xi_0}(u)du\right)\\ &\left(\int_{N_{\{z_2,\dots,z_\ell\}}\times U_{\{z_2,\dots,z_\ell\}}} F(\gamma,\varphi_2)[w_0^H\eta_H w_{\ell-1}^{\#}num]\Lambda_\xi(nu)dndu\right)dm.\end{aligned}$$

Then we finally compute the form

$$\begin{aligned}\langle\langle V_j(F)\mid T(w)\rangle\rangle = &\int_{N_{\{z_2,\dots z_\ell\}}U_{\{z_2,\dots,z_\ell\}}SO(W_1(S)\cap W_1(R_i))\backslash SO((\xi_0)^\perp,q_{W_1})}\\ &\cdot\{V_j(F)(h)\mid T(w)(h)\}_\sigma dh.\end{aligned}$$

Here dh is invariant measure on the space

$$N_{\{z_2,\dots,z_\ell\}}U_{\{z_2,\dots,z_\ell\}}SO(W_1(S)\cap W_1(R_i))\backslash SO((\xi_0)^\perp,q_{W_1})$$

(again we note here that if $j = \dim W_1$ and i odd, then we replace z_ℓ by z_ℓ^*, i.e., $\ell = j - i$).

The first task is to determine when the above integral is convergent. For dh we may replace by $\delta_{P_{\{z_2,\dots,z_\ell\}}}^{-1}(t) d^x_{T_{\{z_2,\dots,z_\ell\}}}(t) \otimes d\nu$ where $d\nu$ is invariant measure on the group $SO(\{z_1^-, y_1, \dots, 1\})$. The relevant domain of integration here is the product variety:

$$T^+_{\{z_2,\dots,z_\ell\}}(C) \times SO(\{z_1^-, y_1, \dots\}).$$

Here $T^+_{\{z_2,\dots,z_\ell\}}(C) = \{D(t_2, \dots, t_\ell) \mid |t_\ell| \leq C|t_{\ell-1}|, \dots, |t_3| \leq C|t_2|, |t_2| \leq C\}$. Then we use the estimates on $V_j(F'(\tau, \sigma, s))$ (given above) and the explicit formula of $\langle \mid \rangle_\sigma$ to deduce that $\langle\langle \Lambda_j(F) \mid T(w) \rangle\rangle$ is majorized by an integral of the form

$$\int_{T^+_{\{z_2,\dots z_\ell\}}(C) \times SO(\{z_1^-, y_1, \dots\})} |t_2 \dots t_\ell|^{Re(s)} \left(\prod_{i=2}^{i=\ell} \widetilde{\rho}_i(t_i) \right)$$
$$\left(\int_{\Sigma_{ij}} F'(|\chi| + |\boldsymbol{s}| \otimes |\chi'|, |\varphi|)[w_{(i,j)} w_S u g_1] du \right)$$
$$(F(|\gamma|, |\varphi_2|)[w_0^H \eta_H w^\#_{\ell-1} g_1]) d^x_{T_{\{z_2,\dots,z_\ell\}}}(t) \otimes d\nu(g).$$

Then assuming $Re(s)$ is sufficiently large, we must determine the convergence of

$$\int_{SO(\{z_1^-, y_1, \dots\})} \left(\int_{\Sigma_{ij}} F'(|\chi| + |\boldsymbol{s}| \otimes \chi', |\varphi|)[w_{(i,j)} w_S u m] du \right)$$
$$F(|\gamma|, |\varphi_2|)[w_0^H \eta_H w^\#_{\ell-1} m] d\nu(m).$$

Indeed following the steps of the proof of **Lemma 1** of §8, we require 2 points for this convergence:

(1) The values of $\chi + \boldsymbol{s} \otimes \chi'$ so that

$$\int_{\Sigma_{ij}} F'(|\chi| + |\boldsymbol{s}| \otimes |\chi'|, |\varphi|)[w_{(i,j)} w_S u] du < \infty.$$

(2) If $F_3((\chi')^{w_S}, \varphi_3)$ lies in the space $\mathrm{ind}_{B_1}^{SO(\{z^-1, y_1, \dots\})}((\chi')^{w_S})$, then the integral

$$\int_{SO(\{z_1^-, y_1, \dots\})} F_3(|(\chi')^{w_S}|, |\varphi_3|)[m] F(|\gamma|, |\varphi_2|)[w_0^H \eta_H m^{w^\#_{\ell-1}} w^\#_{\ell-1}] dm < \infty.$$

We verify these two conditions in **Remark** $(**)$ below (after **Theorem** (A)). In particular for (2) we require that the characters $(\gamma, (\chi')^{w_S})$ relative to the pair of Borels (B_1, B_2) satisfy the hypotheses of **Lemma 2** of §7.

The two Borel subgroups $B_{\mathbb{G}}$ and $B'_{\mathbb{G}}$. We know that $B_{\mathbb{G}}$ and $B'_{\mathbb{G}}$ are $\mathbb{G}$ conjugate. In fact, we let $\widetilde{w}$ be the element in the $N_{K_{\mathbb{G}}}(T)$ (= normalizer in $K_{\mathbb{G}}$ which stabilizes the torus T associated to the basis $\{z_1, \ldots, z_\nu, v_1, \ldots, v_i\}$ (defined above) which conjugates $B_{\mathbb{G}}$ to $B'_{\mathbb{G}}$, i.e., $\widetilde{w} B_{\mathbb{G}} \widetilde{w}^{-1} = B'_{\mathbb{G}}$.

We clarify first the relation between $\widetilde{w}$ and w_{ij} given above. We note that w_{ij}^{-1} is characterized as the ' element which sends the flag $\{z_{j-i}, \ldots, z_1, v_i, \ldots, v_1\} \supseteq \cdots \supseteq \{v_1\}$ into the flag $\{v_i, \ldots, v_1, z_1, \ldots, z_{j-i}\} \supseteq \cdots \supseteq \{z_{j-i}\}$. Indeed $w_{(i,j)}$ satisfies the property that $w_{ij} N^j_{+,-} w_{ij}^{-1} \subseteq N^j_- =$ the negative Borel relative to N^j (defined by the Borel B_{R_j}). In any case, just a simple restatement of this fact determines the property of w_{ij}^{-1} above. On the other hand $\widetilde{w}$ sends the flag $\{z_{j-i}, \ldots, z_1, v_1, \ldots, v_i\} \supseteq \cdots \supseteq \{v_i\}$ into the flag $\{v_1, \ldots, v_i, z_1, \ldots z_{j-i}\} \supseteq \cdots \supseteq \{z_{j-i}\}$. Thus we can tie w and $w_{i,j}$ together. We let σ_0 be the map on R_i which satisfies $\sigma_0(v_i) = v_1$, $\sigma_0(v_{i-1}) = v_2$, $\ldots, \sigma_0(v_1) = v_i$. Then we deduce that $\sigma_0 \widetilde{w} = w_{ij}^{-1} \sigma_0$ or $\widetilde{w} = \sigma_0 w_{ij}^{-1} \sigma_0$. Also we have that $\widetilde{w}^{-1} w_{ij} = \sigma_0 w_{ij} \sigma_0 w_{ij} = (\sigma_0 w_{ij})^2$. In fact, since w_{ij} and $\widetilde{w}$ belong to $GL(R_j)$ we, in fact, can choose w_{ij} specifically to match the basis elements between $\{z_{j-i}, \ldots, z_1, v_i, \ldots, v_1\}$ and $\{v_i, \ldots, v_1, z_1, \ldots, z_{j-i}\}$. In any case, we then deduce that $(\sigma_0 w_{ij})^2 = 1$. This implies that $\widetilde{w}^{-1} w_{ij} = e_i$.

Then we have the following relationship between F'_1 and F_1. Here

$$F'_1(\chi, \varphi_1)(x) = \int_{B'_{\mathbb{G}}} \varphi_1(b'_{\mathbb{G}} x) \chi^{-1} \delta_B^{1/2}(b'_{\mathbb{G}}) d_\ell(b'_{\mathbb{G}})$$

and

$$F_1(\chi, \varphi_2)(x) = \int_{B_{\mathbb{G}}} \varphi_2(b_{\mathbb{G}} x) \chi^{-1} \delta_B^{1/2}(b_{\mathbb{G}}) d_\ell(b_{\mathbb{G}}) .$$

Now by **Remark** 2 of §7 we can identify $\chi^{\widetilde{w}}$ with χ and thus we have $F'_1(\chi, \varphi_1)(x) = F_1(\chi, {}^{\widetilde{w}^{-1}} \varphi_1)(\widetilde{w}^{-1} x)$. Thus we have that

$$\begin{aligned} &\int_{\Sigma_{ij}} F'_1(\chi + \boldsymbol{s} \otimes \chi', \varphi)[w_{(i,j)} w_S u x] \psi_{i,j,\xi_0}(u) du \\ &\quad = \int_{\Sigma_{ij}} F_1((\chi + \boldsymbol{s} \otimes \chi'), {}^{\widetilde{w}^{-1}} \varphi)[w_S u x] \psi_{i,j,\xi_0}(u) du] \\ &\int_{\Sigma_{ij}} F_1((\chi + \boldsymbol{s} \otimes \chi'), {}^{\widetilde{w}^{-1}} \varphi)[w_S u x] \psi_{i,j,\xi_0}(u) du . \end{aligned}$$

In particular the form $\langle\langle V_j(F) \mid T(w) \rangle\rangle$ can be expressed in the form

$$\begin{aligned} &\int_{N_{\{z_2,\ldots,z_\ell\}} U_{\{z_2,\ldots,z_\ell\}} SO(W_1(S) \cap W_1(R_i)) \backslash SO((\xi_0)^\perp, q_{W_1}} \{V_j(F)(h) \mid T(w)(h)\}_\sigma dh \\ &= \int_{SO((\xi_0)^\perp, q_{W_1})} \left(\int_{\Sigma_{ij}} F_1((\chi + \boldsymbol{s} \otimes \chi'), {}^{\widetilde{w}^{-1}} \varphi)[w_S u h] \psi_{i,j,\xi_0}(u) du \right) \\ &F(\gamma, \varphi_2)[w_0^H \eta_H w_{\ell-1}^{\#} h] dh . \end{aligned}$$

Again we emphasize here that F_1 is constructed relative to the Borel $B_{\mathbb{G}}$ and $F(\gamma, \varphi_2)$ is constructed relative to the Borel B_1. Moreover $B_2 \subseteq B_{\mathbb{G}}$ where B_2 is the Borel subgroup of $SO(W_1(S) \cap W_1(R_i))$ so that the $B_1 \times B_2$ open orbit in $SO(\{z_1^-, y_1, \dots\})$ has representative $w_0^H \eta_H$. Here $w_0^H \eta_H$ is the representative of the pair $B_1 \times B_2$ acting in $SO(\{z_1^-, y_1, \dots\})$ (using **Lemma 2** of §6).

We first note that in the construction of w_S given in §5 (above) we have $w_S(z_1) = z_1$ and $w_S(z_1^*) = z_1^*$. This implies that $w_S(\xi_0) = \xi_0$ ($\xi_0 = z_1 + \gamma z_1^*$ for appropriate $\gamma \neq 0$) and hence $w_S SO((\xi_0)^\perp, q_{W_1}) w_S = SO((\xi_0)^\perp, q_{W_1})$.

In fact, if we let $Z^{-1} = w_0^H \eta_H w_{\ell-1}^\#$, then by change of variables in the formula above we obtain

$$\int_{SO((\xi_0)^\perp, q_{W_1})} \left(\int_{\Sigma_{ij}} F_1([(\chi + \boldsymbol{s}) \otimes \chi'], ^{\widetilde{w}^{-1}} \varphi)[w_S u Z h] \psi_{i,j,\xi_0}(u) du \right) F(\gamma, \varphi_2)(h) dh .$$

Remark. In the case where $j = i+1$ we make this more precise. Since $Z \in SO((\xi_0)^\perp, q_{W_1})$ and $\Sigma_{ij} = N^i \mathbb{U}_i$ we deduce that the above integral equals

$$\int_{SO((\xi_0)^\perp, q_{W_1})} \left(\int_{\Sigma_{ij}} F_1([(\chi + \boldsymbol{s}) \otimes \chi'], ^{\widetilde{w}^{-1}} \varphi)[(w_S Z w_S) w_S u h] \psi_{i,j,\xi_0}(u) du \right) \cdot F(\gamma, \varphi_2)(h) dh .$$

In the more general case, $j \geq i+1$ we note that $\Sigma_{ij} = N^{j-i}(N^i \mathbb{U}_i)$. We note here N^{j-i} is taken in $GL(\{z_1, \dots, z_{j-i}\})$ and, in fact, belongs to the Borel subgroup of $GL(\{z_1, \dots, z_{j-i}\})$ which stabilizes the flag $\{z_1, \dots z_{j-i}\} \supseteq \cdots \supseteq \{z_{j-i}\}$. In any case, $N^{j-i} \subseteq SO(W_1(R_i))$; moreover w_S commutes pointwise with elements in $GL(\{z_1, \dots, z_{j-i}\})$. Thus using the decomposition $u = u_1 \cdot u_2$ with $u_1 \in N^{j-i}$ and $u_2 \in N^i \mathbb{U}_i$ we have that the above integral has the form

$$\int_{SO((\xi_0)^\perp, q_{W_1})} \left(\int_{N^{j-i} \times N^i \mathbb{U}_i} F_1((\chi + \boldsymbol{s} \otimes \chi'), ^{\widetilde{w}^{-1}} \varphi)[u_1 (w_S Z w_S) w_S u_2 h] \cdot \psi_{i,j,\xi_0}(u_1 u_2) du_1 du_2 \right) F(\gamma, \varphi_2)(h) dh .$$

Now we restrict to the case where $j = i+1$. We note here the definition of w_i in §8, which is taken relative to the choice of basis associated to the flags adapted to B_G and $\widetilde{B}_{w(R_i)}$ below *(we note here very specifically how w_i is constructed in* **Appendix** 3 *to this section.)* Thus the integral in the **Remark** above has the form

$$\int_{SO((\xi_0)^\perp, q_{W_1})} \left(\int_{N^i \mathbb{U}_i} F_1((\xi + \boldsymbol{s} \otimes \chi'), ^{\widetilde{w}^{-1}} \varphi)[(w_S Z w_S) w_i^S w_i u h] \psi_{i,j,\xi_0}(u) du \right) \cdot F(\gamma, \varphi_2)(h) dh .$$

We note here that F_1 is constructed relative to the Borel $B_{\mathbb{G}}$ and $F(\gamma, \varphi_2)$ relative to the Borel B_1. We note that the pair $(B_{W_1(R_i)}), B_1)$ do not necessarily satisfy the hypothesis of **Lemma 1** of §7 (here $B_{W_1(R_i)}$ defined relative to the flag $\{z_\nu, \ldots, z_1\} \supseteq \cdots \supseteq \{z_1\}$ in the space $W_1(R_i)$).

However, what is possible immediately is the following type of conclusion. First assuming the issue of convergence, we define the function

$$x \rightsquigarrow H(\varphi, \chi, \chi', s)(x) = \int_{N^i \mathbb{U}_i} F_1(\chi + \boldsymbol{s} \otimes \chi', \varphi)[x w_i u] \psi_{i,j,\xi_0}(u) du \,.$$

Then the basic integral above has the form

$$\int_{SO(\xi_0)^\perp, q_{W_1})} H(\varphi, \chi, \chi', s)(w_S Z w_S w_S^i h) F(\gamma, \varphi_2)(h) dh \,.$$

Thus the above integral is similar to the integral in §7 defining $T^{\mathbb{G}}_{\chi \otimes \gamma}$. In fact, from the argument in the proof of **Lemma 2** of §7 we deduce that if the above integral is absolutely convergent, then the point $w_S Z w_S w_S^i$ is a representative of the open $B_{W_1(R_i)} \times B_1$ orbit in $SO(W_1(R_i))$. The relation between the integral above (relative to the pair $(B_{W_1(R_i)} \times B_1)$ and the canonical $T_{\chi \otimes \gamma}$ is given in **Remark 2** of §7. Indeed to make this point more precise we need to determine $\widetilde{B}_{W_1(R_i)}$ and $\widetilde{B}_1$ (Borel subgroups of $SO(W_1(R_i))$ and $SO((\xi_0)^\perp, q_{W_1}))$ which satisfy the hypotheses of **Remark 1** of §7. Moreover, we must determine γ_1 and γ_2 so that $\gamma_1 \widetilde{B}_{W_1(R_i)} \gamma_1^{-1} = B_{W_1(R_i)}$, $\gamma_2 \widetilde{B}_1 \gamma_2^{-1} = B_1$ and determine $(b_H)_0$ and $(b_M)_0$ which satisfy $\gamma_1^{-1} w_S Z w_S w_S^i \gamma_2 = (b_H)_0 w_H^0 \eta_H (b_M)_0$ (with $w_H^0 \eta_H$ the representative given in **Remark 1** of §7 for the $\widetilde{B}_{W_1(R_i)}$ and $\widetilde{B_1}$ action).

This now requires a *detailed case by case analysis* following from the cases given in **Appendix 1** of §7 (specifically cases (1), (2) and (3)).

(a) Let $W_1(R_i) = \mathrm{Sp}\{E_r^*, \ldots E_1^*\} \oplus \mathrm{Sp}\{E_1, \ldots, E_r\}$, and $\xi_0 = E_1 + \gamma E_1^*$ and $\xi_1 = E_1 - \gamma E_1^*$. Then

$$(\xi_0)^\perp = \mathrm{Sp}\{E_r^*, \ldots, E_2^*\} \oplus (\xi_1) \oplus \mathrm{Sp}\{E_2, \ldots, E_r\}$$

and $(\xi_0)^\perp \cap (\xi_1)^\perp = \mathrm{Sp}\{E_r^*, \ldots, E_2^*\} \oplus \mathrm{Sp}\{E_2, \ldots, E_r\}$.

Then we choose $B_1 = \mathrm{Stab}(\{E_2, \ldots, E_r\} \supseteq \cdots \supseteq \{E_r\})$ and $B_2 = \mathrm{Stab}(\{E_2, \ldots, E_r\} \supseteq \cdots \supseteq \{E_r\})$. We note Stab is taken in the groups $SO((\xi_0)^\perp, q_{W_1})$ and $SO((\xi_0)^\perp \cap (\xi_1)^\perp, q_{W_1})$, respectively.

We let $\widetilde{B}_{W_1(R_i)} = \mathrm{Stab}(\{E_1, \ldots, E_r\} \supseteq \cdots \supseteq \{E_r\})$. In this case, $\widetilde{B}_{W_1(R_i)} \times B_1$ and $B_1 \times B_2$ satisfy the hypotheses of **Remark 1** of §7.

Then to be compatible with the data above, we have that $z_\nu = E_2, \ldots, z_2 = E_\nu$ and $z_1 = E_1$. Thus we must determine the $\gamma_1 \in SO(W_1(R_i))$ which maps the Borel group $B_{W_1(R_i)} = \mathrm{Stab}(\{z_\nu, \ldots, z_1\} \supseteq \cdots \supseteq \{z_1\}) = \mathrm{Stab}(\{E_2, \ldots, E_\nu, E_1\} \supseteq \cdots \supseteq \{E_1\})$ to $\widetilde{B}_{W_1(R_i)} = \mathrm{Stab}\{\{z_1, z_\nu, \ldots, z_2\} \supseteq \cdots \supseteq \{z_2\}\} = \mathrm{Stab}\{\{E_1, \ldots, E_\nu\} \supseteq \cdots \supseteq \{E_\nu\}\}$.

(b) Let $W_1(R_i) = \mathrm{Sp}\{E_r^*, \dots, E_1^*\} \oplus \mathrm{Sp}\{Z_0, Z_1\} \oplus \mathrm{Sp}\{E_1, \dots, E_r\}$ and $\xi_0 = E_1 + \gamma E_1^*$, $\xi_1 = E_1 - \gamma E_1^*$; here we choose γ so that $q_{W_1}(Z_0, Z_0) = 2\gamma$ (i.e., $q_{W_1}(Z_0, Z_0) = -q_{W_1}(\xi_1, \xi_1)$). Here $\{Z_0, Z_1\}$ span a 2 dimensional anisotropic space isomorphic to a multiple of a norm form of a quadratic extension K/k. Then

$$(\xi_0)^\perp = \mathrm{Sp}\{E_r^*, \dots, E_2^*\} \oplus \mathrm{Sp}\{\xi_1, Z_0, Z_1\} \oplus \mathrm{Sp}\{E_2, \dots, E_r\}$$

and

$$(\xi_0)^\perp \cap (\xi_1)^\perp = \mathrm{Sp}\{E_r^*, \dots, E_2^*\} \oplus \mathrm{Sp}\{Z_0, Z_1\} \oplus \mathrm{Sp}\{E_2, \dots, E_r\}.$$

Now $\mathrm{Sp}\{\xi_1, Z_0, Z_1\}$ is a 3 dimensional undergenerate space with Witt index = 1. In fact $\mathrm{Sp}\{\xi_1, Z_0, Z_1\} = (X^*) \oplus (W_0) \oplus (X)$ ($\{X, X^*\}$ hyperbolic plane).

Then we choose $B_1 = \mathrm{Stab}(\{X, E_2, \dots, E_r\} \supseteq \cdots \supseteq \{E_r\})$ and $B_2 = \mathrm{Stab}(\{E_2, \dots, E_r\} \supseteq \cdots \supseteq \{E_r\})$.

We let $\widetilde{B}_{W_1(R_i)} = \mathrm{Stab}(\{X, E_2, \dots, E_r\} \supseteq \cdots \supseteq \{E_r\})$. In this case $\widetilde{B}_{W_1(R_i)} \times B_1$ and $B_1 \times B_2$ satisfy the hypotheses of **Remark 1** of §7.

Then to be compatible with the data above, we have that $z_\nu = E_2, \dots, z_2 = E_\nu$ and $z_1 = E_1$. Thus we must determine the $\gamma \in SO(W_1(R_i))$ which maps the Borel group $B_{W_1(R_i)} = \mathrm{Stab})(\{z_\nu, \dots, z_1\} \supseteq \cdots \supseteq \{z_1\})$ to the Borel group $\widetilde{B}_{W_1(R_i)} = \mathrm{Stab}(\{X, z_\nu, \dots, z_2\} \supseteq \cdots \supseteq \{z_2\})$.

(c) Let $W_1(R_i) = \mathrm{Sp}\{E_r^*, \dots, E_1^*\} \oplus \mathrm{Sp}(Z_0) \oplus \mathrm{Sp}\{E_1, \dots, E_r\}$ and $\xi_0 = E_1 + \gamma E_1^*$, $\xi_1 = E_1 - \gamma E_1^*$; here $q_{W_1}(Z_0, Z_0) = 2\gamma$. Here $\{Z_0\}$ spans a one dimensional anisotropic space. Then

$$(\xi_0)^\perp = \mathrm{Sp}\{E_r^*, \dots, E_2^*\} \oplus \mathrm{Sp}\{\xi_1, Z_0\} \oplus \mathrm{Sp}\{E_2, \dots, E_r\}$$

and

$$(\xi_0)^\perp \cap (\xi_1)^\perp = \mathrm{Sp}\{E_r^*, \dots, E_2^*\} \oplus \mathrm{Sp}(Z_0) \oplus \mathrm{Sp}\{E_2, \dots, E_r\}.$$

Now $\mathrm{Sp}\{\xi_1, Z_0\}$ is a nondegenerate space which is not anisotropic (i.e., $q_{W_1}(Z_0, Z_0) = -q_{W_1}(\xi_1, \xi_1)$). Thus we can write $\mathrm{Sp}\{\xi_1, Z_0\} = (X^*) \oplus (X)$ ($\{X, X^*\}$ hyperbolic plane).

Then we choose $B_1 = \mathrm{Stab}(\{X, E_2, \dots, E_r\} \supseteq \cdots \supseteq \{E_r\})$ and $B_2 = \mathrm{Stab}(\{E_2, \dots, E_r\} \supseteq \cdots \supseteq \{E_r\})$.

We let $\widetilde{B}_{W_1(R_i)} = \mathrm{Stab}(\{X, E_2, \dots, E_r\} \supseteq \cdots \supseteq \{E_r\})$. Thus in this case $\widetilde{B}_{W_1(R_i)} \times B_1$ and $B_1 \times B_2$ satisfy the hypotheses of **Remark 1** of §7.

Then to be compatible with the data above, we have that $z_\nu = E_2, \dots, z_2 = E_\nu$ and $z_1 = E_1$. Thus we must determine the $\gamma \in SO(W_1(R_i))$ which maps the Borel group $B_{W_1(R_i)} = \mathrm{Stab}(\{z_\nu, \dots, z_1\} \supseteq \cdots \supseteq \{z_1\})$ to $\widetilde{B}_{W_1(R_i)} = \mathrm{Stab}(\{X, z_\nu, \dots, z_2\} \supseteq \cdots \supseteq \{z_2\})$.

(d) Let $W_1(R_i) = \mathrm{Sp}\{E_r^*, \dots, E_2^*\} \oplus \mathrm{Sp}\{E_1^*, Z_0, E_1\} \oplus \mathrm{Sp}\{E_2, \dots E_r\}$ and $\xi_0 = E_1 + \gamma E_1^*$, $\xi_1 = E_1 - \gamma E_1^*$, here $q_{W_1}(Z_0, Z_0) \neq 2\gamma$. Here $\mathrm{Sp}\{E_1, Z_0, E_1^*\}$ spans a 3 dimensional space which has Witt index 1. Then

$$(\xi_0)^\perp = \mathrm{Sp}\{E_r^*, \dots, E_2^*\} \oplus \mathrm{Sp}\{\xi_1, Z_0\} \oplus \mathrm{Sp}\{E_2, \dots, E_r\}$$

and

$$(\xi_0)^\perp \cap (\xi_1)^\perp = \mathrm{Sp}\{E_r^*, \dots, E_2^*\} \oplus \mathrm{Sp}\{Z_0\} \oplus \mathrm{Sp}\{E_2, \dots, E_r\}.$$

Here $\mathrm{Sp}\{\xi_1, Z_0\}$ is an anisotropic 2 dimensional space.

Then we choose $B_1 = \mathrm{Stab}(\{E_2, \dots, E_r\} \supseteq \cdots \supseteq \{E_r\})$ and $B_2 = \mathrm{Stab}(\{E_2, \dots, E_r\} \supseteq \cdots \supseteq \{E_r\})$.

We let $\widetilde{B}_{W_1(R_i)} = \mathrm{Stab}(\{E_1, \dots, E_r\} \supseteq \cdots \supseteq \{E_r\})$.

Then in this case $\widetilde{B}_{W_1(R_i)} \times B_1$ and $B_1 \times B_2$ satisfy the hypotheses of **Remark 1** of §7.

To be compatible with the data above, we have that $z_\nu = E_2, \dots, z_2 = E_\nu$ and $z_1 = E_1$. Thus we must determine $\gamma \in SO(W_1(R_i))$ which maps the Borel group $B_{W_1(R_i)} = \mathrm{Stab}(\{z_\nu, \dots z_1\} \supseteq \cdots \supseteq \{z_1\})$ to the Borel group $\widetilde{B}_{W_1(R_j)} = \mathrm{Stab}(\{z_1, z_\ell, \dots, z_2\} \supseteq \cdots \supseteq \{z_2\})$.

Summarizing the above 4 cases, we have the following facts. In (a) and (d) the element $\gamma \in SO(W_1(R_i))$ so that $\gamma \widetilde{B}_{W_1(R_i)} \gamma^{-1} = B_{W_1(R_i)}$ is given by $\gamma^{-1} \in GL(\{z_1, \dots, z_\nu\})$ so that $\gamma^{-1}(z_1) = z_2, \gamma^{-1}(z_2) = z_3, \dots, \gamma^{-1}(z_\nu) = z_1$. In case (b) (case c) resp.) we consider the 4 dimensional space $\mathrm{Sp}\{E_1, E_1^*, Z_0, Z_1\} = \mathrm{Sp}\{\xi_0, X, X^*, W_0\}$ (3 dimensional space $\mathrm{Sp}\{E_1, E_1^*, Z_0\} = \mathrm{Sp}\{\xi_0, X, X^*\}$ resp.). Then there exists an element $\widetilde{\gamma} \in SO(\{E_1, E_1^*, Z_0, Z_1\})$ ($\widetilde{\gamma} \in SO(\{E_1, E_1^*, Z_0\})$ resp.) so that $\widetilde{\gamma}(E_1) = X$, $\widetilde{\gamma}(E_1^*) = X^*$ ($\widetilde{\gamma}(E_1) = X$, $\widetilde{\gamma}(E_1^*) = X^*$ resp.). In fact $\widetilde{\gamma}$ also satisfies $\widetilde{\gamma}(Z_0) = \pm\xi_0$, and $\widetilde{\gamma}(Z_1) = \pm W_0$ ($\widetilde{\gamma}(Z_0) = \pm\xi_0$ resp.). We note the $\pm$ sign is chosen above so that $\det(\widetilde{\gamma}) = 1$ (i.e., $\widetilde{\gamma} \in SO(\{E_1, E_1^*, Z_0, Z_1\})$ or $SO(\{E_1, E_1^*, Z_0\})$). In particular, we note that $\widetilde{\gamma}$ fixes pointwise E_i and E_j^* where $i, j \geq 2$. Thus we deduce first that $\widetilde{\gamma} B_{W_1(R_i)} \widetilde{\gamma}^{-1} = \boldsymbol{B}_{W_1(R_i)} =$ the Borel subgroup stabilizing the flag $\{z_\nu, \dots, z_2, X\} \supseteq \cdots \supseteq \{X\}$. Then in cases (b) and (c) the element $\Gamma \in GL(\{X, z_\nu, \dots, z_2\})$ so that $\Gamma \boldsymbol{B}_{W_1(R_i)} \Gamma^{-1} = \widetilde{B}_{W_1(R_i)}$ has the formula $\Gamma(X) = z_\nu, \Gamma(z_\nu) = z_{\nu-1}, \dots, \Gamma(z_2) = X$. Thus we deduce in cases (b) and (c) above that $B_{W_1(R_i)} = \widetilde{\gamma}^{-1} \Gamma \widetilde{B}_{W_1(R_i)} \Gamma^{-1} \widetilde{\gamma}$.

Remark. The standard maximal compact subgroup $K_{\mathbb{G}}$ of $\mathbb{G}$ is chosen as the stabilizer of the $\mathcal{O}$-lattices in W_1 given in the following way. First we choose $\mathcal{O}$ splitting of $W_1 = R_i \oplus W_1(R_i) \oplus R_i^*$ so that $R_i = \mathcal{O}$ span of $\{v_1, \dots, v_i\}$, $R_i^* = \mathcal{O}$ span of $\{v_1^*, \dots, v_i^*\}$ and $W_1(R_i)$ is given in cases (a), (b), (c), (d) as follows:

(a) $\mathcal{O}$ span of $\{E_1, \dots, E_r, E_r^*, \dots, E_1^*\}$

(b) $\mathcal{O}$ span of $\{E_1, \dots, E_r, Z_0, Z_1, E_r^*, \dots, E_1^*\}$

(c) $\mathcal{O}$ span of $\{E_1, \dots, E_r, Z_0, E_r^*, \dots, E_1^*\}$

(d) $\mathcal{O}$ span of $\{E_1, \dots, E_r, Z_0, E_r^*, \dots, E_1^*\}$

Then $K_{SO(W_1(R_i))} =$ stabilizer in $SO(W_1(R_i))$ of the $\mathcal{O}$ lattices (a)-(d) above $= K_{\mathbb{G}} \cap SO(W_1, (R_i))$. Also $K_{SO((\xi_0)^\perp, q_{W_1})} = K_{\mathbb{G}} \cap SO((\xi_0)^\perp, q_{W_1}) =$ stabilizer of $\mathcal{O}$-lattice (a') $\mathcal{O}$ span of $\{E_2, \ldots, E_r, \xi_1, E_r^*, \ldots, E_2^*\}$

(b') $\mathcal{O}$ span of $\{E_2, \ldots, E_r, \xi_1, Z_0, Z_1, E_r^*, \ldots, E_2^*\} = \mathcal{O}$ span of $\{E_2, \ldots, E_r, X, W_0, X^*, E_r^*, \ldots, E_2^*\}$

(c') $\mathcal{O}$ span of $\{E_2, \ldots, E_r, \xi_1, Z_0, E_r^*, \ldots, E_2^*\} = \mathcal{O}$ span of $\{E_2, \ldots, E_r, X, X^*, E_r^*, \ldots, E_2^*\}$

(d') $\mathcal{O}$ span of $\{E_2, \ldots, E_r, \xi_1, Z_0, E_r^*, \ldots, E_2^*\}$

We note here that the elements γ and $\tilde{\gamma}$ given above belong to $K_{GL\{z_1,\ldots,z_\nu\}}$, $K_{SO(\{E_1,E_1^*,Z_0,Z_1\})}$, $K_{SO(\{E_1,E_1^*,Z_0\})}$ (the last two correspond to cases (b) and (c) above). Moreover, Γ is also chosen in $K_{\mathbb{G}}$! (See here specifically **Remark 6 of §7** and **Appendix 3 of §7**.)

We then prove in **Appendix III** to this section that the $(b_H)_0$ and $(b_M)_0$ components of the element $\gamma_1^{-1} w_S z w_S w_S^i$ satisfy the condition that the $t_1, \ldots, t_\ell$ $((t'_1, \ldots, t'_\ell)$ resp.) components of T_H (T_M resp.) are in fact *units*! This proof is contained in the analysis of the 4 cases above give in **Appendix III** to this section.

Then we consider the flag in W_1 which is obtained by $\{L_1, \ldots, L_t, v_1, \ldots, v_i\} \supseteq \cdots \supseteq \{v_i\}$ where $\{L_1, \ldots, L_t\} \supseteq \cdots \supseteq \{L_t\}$ is the flag in $W_1(R_i)$ which is constructed in the 4 cases above (so that $\widetilde{B}_{W_1(R_i)}$ is the Borel subgroup of $SO(W_1(R_i))$ stabilizing this flag). We then consider the Borel group $`B_{\mathbb{G}}$ which is defined by this flag. In particular we note that $`B_{\mathbb{G}} \times B_1$ satisfies the hypotheses in §8 that is used in the construction of the distribution $T^{\mathbb{G}}_{\chi\otimes\gamma}$. In fact, the above discussion implies that the integral

$$\int_{SO((\xi_0)^\perp, q_{W_1})} \left(\int_{N^i \mathbb{U}_i} F_1(\chi + \boldsymbol{s} \otimes \chi', {}^{\tilde{w}^{-1}} \varphi)[w_S Z w_S w_i^S w_i u h] \psi_{i,j,\xi_0}(u) du \right) \cdot F(\gamma, \varphi_2)(h) dh$$

equals upto a *nonzero scalar* (dependent of χ, s and γ given below) the integral given in §8 by

$$T^{\mathbb{G}}_{(\chi+\boldsymbol{s}\otimes\chi')\otimes\gamma}({}^{(w\gamma_1)^{-1}}\varphi \otimes \varphi_2) .$$

Again we emphasize here that $T^{\mathbb{G}}_{(\chi+\boldsymbol{s}\otimes\chi')\otimes\gamma}$ is constructed relative to the pair $`B_{\mathbb{G}} \times B_1$ defined above; $\chi + \boldsymbol{s} \otimes \chi'$ ($\gamma$ resp.) is a character on $`B_{\mathbb{G}}$ (B_1 resp.).

We note that the nonzero scalar in this case is given by

$$(\chi + \boldsymbol{s} \otimes \chi')\delta^{1/2}_{`B_{\mathbb{G}}}(T_1^0)(\gamma\delta_{B_1}^{+1/2})([(T_2^0)^{w_S}]^{-1}) .$$

Here T_1^0, T_2^0 represent the toral parts of $(b_H)_0$ and $(b_M)_0$. We know that the components of T_1^0 and T_2^0 are units (from **Appendix III** of this section). This implies that the above product is, in fact, equal to one (the characters χ, χ' and γ' are assumed to be unramified). Hence we have that the integral

$$\int_{SO((\xi_0)^\perp, q_{W_1})} \left(\int_{N^i \mathbb{U}_i} F_1(\chi + \boldsymbol{s} \otimes \chi', {}^{\tilde{w}^{-1}} \varphi)[w_S Z w_S w_S^i w_i u h] \psi_{i,j,\xi_0}(u) du \right) \cdot F(\gamma, \varphi_2)(h) dh$$

equals the integral

$$T^{\mathbb{G}}_{(\chi+\boldsymbol{s}\otimes\chi')\otimes\gamma}({}^{(w\gamma_1)^{-1}}\varphi\otimes\varphi_2)\,.$$

Thus we can state **Theorem (A)** describing the properties of the pairing:

$$\langle\langle V_j(F)\mid T(w)\rangle\rangle\,.$$

Theorem (A). *We let $F_1'(\chi+\boldsymbol{s}\otimes\chi',\varphi)$ belong to $I_j(\tau,\sigma,s)$ where $\tau\otimes|\;|^s = ind_{B_{R_j}}^{GL(R_j)}(\chi+\boldsymbol{s})$ and $\sigma = ind_{B_{W_1(R_j)}}^{SO(W_1(R_j))}(\chi'\,|_{T_{W_1(R_j)}})$. We let $T\in Hom_{SO((\xi_0)^\perp,qw_1)}$ $(\Pi,R_{\ell-1}(\sigma)^*)$ where $\Pi = ind_{B'}^{SO((\xi_0)^\perp,qw_1)}(\gamma)$ is given by the integral formula*

$$T(F(\gamma,\varphi_2))[x] = \int_{B_2\times N_{\{z_2,\ldots,z_\ell\}}} F(\gamma,\varphi_2)[w_0^H\eta_H w_{\ell-1}^{\#}nub_2x]\chi^{**}\delta_{B_2}^{-1/2}(b_2)$$
$$\psi_{\ell-1}\otimes\psi_{Z_\beta}(nu)d_\ell(b_2)dndu$$

where χ^{**} is defined as above, etc. The pairing

$$\langle\langle V_j(F_1'(\chi+\boldsymbol{s}\otimes\chi',\varphi)\mid T(F(\gamma,\varphi_2))\rangle\rangle$$

is given by a convergent integral of the form

$$\int_{SO((\xi_0)^\perp,qw_1)}\left(\int_{N^{j-i}\times N^i\mathbb{U}_i} F_1((\chi+\boldsymbol{s}\otimes\chi',{}^{\widetilde{w}^{-1}}\varphi)[u_1w_SZw_Sw_Su_2h]\right.$$
$$\left.\cdot\,\psi_{i,j,\xi_0}(u_1u_2)du_1du_2\right)F(\gamma,\varphi_2)(h)dh\,.$$

The domain of convergence is given precisely in **Remark (**)** *below.*

In the case $j=i+1$, then the above integral equals

$$T^{\mathbb{G}}_{(\chi+\boldsymbol{s}\otimes\chi')\otimes\gamma}({}^{(w\gamma_1)^{-1}}\varphi\otimes\varphi_2)$$

which is defined in §8.

We note here that the basic calculation in §6 to §8 is involved with the integral

$$T_{(\chi+\boldsymbol{s}\otimes\chi')\otimes\gamma}(\varphi_1\otimes\varphi_2)\,.$$

In particular we are interested in the case where φ_1 and φ_2 are fixed by $K_{\mathbb{G}}$ and $K_{SO((\xi_0)^\perp,qw_1)}$.

Remark (**). At this point we determine the convergence of the integral

$$\int_{N_+^{j-i}\times N^i\mathbb{U}_i} |F_1(\chi + \boldsymbol{s}\otimes\chi', {}^{\tilde{w}^{-1}}\varphi)[u_1 w_S u_2 x]| du_2 du_2 \,.$$

We note that $w_S N^i \mathbb{U}_i w_S = N_-^i \mathbb{U}_{i,-}$. here N_-^i = the unipotent radical of the Borel subgroup in $GL(R_i)$ stabilizing the flag $\{v_i, v_{i-1}, \ldots, v_1\} \supseteq \cdots \supseteq \{v_1\}$ and $\mathbb{U}_{i,-}$ = the opposed unipotent radical to $\mathbb{U}_i$ = the unipotent radical of the parabolic group $(\mathbb{P}_i)_-$ which stabilizes the subspace $\mathrm{Sp}\{v_1^*, \ldots, v_i^*\}$. Thus the above integral becomes

$$\int_{N_+^{j-i}\cdot N_-^i\cdot\mathbb{U}_{i,-}} |F_1(\chi + \boldsymbol{s}\otimes\chi', {}^{\tilde{w}^{-1}}\varphi)[u_1 \bar{u}_2 x]| du_1 d\bar{u}_2$$

$d\bar{u}_2$ = invariant measure on $N_-^i\mathbb{U}_{i,-}$.

Then taking the positive roots Δ_+ of $\mathbb{G}$ defined relative to the Borel $B_{\mathbb{G}}$ (given above) we note that the set $S_{i,j} = \{\beta \mid$ the root subgroup x_β in $\mathbb{G}$ belongs to $N^{j-i}N_-^i\mathbb{U}_{i,-}\}$. Now we note that $S_{i,j}$ is given as the following type of set: $-\{\gamma \in \Delta_+ \mid w(\gamma) < 0\}$ for an appropriate $w \in$ Weyl group $SO(W_1, q_{W_1})$. In any case, applying the criterion for convergence of intertwining integrals we have that the above integral converges under the following conditions. Let $\chi + \boldsymbol{s}\otimes\chi' = \left(\sum_{t=1}^{t=j} (\chi_t + s)\varepsilon_t\right) + \sum_{t>j} \chi_t'\varepsilon_t$. Then the integral above converges provided

(1) $\chi_{\ell_1} - \chi_{\ell_2} >> 0$ with $1 \le \ell_1 < \ell_2 \le i$

(1') $\chi_{\ell_1} + \chi_{\ell_2} + 2s \gg 0$ with $1 \le \ell_1 < \ell_2 \le i$

(2) $\chi_\ell \pm \chi_{\ell'} + \varepsilon_{\ell'} s >> 0$ with $1 \le \ell \le i$ and $i+1 \le \ell'$.

$$\varepsilon_{\ell'} = \begin{cases} 0 & \text{if } i+1 \le \ell' \le j & \text{with} & -\chi_{\ell'} \\ 1 & \text{if } j < \ell' & \text{with} & \pm\chi_{\ell'} \\ 2 & \text{if } i+1 \le \ell \le j & \text{with} & +\chi_{\ell'} \end{cases}$$

(2') if ε_ℓ a root $(1 \le \ell \le i)$, then $\chi_\ell + s \gg 0$.

(3) $\chi_{\nu_1} - \chi_{\nu_2} >> 0$ with $i+1 \le \nu_1 < \nu_2 \le j$. Here ">> 0" means sufficiently large.

Thus, in fact, if $Re(s)$ is sufficiently large, $(\chi_1, \ldots, \chi_j)$ satisfies, $Re(\chi_{\ell_1}) - Re(\chi_{\ell_2}) \gg 0$ $(1 \le \ell_1 < \ell_2 \le j)$ and $(\chi_{j+1}, \ldots, \chi_\nu)$ remain in a bounded set, then the integral

$$\int_{N^{j-i}\times N^i\mathbb{U}_i} |F_1(\chi + \boldsymbol{s}\otimes\chi', {}^{\tilde{w}^{-1}}\varphi)[u_1 w_S u_2 x]| du_1 du_2$$

is absolutely convergent.

If now $\gamma\,|_{T_{B_1}} = (\gamma_1, \ldots, \gamma_t)$ (t suitably chosen) and $(\chi')^{W_S}\,|_{T_{B_2}} = (\chi_{j+1}, \ldots, \chi_\nu)$ then we let Ω be chosen as in **Lemma 7.2**. We note then that we have $(\chi_{j+1}, \ldots, \chi_\nu)$ lies in a bounded set and hence we have the convergence of the basic integral in **Theorem (A)**.

Appendix II.

There is, however, even a better way to insure convergence of the integral

$$V_j(F) = \int_{N_\beta\backslash\mathbb{U}_i} L_j[F][w_S ux]\Lambda_{xio}(u)du\,.$$

There are two issues here. How does the convergence of the above integral depend on τ, σ and s? Also how does the choice of L_j effect the convergence? Here $j = i+1$.

We recall above we have used the Whittaker-Jacquet integral defining L_j (and then its analytic continuation). In fact, we can normalize L_j by the factor $\zeta_{GL_j}(\chi, 1)^{-1}$ to form L_j^*. In such a case $L_j^*(F_{\chi\otimes\chi'})(g)$ (for fixed g) is an analytic function in (χ, χ'). In such a case we first estimate $L_j^*(F)$ in the following way.

Recall $F \in \mathrm{ind}_{GL_j\times SO(W_1(R_j))\times\mathbb{U}_j}^{\mathbb{G}}(\tau\otimes|\ |^s\otimes\sigma)$. Now $\tau = \mathrm{ind}_{B_{R_j}}^{GL(R_j)}(\chi)$ and $\sigma = \mathrm{ind}_{B_{W_1(R_j)}}^{SO(W_1(R_j))}(\chi')$.

We recall that a gauge form ξ on GL_j is a function on GL_j which is right invariant under K_{GL_j}, left invariant under $N_{+,-}^j$ and on diagonal matrices

$$D(a_1\dots a_j, a_2\dots a_j, \dots, a_j)$$

has the form $\varphi(a_1,\dots,a_{j-1})|a_1\dots a_{j-1}|^{-r}$ where φ is a Schwartz function on k^{j-1}, $\varphi\geq 0$ and $r>0$.

Now let $g_1\in GL_j$, $g_2\in SO(W_1(R_j))$, $u\in\mathbb{U}_j$ and $k\in K_{\mathbb{G}}$. Then we deduce simply the following estimate for $L_j^*(F)$:

$$|L_j^*(F)[g_1g_2uk]\,|\leq|\det g_1|^{s_0+s}\xi(g_1)h_{|\chi'|}(g_2)\,.$$

Here $h_{|\chi'|}\in\mathrm{ind}_{B_{W_1(R_j)}}^{SO(W_1(R_j))}(|\chi'|)$ and the choice of s_0, ξ and $h_{|\chi'|}$ depends on (χ,χ') and F.

A more precise estimate can be obtained in the case where F is a $K_{\mathbb{G}}$ fixed vector. In fact, we have

$$|L_j^*(F)(g_1g_2uk)|\leq\delta_{B_{GL_j}}^{\frac{1}{2}}(H_j(g_1))\lambda(H_j(g_1))|\det g_1|^{s-s_0}\phi_{|\chi'|}(g_2)\,.$$

where H_j is the diagonal part in the Iwasawa decomposition relative to $GL_j = N_{+,-}^jDK_{\mathbb{G}L_j}$, and $\phi_{|\chi'|}$ is the normalized spherical vector in the space $\mathrm{ind}_{B_{W_1(R_j)}}^{SO(W_1(R_j))}(|\chi'|)$. Moreover, for any $T>0$ so that $|Re(\chi_i)|<T(i=1,\dots,j)$ there is an $s_0>0$ so that the above inequality is valid. Here λ is some positive character on D_j (independent of (χ,χ')).

We let $\tilde{\tilde{B}}_G$ be the Borel Subgroup of $\mathbb{G}$ associated to the flag in W_1 given by

$$\{z_\nu,\dots,z_{j-i+1}v_i,\dots,v_1,z_1,\dots,z_{j-i}\}\subseteq\dots\subseteq\{z_{j-i}\}\,.$$

Then let $\hat{\Omega}$ be the function on the right hand side of the inequalities above. (In the case of the gauge ξ, we replace the Schwartz function φ by the constant function here.) In any case, $\hat{\Omega}$ determines an element in the space

$$\mathrm{ind}_{\tilde{\tilde{B}}_{\mathbb{G}}}^{\mathbb{G}}(\alpha_j + \underset{\sim}{s}) \otimes |\chi'|)$$

where α_j is a fixed positive character.

The goal is to show that the integral

$$\int_{\mathbb{U}_i'} \hat{\Omega}(w_S u_i w_S x) du < \infty$$

where $\mathbb{U}_i'$ = the subgroup of $\mathbb{U}_i$ given by

$$\{N_\ell((\mu_1', \ldots, \mu_i'), s) \mid \mu_t' \in Sp\{z_1^*, \ldots, z_{j-i}^*\} \oplus (W_1(S) \cap W_1(R_i))\}$$

(note in the case where $j = \dim W_i$ and i odd, then in the above formula replace $z_{j=i}^*$ by z_{j-i}, $\ell = j - i$).

Choosing the root system adapted to the Borel $\tilde{\tilde{B}}_{\mathbb{G}}$ we see that $w_S \mathbb{U}_i' w_S$ is contained in the unipotent group $\mathbb{U}_j^-$. Then using the standard theory of convergence of intertwining operators (see **Appendix 1 of** §**5**) we deduce that the above integral *converges* if

$$Re(s) \pm s_0 \pm Re(\chi_t) \gg 0$$

where $j \leq t \leq$ rank $(\mathbb{G})$. In any case, when $Re(s)$ is large enough this is true. We thus deduce that for a general section F, the integral defining $V_j(F)$ is convergent for $Re(s)$ sufficiently large. Hence the integral defining $V_j(F)$ for $Re(s)$ sufficiently large is *analytic* in s. (The size of $Re(s)$ depends on τ, σ and s_0.) On the other hand, if F is unramified, we get much more information. In fact, for any number $T' > 0$ so that $|Re(\chi_i)| < T'$ for $i = 1$, rank$(\mathbb{G})$, then there exists s_1 sufficiently large so that with $Re(s) > s_1$, then the integral defining $V_j(F)$ (for this family of χ) is convergent. In fact, this shows that for *any* $g_1 \in \mathbb{G}$ the function $V_j(F^{g_1})$ admits exactly the same properties of convergence! Thus for F the unramified spherical vector in $\mathrm{ind}_{B_{\mathbb{G}}'}^{\mathbb{G}}(\chi + s \otimes \chi')$ (normalized so that $F(e) = 1$) we have that $V_j(F * \phi_0)$ is convergent for the same family $(\chi + \underset{\sim}{s}) \otimes \chi'$. This implies that the integral defining $V_j(F * \phi_0)$, in fact, extends the integral defining $V_j(F * \phi_0)$ given by

$$(\int_{\Sigma_{ij}} (F_1(\chi + s \otimes \chi')[^{\tilde{w}^{-1}} \varphi_1] * \phi_0)(ux) \psi_{i,j,\xi_0}(u) du)$$

(defined in its given domain (χ, χ', s) used above, see **Remark** $(**)$ in **Appendix 1 of** §**5**).

Appendix III.

At this point we let V be a vector space endowed with a nondegenerate quadratic form q_V.

We form the r-th exterior power $\Lambda^r(V)$ of V.

Then we know that $\Lambda^r(V)$ possesses a nondegenerate symplectic or quadratic form q_{Λ^r} which satisfies

$$q_{\Lambda^r}(g_1 w, g_2 w') = q_{\Lambda^r}(w, w')$$

for all $g_1 \in SO(V, q_V) =$ isometry group of q_V and $w, w' \in \Lambda^r V$.

In fact we know that $\Lambda^r(V)$ is an irreducible module for $SO(V, q_V)$ except in the case where $r = 1/2 dimV$. Moreover $\Lambda^r(V)$ and $\Lambda^{dimV-r}(V)$ are equivalent representatives for $SO(V, q_V)$. In the case when $r = 1/2dimV$ then $\Lambda^r(V)$ is either irreducible or is a direct sum of 2 inequivalent $SO(V, q_V)$ modules. We note that if q_V is totally split, then the latter condition holds.

We recall the decomposition

$$V = W \oplus V_{\text{anis}} \oplus W^*$$

where W, W^* are maximal isotropic subspaces in V which are nonsingularly paired. V_{anis} is an anisotropic space relative to q_V.

We let $V = (\xi) \oplus V'$ be an orthogonal direct sum with $q_V(\xi) \neq 0$. Then we have a decomposition as spaces: $\Lambda^r(V) = \Lambda^r(V') \oplus \Lambda^{r-1}(V') \wedge (\xi)$.

We let Ω^r ($\Omega_r^{'}$ *resp.*) be highest weight vectors in $\Lambda^r(V)(\Lambda^r(V')$ *resp.*) of Borel subgroups B and B' in $SO(V, q_V)$ and $SO(V', q_{V'})$. Then we let $g = b_1 m b_2$ with $b_1 \in B$ and $b_2 \in B'$. We have the formula

$$q_{\Lambda^r}(g\Omega_r^{'}, \Omega_r) = \omega_r(b_1)^{-1}\omega_r^{'}(b_2) q_{\Lambda^r}(m\Omega_r^{'}, \Omega_r)$$

with ω_r ($\omega_r^{'}$ *resp.*) the character of $B(B'$ *resp.*) according to which Ω_r ($\Omega_r^{'}$ *resp.*) transforms. We also have the formula:

$$q_{\Lambda^r}(g(\Omega_{r-1}^{'} \wedge \xi), \Omega_r) = w_r(b_1)^{-1} w_{r-1}^{'}(b_2) q_{\Lambda^r}(m(\Omega_{r-1}^{'} \wedge \xi), \Omega_r)$$

From this data above we deduce that

$$\frac{q_{\Lambda^r}(g\Omega_r^{'}, \Omega_r)}{q_{\Lambda^r}(g(\Omega_{r-1}^{'} \wedge \xi), \Omega_r)} = \left(\frac{\omega_r^{'}(b_2)}{\omega_{r-1}^{'}(b_2)}\right)\left(\frac{q_{\Lambda^r}(m\Omega_r^{'}, \Omega_r)}{q_{\Lambda^r}(m\Omega_{r-1}^{'} \wedge \xi), \Omega_r)}\right)$$

provided we know that $q_{\Lambda^r}(g(\Omega_r^{'} \wedge \xi), \Omega_r) \neq 0$. We can invert the above formula as long as $q_{\Lambda^r}(g\Omega_r^{'}, \Omega_r) \neq 0$.

We analyze below when $q_{\Lambda^r}(m\Omega_r^{'}, \Omega_r) \neq 0$ and when $q_{\Lambda^r}(m\Omega_{r-1}^{'} \wedge \xi), \Omega_r) \neq 0$.

Now we must analyze the data $w_S, w_0^{\mathbb{G}}, w_0^H, w_0^M, w_i, w_i^S$, and γ in the 4 cases (a), (b), (c), and (d) given in **Appendix** 1.

Our basic objective here is to compute for the element

$$\gamma^{-1} w_S Z w_S w_S^i = (b_H)_0 w_0^H \eta_H (b_M)_0$$

the components $(b_H)_0$ *and* $(b_M)_0$ *or at least the toral parts.*

We do now this problem in the 4 cases mentioned above. In fact we collect all the data in each of the cases.

We use here the pairs $(\tilde{B}_{W_1(R_i)}, B_1)$ and (B_1, B_2) given in the construction of the cases (a) ... (d) in **Appendix 1** of §5.

(a) Here $W_1(R_1) = \mathrm{Sp}\{z_1^*, \dots z_\nu^*\} \oplus \mathrm{Sp}\{z_1, \dots z_\nu\}$, $\xi_0 = z_1 + \gamma z_1^*$, $\xi_1 = z_1 - \gamma z_1^*$. Now $(\xi_0)^\perp = \mathrm{Sp}\{z_2^*, \dots z_\nu^*\} \oplus (\xi_i) \oplus \mathrm{Sp}\{z_2, \dots z_\nu\}$ and $(\xi_0)^\perp \cap (\xi_1)^\perp = \mathrm{Sp}\{z_2^*, \dots z_\nu^*\} \oplus \mathrm{Sp}\{z_2, \dots z_\nu\}$. The group $H = SO(W_1)$, $M = SO((\xi_0)^\perp)$ and $H_1 = SO((\xi_0)^\perp \cap (\xi_1)^\perp)$. Then we let B_H = the Borel subgroup of H stabilizing the flag $\{z_1, z_\nu, \dots, z_2\} \supseteq \dots \supseteq \{z_2\}$, B_M = the Borel subgroup of M stabilizing the flag $\{z_\nu, \dots, z_2\} \supseteq \dots \supseteq \{z_2\}$. and B_{H_1} = the Borel subgroup of H_1 stabilizing the flag $\{z_\nu, \dots, z_2\} \supseteq \dots \supseteq \{z_2\}$. Here $\tilde{B}_{W_1(R_L)} = B_H$, $B_1 = B_M$ and $B_2 = H_1$. Now we let T_H be the torus $T(t_1, \dots, t_\nu)$ which is given by $T(t_1, \dots, t_\nu)(z_j) = t_j z_j$ and $T(t_1, \dots, t_\nu)(z_j^*) = t_j^{-1} z_j^*$ for $j = 1, \dots \nu$.

Now we have the following formulae:

(1) $w_S =$

$$\begin{cases} v_\alpha \to v_\alpha^*,\ v_\alpha^* \to v_\alpha & (\alpha = 1, \dots i) \\ z_i \to z_i,\ z_i^* \to z_i^* & (i = 1, \dots \nu) \quad \text{except where } i \text{ is odd;} \\ \text{then} & \\ \quad z_i \to z_i,\ z_i^* \to z_i^* & (i = 1, \dots \nu - 1) \\ \text{and } z_\nu \to z_\nu^*,\ z_\nu^* \to z_\nu & \end{cases}$$

(2) $w_0^{\mathbb{G}} =$

$$\begin{cases} v_\alpha \to v_\alpha^*,\ v_\alpha^* \to v_\alpha & (\alpha = 1, \dots i) \\ z_i \to z_i^*,\ z_i^* \to z_i & (i = 1, \dots \nu) \quad \text{except where } i + \nu \text{ is odd} \\ \text{then} & \\ \quad z_i \to z_i^*,\ z_i^* \to z_i & (i = 2, \dots \nu) \\ \text{and } z_1 \to z_1,\ z_1^* \to z_1^* & \end{cases}$$

(3) $w_0^H =$

$$\begin{cases} z_i \to z_i^*,\ z_i^* \to z_i \qquad (i = 1, \dots \nu), \quad \nu \text{ even} \\ \begin{cases} z_i \to z_i^*,\ z_i^* \to z_i & (i = 2, \dots \nu), \quad \nu \text{ odd} \\ z_1 \to z_1,\ z_1^* \to z_1^* & \end{cases} \end{cases}$$

(4) $w_i = w_0^{\mathbb{G}} w_0^H =$

$$\begin{cases} v_\alpha \to v_\alpha^* \, , \; v_\alpha^* \to v_\alpha & (\alpha = 1, \ldots i) \\ I \text{ on } W_1(R_i) \text{ if} & (i+\nu, \nu) \text{ is (even, even) or (odd, odd)} \\ z_1 \to z_1^* \, , \; z_1^* \to z_1 & \text{if } (i+\nu,\nu) \text{is (odd, even) or (even, odd)} \\ \text{and } I \text{ in } < z_1, z_1^* >^\perp & \end{cases}$$

(5) $w_0^M =$

$$\begin{cases} z_t \to z_t^* \, , \; z_t^* \to z_t & (t = 2, \ldots \nu) \\ \xi_1 \to & \begin{cases} +\xi_1, \text{ if } \nu \text{ odd} \\ -\xi_1, \text{ if } \nu \text{ even} \end{cases} \end{cases}$$

Remark. We note in the above choice of $w_0^{\mathbb{G}}$ and w_0^H (with $w_i = w_0^{\mathbb{G}} w_0^H$) we do not necessarily have the condition that $w_i(\xi_0) = \xi_0$ (see **Appendix** 1 to §8). However we note that we can adjust the choice of $w_0^{\mathbb{G}}$ and w_0^H minimally so that the above condition is valid. For instance if $(i+\nu, \nu)$ is (odd, even) we alter $w_0^{\mathbb{G}}$ by requiring $z_1 \to \gamma z_1, z_1^* \to \gamma^{-1} z_1^*$ and the rest remains the same. Then w_i is changed so that $z_1 \to \gamma z_1^*$ and $z_1^* \to \gamma^{-1} z_1$. In any case we emphasize here that the choices of $w_0^{\mathbb{G}}$ and w_0^H are tied to one another as indicated in the above example. We note that the new w_i has the exact form as in (4) except the condition on the space z_1, z_1^* becomes:

$$\begin{cases} I & \text{if } (i+\nu, \nu) \text{ have same parity} \\ z_1 \to \gamma z_1^* \, , \; z_1^* \to \gamma^{-1} z_1 & \text{if } (i+\nu,\nu) \text{ have opposite parity} \end{cases}$$

We note then $w_1(\xi_0) = \xi_0$ and $w_1(\xi_1) = \pm\xi_1$ ($-$ in the case of opposite parity). However we note here then that $w_i B_H w_i^{-1} = B_H$ and $w_i B_M w_i^{-1} = B_M$.

In any case, we note that $w_i = w_0^{\mathbb{G}} w_0^H = t_1 w_0^H w_0^{\mathbb{G}}$ where $t \in T_H \cap K_H$.

We note that the highest weight vectors of B_H in the various Λ^r have the form $z_2, z_3 \wedge z_2, \ldots z_\nu \wedge \cdots \wedge z_2, z_1 \wedge z_\nu \wedge \ldots z_2$ and $z_1^* \wedge z_\nu \wedge \cdots \wedge z_2$. Also the highest weight vectors for B_M in the various Λ^r also have the form $z_2, z_3 \wedge z_2, \ldots, z_\nu \wedge \cdots \wedge z_2$. Thus we compute that for $g = b_H w_0^H \eta_H b_M$ $(m \leq \nu)$

(i) $q_{\Lambda^{m-1}}(g(z_m \wedge \cdots \wedge z_2) \, , \; z_m \wedge \cdots \wedge z_2) =$

$$(t_m \ldots t_2)^{-1} \, (t_m^{'} \ldots t_2^{'}) \; q_{\Lambda^{m-1}}(\eta_H(z_m \wedge \cdots \wedge z_2) \, , \; z_m^* \wedge \cdots \wedge z_2^*)$$

(ii) $q_{\Lambda^\nu}(g(z_1 \wedge z_\nu \wedge \cdots \wedge z_2) \, , \; z_1 \wedge z_\nu \wedge \cdots \wedge z_2) \;=\; (t_\nu \ldots t_1)^{-1} \, t_\nu^{'} \ldots t_2^{'}$

$$q_{\Lambda^\nu}(\eta_H(z_1 \wedge z_\nu \wedge \cdots \wedge z_2) \, , \; z_1^* \wedge z_\nu^* \wedge \cdots \wedge z_2^*) \quad (\text{ for } \nu \text{ even})$$

(iii) $q_{\Lambda^\nu}(g(z_1 \wedge z_\nu \wedge \cdots \wedge z_2) \, , \; z_1^* \wedge z_\nu \wedge \cdots \wedge z_2)) =$

$$(t_\nu \ldots t_2 t_1^{-1})^{-1} \, (t_\nu^{'} \ldots t_2^{'}) \, q_{\Lambda^\nu}(\eta_H(z_1 \wedge z_\nu \wedge \cdots \wedge z_2) \, , \; z_1^* \wedge z_\nu^* \wedge \cdots \wedge z_2^*) \; (\text{for } \nu \text{ odd})$$

(iv) $(m \leq \nu)$ $q_{\Lambda^{m-1}}(g(\xi_0 \wedge z_{m-1} \wedge \cdots \wedge z_2)\ ,\ z_m \wedge \cdots \wedge z_2) =$

$$(t_m \ldots t_2)^{-1}\ (t_m^{'} \ldots t_2^{'})\ q_{\Lambda^{m-1}}(\eta_H(\xi_0 \wedge z_{m-1} \wedge \cdots \wedge z_2)\ , z_m^* \wedge \cdots \wedge z_2^*)$$

We also take the case here where $m = 2$, i.e.

$$q_{\Lambda^1}(g\xi_0, z_2) = t_2^{-1} q_{\Lambda^1}(\eta_H(\xi_0), z_2^*)\,.$$

We recall here that the element γ given by $\gamma B_H \gamma^{-1} = B_{W_{1(R_i)}}$ is defined by the formula $\gamma^{-1}(z_1) = z_2, \gamma^{-1}(z_2) = z_3, \ldots \gamma^{-1}(z_\nu) = z_1$. Then we compute $\gamma(z_2) = z_1, \gamma(z_3 \wedge z_2) = z_2 \wedge z_1, \ldots, \gamma(z_\nu \wedge \cdots \wedge z_2) = z_{\nu-1} \wedge \cdots \wedge z_1$ and $\gamma(z_1 \wedge z_\nu \wedge \cdots \wedge z_2) = z_\nu \wedge z_{\nu-1} \wedge \cdots \wedge z_1$.

Thus we have, for $g = \gamma^{-1} w_S Z w_S w_S^i$ with $Z = \eta_M^{-1}(w_0^M)^{-1}$,

$$q_{\Lambda^{r-1}}(g(z_r \wedge \cdots \wedge z_2), z_r \wedge \cdots \wedge z_2) = q_{\Lambda^{r-1}}(\eta_M^{-1}(z_r^* \wedge \cdots \wedge z_2^*)\ ,\ z_{r-1} \wedge \cdots \wedge z_1).$$

Similarly $q_{\Lambda^\nu}(g(z_1 \wedge z_\nu \wedge \cdots \wedge z_2)\ ,\ z_1 \wedge z_\nu \wedge \cdots \wedge z_2)$

$$= \begin{cases} q_{\Lambda^\nu}(\eta_M^{-1}(z_1^* \wedge z_\nu^* \wedge \cdots \wedge z_2^*), z_1 \wedge z_\nu \wedge \cdots \wedge z_2)\gamma \text{ where } \xi_0 = z_1 + \gamma z_1^* \ (i \text{ even}) \\ q_{\Lambda^\nu}(\eta_M^{-1}(z_1 \wedge z_\nu^* \wedge z_{\nu-1}^* \wedge \cdots \wedge z_2^*), z_1^* \wedge z_\nu \wedge \cdots \wedge z_2)\ (i \text{ odd}) \end{cases}$$

Also (for ν odd) $q_{\Lambda^\nu}(g(z_1 \wedge z_\nu \wedge \cdots \wedge z_2), z_1^* \wedge z_\nu \wedge \cdots \wedge z_2)$

$$= \begin{cases} q_{\Lambda^\nu}(\eta_M^{-1}(z_1 \wedge z_\nu^* \wedge \cdots \wedge z_2^*), z_1^* \wedge z_\nu \wedge \cdots \wedge z_2)\ (i \text{ even}) \\ q_{\Lambda^\nu}(\eta_M^{-1}(z_1^* \wedge z_\nu^* \wedge \cdots \wedge z_2^*), z_1 \wedge z_\nu \wedge \cdots \wedge z_2)\ (i \text{ odd}) \end{cases}$$

But we also have that $(m \leq \nu)$ $q_{\Lambda^r}(g(\xi_0 \wedge z_{m-1} \wedge \cdots \wedge z_2), z_m \wedge \cdots \wedge z_2) =$

$$q_{\Lambda^m}(\xi_0 \wedge z_m^* \wedge \cdots \wedge z_2^*, z_m \wedge \cdots \wedge \eta_M\left(\begin{Bmatrix} z_1 \\ z_1^* \end{Bmatrix}\right)\,.$$

Here $\{\ \}$ is chosen to depend what parity $(i + \nu, \nu)$ has.

We note that it is possible to choose η_M so that

$$\eta_M(\xi_1) = \xi_1 + \sum_{i \geq 2} \tilde{\beta}_i z_i$$

where the β_i are nonzero *units*.

Then we have that $\eta_M(\xi_0) = \xi_0$ and $\eta_M(\xi_1)$ as above implying that

$$\eta_M(z_1) = z_1 + 1/2 \sum_{i \geq 2} \tilde{\beta}_i z_i$$

and

$$\eta_M(z_1^*) = z_1^* - \frac{1}{2\gamma} \sum \tilde{\beta}_i z_i$$

Hence we have

$$\eta_M(z_{r-1}\wedge\cdots\wedge z_1) = (z_{r-1}\wedge\cdots\wedge z_2)\wedge(z_1+1/2\sum_{i\geq 2}\tilde{\beta}_i z_i)$$
$$= z_{r-1}\wedge\cdots\wedge z_1 + 1/2\left(\sum_{t\geq r}\tilde{\beta}_t(z_{r-1}\wedge\cdots\wedge z_2\wedge z_t)\right).$$

Hence $q_{\Lambda^{r-1}}(\eta_M^{-1}(z_r^*\wedge\cdots\wedge z_2^*), z_{r-1}\wedge\cdots\wedge z_1) = 1/2\beta_r q_{\Lambda^{r-1}}(z_r^*\wedge\cdots\wedge z_2^*), z_{r-1}\wedge\cdots\wedge z_2\wedge z_r)$ (using the fact that $z_r^*\wedge\cdots\wedge z_2^*$ and $z_{r-1}\wedge\cdots\wedge z_2\wedge z_t$ $(t\neq r)$ are not dual to each other relative to the torus action T_H of B_H). Note that $q_{\Lambda^{r-1}}(z_r^*\wedge\cdots\wedge z_2^*, z_{r-1}\wedge\cdots\wedge z_2\wedge z_r)$ can be chosen to be a unit! Thus $q_{\Lambda^{r-1}}(g(z_r\wedge\cdots\wedge z_2), z_r\wedge\cdots\wedge z_2)$ is a unit!

Then for the case (ν even) $g_{\Lambda^\nu}(g(z_1\wedge z_\nu\wedge\cdots\wedge z_2),\ z_1\wedge z_\nu\wedge\cdots\wedge z_2) = g_{\Lambda^\nu}\left(\begin{Bmatrix} z_1 \\ z_1^* \end{Bmatrix}\wedge z_\nu^*\wedge\cdots\wedge z_2^*, \eta_M\left(\begin{Bmatrix} z_1 \\ z_1^* \end{Bmatrix}\right)\wedge z_\nu\wedge\cdots\wedge z_2\right)$. Using the formula of η_M applied to either z_1 or z_1^* and the compatibility of how $\begin{Bmatrix} z_1 \\ z_1^* \end{Bmatrix}$ is chosen on one side or the other we deduce again that the above number is a unit. A similar argument works in the ν odd case.

Next we consider the case of the terms $q_{\Lambda^\nu}(g(\xi_0\wedge z_{m-1}\wedge\cdots\wedge z_2), z_m\wedge\cdots\wedge z_2)$.

First we note that

$$\eta_M\left(z_{m-1}\wedge\cdots\wedge z_2\wedge\begin{Bmatrix} z_1 \\ z_1^* \end{Bmatrix}\right) = z_{m-1}\wedge\cdots\wedge z_2\wedge\eta_M\begin{Bmatrix} z_1 \\ z_1^* \end{Bmatrix}$$
$$= z_{m-1}\wedge\cdots\wedge z_2\wedge\begin{Bmatrix} z_1 \\ z_1^* \end{Bmatrix} + 1/2\sum_{i\geq m}\begin{Bmatrix} \beta_i \\ -\beta_c/\gamma \end{Bmatrix} z_{m-1}\wedge\cdots\wedge z_2\wedge z_i.$$

Hence we have that

$$q_{\Lambda^\nu}\left(\alpha z_1+\beta z_1^*\wedge z_{m-1}^*\wedge\cdots\wedge z_2^*, z_{m-1}\wedge\cdots\wedge z_2\wedge\begin{Bmatrix} z_1 \\ z_1^* \end{Bmatrix}\right)$$

is a unit provided that α and β are both units.

This implies that $q_{\Lambda^m}(g(\xi_0\wedge z_{m-1}\wedge\cdots\wedge z_2), z_m\wedge\cdots\wedge z_2)$ is a unit!

Finally we compute the effect of η_H on various highest weight vectors; indeed $\eta_H(z_m\wedge\cdots\wedge z_2) = z_m\wedge\cdots\wedge z_2$, $(m\leq\nu)$, $\eta_H(z_1\wedge z_\nu\wedge\cdots\wedge z_2) = z_1\wedge z_\nu\wedge\cdots\wedge z_2$ and $\eta_H(z_1^*\wedge z_\nu\wedge\cdots\wedge z_2) = z_1^*\wedge z_\nu\wedge\cdots\wedge z_2$. We now must check $\eta_H(\xi_0\wedge z_{m-1}\wedge\cdots\wedge z_2) = \eta_H(\xi_0)\wedge z_{m-1}\wedge\cdots\wedge z_2$. We note that η_H can be chosen so that

$$\eta_H(\xi_0) = \xi_0 + \sum_{i\geq 2}\alpha_i z_i$$

with α_i nonzero *units*. Hence $\eta_H(\xi_0)\wedge z_{m-1}\wedge\cdots\wedge z_2 = \xi_0\wedge z_{m-1}\wedge\cdots\wedge z_2 + \sum_{i\geq 2}\alpha_i z_i\wedge z_{m-1}\wedge\cdots\wedge z_2$.

Thus $q_{\Lambda^{m-1}}(\eta_H(\xi_0 \wedge z_{m-1} \wedge \cdots \wedge z_2), z_m^* \wedge \cdots \wedge z_2^*)$ is a unit! Thus we have that in the formulae (i) – (iv) the q dot products on the right hand side are all units.

Thus the above computations prove that the t_i components (t_i^i components *resp.*) of b_H (b_M *resp.*) are all units!

Remark. We note at this point that the same calculations as above show that if $g = Z = \eta_M^{-1}(w_0^M)^{-1}$, then if $g = b_H(w_0^H)\eta_H b_M$, then the toral parts of b_H and b_M are units!

(b) Here $W_1(R_i) = \mathrm{Sp}\{z_1^*, \ldots, z_\nu^*\} \oplus \mathrm{Sp}\{Z_0, Z_1\} \oplus \mathrm{Sp}\{z_1, \ldots, z_\nu\}$. Here $\xi_0 = z_1 + \gamma z_1^*$ and $\xi_1 = z_1 - \gamma z_1^*$. Also recall $q_{W_i}(Z_0, Z_0) = 2\gamma$. Thus $\{Z_0, Z_1\}$ span a 2 dimensional anisotropic space. Moreover $(\xi_0)^\perp = \mathrm{Sp}\{z_2^*, \ldots, z_\nu^*\} \oplus \mathrm{Sp}\{\xi_1, Z_0, Z_1\} \oplus \mathrm{Sp}\{z_2, \ldots, z_\nu\}$, $(\xi_0)^\perp \cap (\xi_1)^\perp = \mathrm{Sp}\{z_2^*, \ldots, z_\nu^*\} \oplus \{Z_0, Z_1\} \oplus \mathrm{Sp}\{z_2, \ldots, z_\nu\}$. Also $\mathrm{Sp}\{\xi_1, Z_0, Z_1\}$ is a 3 dimensional space with Witt index 1. So $\mathrm{Sp}\{\xi_1, Z_0, Z_1\} = (X) \oplus (Z_1) \oplus (X^*)$ with $\{X, X^*\}$ spanning a hyperbolic plane. We note here that we can choose $X = \xi_1 + Z_0$ and $X^* = -\frac{1}{4\gamma}(\xi_1 - Z_0)$. Then let $H = SO(W_1(R_i))$, $M = SO((\xi_0)^\perp)$, and $H_1 = SO((\xi_0)^\perp) \cap (\xi_1)^\perp))$. We define $B_H =$ the Borel group of H stabilizing the flag $\{X, z_\nu, \ldots, z_2\} \supseteq \cdots \supseteq \{z_2\}$, $B_M =$ the Borel subgroup of M stabilizing the flag $\{X, z_\nu, \ldots, z_2\} \supseteq \cdots \supseteq \{z_2\}$ and $B_{H_1} =$ the Borel subgroup of M stabilizing the flag $\{z_\nu, \ldots, z_2\} \supseteq \cdots \supseteq \{z_2\}$. We let $T_H = T(t_i, \ldots, t_\nu)$ be the split torus in B_H defined by $T_H(t_1, \ldots, t_\nu)(X) = t_1 X, T_H(t_1, \ldots, t_\nu)X^* = t_1^{-1}X^*$ and $T_H(t_1, \ldots, t_\nu)(z_i) = t_i z_i$, $T_H(t_i, \ldots, t_\nu)(z_i^*) = t_i^{-1} z_i^*$.

Next we have the following formulae

(1) $w_S =$

$$\begin{cases} v_\alpha \to v_\alpha^* \,,\ v_\alpha^* \to v_\alpha & (\alpha = 1, \ldots i) \\ I \text{ on } W_i(R_i) & \text{except if } i \text{ is odd} \\ I \text{ in } \mathrm{Sp}\{z_1, \ldots z_\nu, Z_0, z_1^*, \ldots z_\nu^* & \\ \text{and } Z_1 \to -Z_1 & \text{where } i \text{ is odd.} \end{cases}$$

(2) $w_0^{\mathbb{G}} =$

$$\begin{cases} v_\alpha \to v_\alpha^*, v_\alpha^* \to v_\alpha & (\alpha = 1, \ldots i) \\ z_i \to z_i^*, z_i^* \to z_i & (i = 2, \ldots \nu) \\ X \to X^*, X^* \to X & \\ \xi_0 \to \xi_0 & \\ Z_1 \to \pm Z_1 & +\text{if } (i+\nu) \text{ even}; \ - \text{ if } (i+\nu) \text{ odd} \end{cases}$$

(3) $w_0^H =$

$$\begin{cases} z_i \to z_i^*, z_i^* \to z_i & (i = 2, \ldots \nu) \\ X \to X^*, X^* \to X & \\ \xi_0 \to \xi_0 & \\ Z_1 \to \pm Z_1 & +\text{if } \nu \text{ even}; \ - \text{ if } \nu \text{ odd} \end{cases}$$

(4) $w_i =$

$$\begin{cases} v_\alpha \to v_\alpha^*, v_\alpha^* \to v_\alpha & (\alpha = 1, \dots i) \\ z_i \to z_i, z_i^* \to z_i^* & (i = 2, \dots \nu) \\ X \to X, X^* \to X^* & \\ \xi_0 \to \xi_0 & \\ Z_1 \to \pm Z_1 & + \text{ if } (i+\nu, \nu) \text{ have same parity; } - \text{ otherwise.} \end{cases}$$

(5) $w_0^M =$

$$\begin{cases} z_\nu \to z_\nu^*, z_\nu^* \to z_\nu & (i = 2, \dots \nu) \\ X \to X^*, X^* \to X & \\ Z_1 \to \pm Z_1 & + \text{ if } \nu \text{ even; } - \text{ otherwise.} \end{cases}$$

Remark. The element w_i given above satisfies $w_i(\xi_0) = \xi_0$ and $w_1(\xi_1) = \xi_1$. Moreover it is clear that w_1 preserves the flags defining B_H and B_M.

Then the highest weight vectors of B_H in the various Λ^r have the form $z_2, z_3 \wedge z_2 \dots, z_\nu \wedge \dots \wedge z_2$ and $X \wedge z_\nu \wedge \dots \wedge z_2$. Similarly the highest weight vectors of B_M are $z_2, z_3, \wedge z_2, \dots,$ and $X \wedge z_\nu \wedge \dots \wedge z_2$. Then for an element $g = b_H w_0^H \eta_H b_M$ we have the formulae:

(i) $q_{\Lambda^{m-1}}(g(z_m \wedge \dots \wedge z_2)\ ,\ z_m \wedge \dots \wedge z_2) =$

$$(t_m \dots t_2)^{-1}\ (t_m^{'} \dots t_2^{'})\ q_{\Lambda^{m-1}}(\eta_H(z_m \wedge \dots \wedge z_2)\ ,\ z_m^* \wedge \dots \wedge z_2^*).$$

(ii) $q_{\Lambda^{\nu-1}}(g(X \wedge z_\nu \wedge \dots \wedge z_2), X \wedge z_\nu \wedge \dots \wedge z_2) =$

$$(t_\nu \dots t_1)^{-1}(t_\nu^{'} \dots t_1^{'}) q_{\Lambda^{\nu-1}}(\eta_H(X \wedge z_\nu \wedge \dots \wedge z_2\ ,\ X^* \wedge z_\nu^* \wedge \dots \wedge z_2^*),$$

(iii) $(m \le \nu)\quad q_{\Lambda^{m-1}}(g(\xi_0 \wedge z_{m-1} \wedge \dots \wedge z_2)\ ,\ z_m \wedge \dots \wedge z_2) =$

$$(t_m \dots t_2)^{-1}\ (t_m^{'} \dots t_2^{'})\ q_{\Lambda^{m-1}}(\eta_H(\xi_0 \wedge z_{m-1} \wedge \dots \wedge z_2\ ,\ z_m^* \wedge \dots \wedge z_2^*).$$

(iv) $q_{\Lambda^{\nu-1}}(g(\xi_0 \wedge z_\nu \wedge \dots \wedge z_2)\ ,\ X \wedge z_\nu \wedge \dots \wedge z_2) =$

$$q_{\Lambda^{\nu-1}}(\eta_H(\xi_0 \wedge z_\nu \wedge \dots \wedge z_2)\ ,\ X^* \wedge z_\nu^* \wedge \dots \wedge z_2^*).$$

We recall that the element ζ so that $\zeta B_H \zeta^{-1} = B_{W_1(R_i)}$ satisfies $\zeta^{-1}(z_\nu) = X, \zeta^{-1}(z_{\nu-1}) = z_\nu, \dots, \zeta^{-1}(z_1) = z_2$. Thus we have $\zeta(z_2) = z_1, \dots, \zeta(z_\nu \wedge \dots \wedge z_2) = z_{\nu-1} \wedge \dots \wedge z_1$ and $\zeta(X \wedge z_\nu \wedge \dots \wedge z_2) = z_\nu \wedge z_{\nu-1} \wedge \dots \wedge z_1$.

Then we have for $g = \zeta^{-1} w_S Z w_S w_S^i$ with $Z = \eta_M^{-1}(w_0^M)^{-1}$

$$q_{\Lambda^{r-1}}(g(z_r \wedge \dots \wedge z_2)\ ,\ z_r \wedge \dots \wedge z_2) = q_{\Lambda^{r-1}}(\eta_M^{-1}(z_r^* \wedge \dots \wedge z_2^*)\ ,\ z_{r-1} \wedge \dots \wedge z_1)$$

and $q_{\Lambda^{\nu-1}}(g(X \wedge z_\nu \wedge \cdots \wedge z_2)\,,\, X \wedge z_\nu \wedge \cdots \wedge z_2) =$

$$q_{\Lambda^{\nu-1}}(\eta_M^{-1}(X^* \wedge z_\nu^* \wedge \cdots \wedge z_2^*)\,,\, z_\nu \wedge \cdots \wedge z_1)$$

Also we note that

$$q_{\Lambda^{m-1}}(g(\xi_0 \wedge z_{m-1} \wedge \cdots \wedge z_2)\,,\, z_m \wedge \cdots \wedge z_2)$$
$$= q_{\Lambda^{m-1}}(\eta_M^{-1}(\xi_0 \wedge z_{m-1}^* \wedge \cdots \wedge z_2^*)\,,\, z_{m-1} \wedge \cdots \wedge z_1)$$

Next we choose η_M to fix the vectors $\xi_0, z_2, \ldots, z_m$, and x.

Then we also choose η_M to satisfy

$$\eta_M(X^*) = X^* + \sum_{i \geq 2} \beta_i z_i + \alpha\xi_0 + \delta Z_1$$

with β_i, α and δ units. Thus we have that $\eta_M(\xi_1) = \eta_M(X - 4\gamma_1 X^*) =$

$$= (X - 4\gamma X^*) - 4\gamma \sum_{i \geq 2} \beta_i z_i - 4\gamma\alpha\xi_0 - 4\gamma\delta Z_1\,.$$

Thus we deduce that

$$\eta_M(z_1) = z_1 - \gamma \sum \beta_i z_i - \gamma\alpha\xi_0 - \gamma\delta Z_1\,.$$

Thus we deduce that

$$q_{\Lambda^{r-1}}(\eta_M^{-1}(z_r^* \wedge \cdots \wedge z_2^*)\,,\, z_{r-1} \wedge \cdots \wedge z_1)$$
$$q_{\Lambda^{r-1}}(z_r^* \wedge \cdots \wedge z_2^*\,,\, z_{r-1} \wedge \cdots \wedge z_r) \cdot 2\gamma\beta_r$$

Since β_r is a nonzero unit we deduce that $q_{\Lambda^{r-1}}(\eta_M^{-1}(z_r^* \wedge \cdots \wedge z_2^*)\,,\, z_{r-1} \wedge \cdots \wedge z_1)$ is a nonzero unit.

Also from the above calculations we deduce that

$$q_{\Lambda^{\nu-1}}(\eta_M^{-1}(X^* \wedge z_\nu^* \wedge \cdots \wedge z_2^*)\,,\, z_\nu \wedge z_{\nu-1} \wedge \cdots \wedge z_1) =$$
$$q_{\Lambda^{\nu-1}}(X^* \wedge z_\nu^* \wedge \cdots \wedge z_2^*\,,\, z_\nu \wedge z_{\nu-1} \wedge \cdots \wedge z_1) =$$
$$q_{\Lambda^{\nu-1}}(\xi_1 \wedge z_\nu^* \wedge \cdots \wedge z_2^*\,,\, z_\nu \wedge z_{\nu-1} \cdots \wedge z_1) =$$
$$(-\gamma) q_{\Lambda^{\nu-1}}(z_1^* \wedge z_\nu^* \wedge \cdots \wedge z_2^*\,,\, z_\nu \wedge \cdots \wedge z_1)$$

We note this last term becomes a unit and is nonzero!

On the other hand

$$q_{\Lambda^{\nu-1}}(\eta_M^{-1}(\xi_0 \wedge z_{m-1}^* \wedge \cdots \wedge z_2^*),\, z_{m-1} \wedge \cdots \wedge z_1)$$
$$= q(\xi_0 \wedge z_{m-1}^* \wedge \cdots \wedge z_2^*, z_{m-1} \wedge \cdots \wedge z_1)$$
$$-\gamma\alpha q(\xi_0 \wedge z_{m-1}^* \wedge \cdots \wedge z_2^*, z_{m-1} \wedge \cdots \wedge z_2 \wedge \xi_0)$$
$$= \pm q(\xi_0, z_1 2) \pm \gamma\alpha q(\xi_0, \xi_0)$$
$$= \pm\gamma \pm \gamma\alpha\,.$$

Then we choose α a unit so that $(1 \pm \alpha)$ is a unit (possible if residual characteristic > 2).

We now emphasize here again, as in case (a) above, that we have that if $g = Z = (w_0^M \eta_M)^{-1}$ and $g = b_H w_0^H \eta_H b_M$, then the toral parts of b_H and b_M are units.

Thus we see directly that the terms (with $g = \zeta^{-1} w_S Z w_S w_S^i$) $q_{\Lambda^{r-1}}(g(z_r \wedge \cdots \wedge z_2)\ ,\ z_r \wedge \cdots \wedge z_2)$, $q_{\Lambda^{\nu-1}}(g(X \wedge z_\nu \wedge \cdots \wedge z_2)\ ,\ X \wedge z_\nu \wedge \cdots \wedge z_2)$, $q_{\Lambda^{m-1}}(g(\xi_0 \wedge z_{m-1} \wedge \cdots \wedge z_2)\ ,\ z_m \wedge \cdots \wedge z_2)$, and $q_{\Lambda^{\nu-1}}(g(\xi_0 \wedge z_\nu \wedge \cdots \wedge z_2)\ ,\ X \wedge z_\nu \wedge \cdots \wedge z_2)$ are all units and nonzero!

Finally we analyze the formulae (i), (ii), (iii) and (iv) above. For this we note that η_H fixes pointwise $X, z_\nu, \ldots, z_2$. On the other hand we can choose η_H so that

$$\eta_H(\xi_0) = \xi_0 + rX + \sum_{i \geq 2} \alpha_i z_i \ (r, \alpha_i \text{ nonzero units})$$

Hence we have that $q_{\Lambda^{m-1}}(\eta_H(\xi_0 \wedge z_{m-1} \wedge \cdots \wedge z_2)\ ,\ z_m^* \wedge \cdots \wedge z_2) = q_{\Lambda^{m-1}}(z_m \wedge z_{m-1} \wedge \cdots \wedge z_2\ ,\ z_m^* \wedge \cdots \wedge z_2^*)\alpha_m$, and $q_{\Lambda^{\nu-1}}(\eta_H(\xi_0 \wedge z_\nu \wedge \cdots \wedge z_2), X^* \wedge z_\nu * \wedge \cdots \wedge z_2^*) = r q_{\Lambda^{\nu-1}}(X \wedge z_\nu \wedge \cdots \wedge z_2, X^* \wedge z_\nu^* \wedge \cdots \wedge z_2^*)$; hence this element is a nonzero unit. Thus in formulae (i), (ii) and (iii) the parts $q_{\Lambda^{m-1}}(\eta_H(z_m \wedge \cdots \wedge z_2)\ ,\ z_m^* \wedge \cdots \wedge z_2^*)$, $q_{\Lambda^{\nu-1}}(\eta_H(X \wedge z_\nu \wedge \cdots \wedge z_2)\ ,\ X^* \wedge z_\nu^* \wedge \cdots \wedge z_2^*)$, and $q_{\Lambda^{m-1}}(\eta_H(\xi_0 \wedge z_{m-1} \wedge \cdots \wedge z_2)\ ,\ z_m^* \wedge \cdots \wedge z_2^*)$ are all nonzero units.

The above arguments then imply that all the t_i and t_j' are *units*!

(c) Here $W_1(R_i) = \mathrm{Sp}\{z_1^*, \ldots, z_\nu^*\} \oplus \mathrm{Sp}(Z_0) \oplus \mathrm{Sp}\{z_1, \ldots, z_\nu\}$. Here $\xi_0 = z_1 + \gamma z_1^*\ ,\ \xi_1 = z_1 - \gamma z_1^*$. Moreover $q_{W_1}(Z_0, Z_0) = 2\gamma$. Then $(\xi_0)^\perp = \mathrm{Sp}\{z_2^*, \ldots, z_\nu^*\} \oplus \mathrm{Sp}\{\xi_1, Z_0\} \oplus \mathrm{Sp}\{z_2, \ldots, z_\nu\}$ and

$$(\xi_0)^\perp \cap (\xi_1)^\perp = \mathrm{Sp}\{z_2^*, \ldots, z_\nu^*\} \oplus \mathrm{Sp}(Z_0) \oplus \mathrm{Sp}\{z_2, \ldots, z_\nu\}\,.$$

Now $\mathrm{Sp}\{\xi_1, Z_0\}$ is a hyperbolic plane $= \{X, X_*\}$ where $X = \xi_1 + Z_0$, $X^* = -\frac{1}{4\gamma}(\xi_1 - Z_0)$. We let $H = SO(W_1(R_I))$, $M = SO((\xi_0)^\perp)$ and $H_1 = SO((\xi_0)^\perp) \cap (\xi_0)^\perp)$.

We define $B_H =$ the Borel subgroup of H stabilizing the flag $\{X, z_\nu, \ldots, z_2\} \supseteq \cdots \supseteq \{z_2\}$, $B_M =$ the Borel subgroup of M stabilizing the flag $\{X, z_\nu, \ldots, z_2\} \supseteq \cdots \supseteq \{z_2\}$. and $B_{H_1} =$ the Borel subgroup of H_1 stabilizing the flag $\{z_\nu, \ldots, z_2\} \supseteq \cdots \supseteq \{z_2\}$. We let $T_H = T(t_1, \ldots, t_\nu)$ be the split torus defined by $T(t_1, \ldots, t_\nu)(X) = t_1 X\ ,\ T_H(t_1, \ldots, t_\nu)(X^*) = t_1^{-1} X^*\ ,\ T_H(t_1, \ldots t_\nu)(z_i) = t_i z_i\ ,\ T_H(t_1, \ldots, t_\nu)(z_i^*) = t_i^{-1} z_i^*$.

Now we have the following formulae

(1) $w_S =$

$$\begin{cases} v_\alpha \to v_\alpha^*\ ,\ v_\alpha^* \to v_\alpha & (\alpha = 1, \ldots i) \\ I \text{ on } W_i(R_i) & \text{except if } i \text{ is odd} \\ \text{and then } Z_0 \to -Z_0 & \end{cases}$$

(2) $w_0^{\mathbb{G}} =$

$$\begin{cases} v_\alpha \to v_\alpha^*, v_\alpha^* \to v_\alpha & (\alpha = 1, \dots i) \\ z_i \to z_i^*, z_i^* \to z_i & (i = 2, \dots \nu) \\ X \to X^*, X^* \to X & \\ Z_0 \to \pm Z_0 & + \text{if } (i+\nu) \text{ even}; \ - \text{ if } (i+\nu) \text{ odd} \end{cases}$$

(3) $w_0^H =$

$$\begin{cases} z_i \to z_i^*, z_i^* \to z_i & (i = 2, \dots \nu) \\ X \to X^*, X^* \to X & \\ Z_0 \to \pm Z_0 & + \text{if } \nu \text{ even}; \ - \text{ if } \nu \text{ odd} \end{cases}$$

(4) $w_i =$

$$\begin{cases} v_\alpha \to v_\alpha^*, v_\alpha^* \to v_\alpha & (\alpha = 1, \dots i) \\ z_i \to z_i, z_i^* \to z_i^* & (i = 2, \dots \nu) \\ X \to X, X^* \to X^* & \\ Z_0 \to \pm Z_0 & + \text{ if } (i+\nu, \nu) \text{ have same parity}; \ - \text{ otherwise.} \end{cases}$$

(5) $w_0^M =$

$$\begin{cases} z_i \to z_i^*, z_i^* \to z_i & (i = 2, \dots \nu) \\ X \to X^*, X^* \to X & \text{if } \nu \text{ is even} \\ X \to X, X^* \to X^* & + \text{ if } \nu \text{ is odd.} \end{cases}$$

Remark. First we note that $w_i(\xi_0) = \xi_0$ and $w_i(\xi_1) = \xi_1$. Moreover we have also that $w_i B_H w_i^{-1} = B_H$ and $w_i B_M w_i^{-1} = B_M$.

The highest weight vectors of B_H in the various Λ^r have the form $z_2, z_3 \wedge z_2, \dots, z_\nu \wedge \cdots \wedge z_2$ and $X \wedge z_\nu \wedge \cdots \wedge z_2$. A similar statement is valid for B_M. Then if $g = b_H w_0^H \eta_H b_M$ we have the formulae:

(i) $q_{\Lambda^{m-1}}(g(z_m \wedge \cdots \wedge z_2)\ ,\ z_m \wedge \cdots \wedge z_2) =$

$$(t_m \dots t_2)^{-1}\ (t_m' \dots t_2')\ q_{\Lambda^{m-1}}(\eta_H(z_m \wedge \cdots \wedge z_2)\ ,\ z_m^* \wedge \cdots \wedge z_2^*)$$

(ii) $q_{\Lambda^{\nu-1}}(g(X \wedge z_\nu \wedge \cdots \wedge z_2)\ ,\ X \wedge z_1 \wedge \cdots \wedge z_2) =$

$$(t_m \dots t_1)^{-1} t_m' \dots t_1' q_{\Lambda^{\nu-1}}(\eta_H(X \wedge z_\nu \wedge \cdots \wedge z_2)\ ,\ X^* \wedge z_\nu^* \wedge \cdots \wedge z_2^*)$$

(iii) $q_{\Lambda^{m-1}}(g(\xi_0 \wedge z_{m-1} \wedge \cdots \wedge z_2)\ ,\ z_m \wedge \cdots \wedge z_2)) =$

$$(t_m \dots t_2)^{-1}\ (t_{m-1}' \dots t_1')\ q_{\Lambda^{m-1}}(\eta_H(\xi_0 \wedge z_{m-1} \wedge \cdots \wedge z_2)\ ,\ z_m^* \wedge \cdots \wedge z_2^*)$$

(iv) $q_{\Lambda^{\nu-1}}(g(\xi_0 \wedge z_\nu \wedge \cdots \wedge z_2)\ ,\ X \wedge z_\nu \wedge \cdots \wedge z_2) =$

$$(t_\nu \dots t_1)^{-1}\ (t_\nu' \dots t_2')\ q_{\Lambda^{\nu-1}}(\eta_H(\xi_0 \wedge z_\nu \wedge \cdots \wedge z_2)\ ,\ X^* \wedge z_\nu^* \wedge \cdots \wedge z_2^*)$$

We recall that the element ζ so that $\zeta B_H \zeta^{-1} = B_{W_1(R_i)}$ satisfies $\zeta^{-1}(z_\nu) = X$, $\zeta^{-1}(z_{\nu-1}) = z_\nu, \dots, \zeta^{-1}(z_1) = z_2$. Thus we have $\zeta(z_2) = z_1, \dots, \zeta(z_\nu \wedge \cdots \wedge z_2) = z_{\nu-1} \wedge \cdots \wedge z_1$ and $\zeta(X \wedge z_\nu \wedge \cdots \wedge z_2) = z_\nu \wedge z_{\nu-1} \wedge \cdots \wedge z_1$.

Then we have for $g = \zeta^{-1} w_S Z w_S w_S^i$ with $Z = \eta_M^{-1}(w_0^M)^{-1}$.

$$q_{\Lambda^{r-1}}(g(z_r \wedge \cdots \wedge z_2)\ ,\ z_r \wedge \cdots \wedge z_2) = q_{\Lambda^{r-1}} \eta_M^{-1}(z_r^* \wedge \cdots \wedge z_2^*)\ ,\ z_{r-1} \wedge \cdots \wedge z_1).$$

Secondly we have $q_{\Lambda^{\nu-1}}(g(X \wedge z_\nu \wedge \cdots \wedge z_2)\ ,\ X \wedge z_\nu \wedge \cdots \wedge z_2)$

$$= q_{\Lambda^{\nu-1}} \eta_M^{-1}(X^* \wedge z_\nu^* \wedge \cdots \wedge z_2^*)\ ,\ z_\nu \wedge z_{\nu-1} \wedge \cdots \wedge z_1).$$

Also for $(m \leq \nu)$ $q_{\Lambda^{m-1}}(g(\xi_0 \wedge z_{m-1} \wedge \cdots \wedge z_2)\ ,\ z_m \wedge \cdots \wedge z_2)$

$$= q_{\Lambda^{m-1}}(\eta_M^{-1}(\xi_0 \wedge z_{m-1} \wedge \cdots \wedge z_2)\ ,\ z_{m-1} \wedge \cdots \wedge z_2).$$

Finally we note that $q_{\Lambda^{\nu-1}}(g(\xi_0 \wedge z_\nu \wedge \cdots \wedge z_2), X \wedge z_\nu \wedge \cdots \wedge z_2)$

$$= q_{\Lambda^\nu}(\eta_M^{-1}(g(\xi_0 \wedge z_\nu^* \wedge \cdots \wedge z_2^*)\ ,\ z_1 \wedge z_\nu \wedge \cdots \wedge z_2).$$

Moreover $q_{\Lambda^{\nu-1}}(g(\xi_0 \wedge z_\nu \wedge \cdots \wedge z_2)\ ,\ X \wedge z_{\nu-1} \wedge \cdots \wedge z_2) =$

$$q_{\Lambda^{\nu-1}}(\eta_M^{-1}(\xi_0 \wedge z_\nu^* \wedge \cdots \wedge z_2^*)\ ,\ z_\nu \wedge z_{\nu-1} \wedge \cdots \wedge z_1).$$

The rest of the proof is similar to case (b). Thus we prove here by similar means that for g as above the components of b_H and b_M along the split torus T_H and T_M are units!

(d) Here $W_1(R_i) = \mathrm{Sp}\{z_2^*, \dots, z_\nu^*\} \oplus \mathrm{Sp}\{z_1, Z_0, z_1^*\} \oplus \mathrm{Sp}\{z_2, \dots, z_\nu\}$. We let $\xi_0 = z_1 + \gamma z_1^*$, $\xi_1 = z_1 - \gamma z_1^*$. We assume here $q_{W_1}(Z_0, Z_0) \neq 2\gamma$. Then $(\xi_0)^\perp = \mathrm{Sp}\{z_\nu^*, \dots, z_2^*\} \oplus \mathrm{Sp}\{\xi_1, Z_0\} \oplus \mathrm{Sp}\{z_2, \dots, z_\nu\}$, $(\xi_0)^\perp \cap \xi_1)^\perp = \mathrm{Sp}\{z_\nu^*, \dots, z_2^*\} \oplus \mathrm{Sp}\{Z_0\} \oplus \mathrm{Sp}\{z_\nu, \dots, z_2\}$. Here $\mathrm{Sp}\{Z_0, \xi_1\}$ is an anisotropic 2 dimensional space. We let $H = SO(W_1(R_i))$, $M = SO((\xi_0)^\perp)$, and $H_1 = SO((\xi_0)^\perp \cap (\xi_1)^\perp)$. We define $B_H =$ the Borel subgroup of H stabilizing the flag $\{z_1, z_\nu, \dots, z_2\} \supseteq \cdots \supseteq \{z_2\}$; $B_M =$ the Borel subgroup of M stabilizing the flag $\{z_\nu, \dots, z_2\} \supseteq \cdots \supseteq \{z_2\}$ and $B_{H_1} =$ the Borel subgroup stabilizing the flag $\{z_\nu, \dots, z_2\} \supseteq \cdots \supseteq \{z_2\}$. Now let $T_H = T(t_1, \dots, t_\nu)$ be the torus defined by the conditions $T(t_1, \dots, t_\nu)(z_i) = t_i z_i, T(t_1, \dots, t_\nu)(z_i^*) = t_i^{-1} z_i^*$.

Now we have the following formulae

(1) $w_S =$

$$\begin{cases} v_\alpha \to v_\alpha^*\ ,\ v_\alpha^* \to v_\alpha & (\alpha = 1, \dots i) \\ I \text{ on } W_1(R_i) & \text{unless } i \text{ is odd;} \\ \text{then } Z_0 \to -Z_0 & \end{cases}$$

(2) $w_0^{\mathbb{G}} =$

$$\begin{cases} v_\alpha \to v_\alpha^*\ ,\ v_\alpha^* \to v_\alpha & (\alpha = 1, \dots i) \\ z_i \to z_i^*\ ,\ z_i^* \to z_i & (i = 1, \dots \nu) \\ Z_0 \to \pm Z_0 & + \text{ if } i + \nu \text{ is even; } - \text{ if } i + \nu \text{ is odd.} \end{cases}$$

(3) $w_0^H =$

$$\begin{cases} z_i \to z_i^*, z_i^* \to z_i & (i = 2, \dots \nu) \\ Z_0 \to \pm Z & + \text{ if } \nu \text{ is even; } - \text{ if } \nu \text{ is odd.} \end{cases}$$

(4) $w_i =$

$$\begin{cases} v_\alpha \to v_\alpha^*, v_\alpha^* \to v_\alpha & (\alpha = 1, \dots i) \\ z_i \to z_i, z_i^* \to z_i^* & (i = 2, \dots \nu) \\ Z_0 \to \pm Z_0 & + \text{ if } (i + \nu, \nu) \text{ have same parity; } - \text{ otherwise.} \end{cases}$$

(5) $w_0^M =$

$$\begin{cases} z_i \to z_i^*, z_i^* \to z_i & (i = 2, \dots \nu) \\ \xi_1 \to \xi_i & \\ Z_0 \to \pm Z_0 & + \text{ if } \nu \text{ is odd; } - \text{ if } \nu \text{ is even.} \end{cases}$$

Remark. We note that the element $w_i(\xi_0) = \xi_0$ and $w_i(\xi_1) = \xi_1$. Moreover $w_i B_H w_i = B_H$ and $w_i B_M w_i = B_M$.

Again by similar reasoning as in case (a) we deduce that $g = \gamma^{-1} w_S Z w_S w_S^i$ has components b_H and b_M where the toral parts are *units*!

Appendix IV to §5. Density Principle.

We now adopt the notation used in the previous **Appendices 1, 2**, and **3** to §**5**.

We consider the map T defined in **Appendix 1 of** §**5**. In particular, this determines the $SO(W_1(R_j))$ invariant pairing between $\pi(\gamma)$ and $\sigma(\chi')$. Specifically we assert that "generically" in (γ, χ') this defines such a pairing. This means we look at all the values (γ, χ') which simultaneously satisfy the following:

(i) The family of distribution $T_{(\gamma,\chi')}(\varphi_1 \otimes \varphi_2)$ is analytic in (γ, χ') for all φ_1 and φ_2.

(ii) The modules $\Pi(\gamma)$ and $\sigma(\chi')$ are irreducible.

In particular, one can verify that such a space of (γ, χ') is controlled by a finite number of hyperplanes: there exists a finite number of linear function $F_\nu(\gamma, \chi')$ so that if $F_\nu(\gamma, \chi') \neq 0$, then (γ, χ') satisfies (i) and (ii) above. Specifically, "finite" here means when we understand the domain of (γ, χ') as a product of the form

$$\prod \mathbb{C}/(2\pi i \log(q)\mathbb{Z}).$$

We verify in §**7** that it is always possible to choose φ_1 and φ_2 so that $T_{\gamma\otimes\chi'}(\varphi_1 \otimes \varphi_2) = 1 \neq 0$.

This insures that for such (γ, χ') we have that $T = T_{\gamma\otimes\chi'}$, in fact defines a *nonzero* $SO(W_1(R_j))$ invariant pairing between $\pi(\gamma)$ and $\sigma(\chi')$.

Then following **Appendix 1 of** §**5** we consider the construction of the pairing

$$\langle\langle V_j(\tilde{F}) \mid T(w_0)\rangle\rangle.$$

In particular, using the notation from **Appendix 1** we let $F_1(\chi,\varphi_1)$ be constructed from the Borel subgroup $B_{\mathbb{G}}$ and $F(\gamma,\varphi_2)$ from the Borel subgroup B' (defined in this section). In specific terms, we get a distribution on $S(\mathbb{G}) \otimes S(SO(\xi_0)^{\perp}, q_{W_1})$ defined by the integral (where it is convergent)

$$\int_{N_{\{z_2,\dots,z_\ell\}}U_{\{z_2,\dots,z_\ell\}}\backslash SO((\xi_0)^{\perp},q_{W_1})} \left(\int_{\Sigma_{ij}} F_1(\chi+s\otimes\chi',\varphi)[w_S uh]\psi_{i,j,\xi_0}(u)du\right) \left(\int_{N_{\{z_2,\dots,z_\ell\}}U_{\{z_2,\dots,z_\ell\}}} F(\gamma,\varphi_2)[w_0^H\eta_H w_{\ell-1}^{\#}n'u'h]\Lambda_\xi(n'u')dn'du'\right) dh\,.$$

We note in the specific case where $j = \mathrm{rank}(\mathbb{G})$, then the element $w_0^H\eta_H$ is not present in the above formula.

We recall that $F_1'(\chi+s\otimes\chi',\tilde{\varphi})$ and $F_1(\chi+s\otimes\chi',\tilde{\tilde{\varphi}})$ are defined relative to the Borel subgroups $B'_{\mathbb{G}}$ and $B_{\mathbb{G}}$ given in **Appendix 1 to §5**. Thus we recall that

$$V_j(F_1'(\chi+s\otimes\chi',\varphi))(x) = \int_{\Sigma_{ij}} F_1(\chi+s\otimes\chi,{}^{\tilde{w}^{-1}}\varphi)[w_S ux]\psi_{i,j,\xi_0}(u)du\,.$$

What we deduce from the above is that the pairing defined between (the choice of B'_j given in **Appendix 1 to §5**)

$$\mathrm{ind}_{B_{\mathbb{G}}}^{\mathbb{G}}((\chi+\underset{\sim}{s})\otimes\chi')\otimes\mathrm{ind}_{B'_j}^{SO((\chi_0)^{\perp},q_{W_1})}(\gamma)$$

given by the integral

$$\int_{N_{\{z_2,\dots,z_\ell\}}U_{\{z_2,\dots,z_\ell\}}SO(W_1(S)\cap W_1(R_i))\backslash SO((\xi_0)^{\perp},q_{W_1})} V_j(F_1'(\chi+s\otimes\chi',{}^{\tilde{w}}\varphi_1)[h]$$
$$\Big(\int_{N_{\{z_2,\dots,z_\ell\}}U_{\{z_2,\dots,z_\ell\}}} F(\gamma,\varphi_2)[w_0^H\eta_H w_{\ell-1}n'u'h]\Lambda_\xi(n'u')dn'du'\Big)dh$$

determines $N^i\mathbb{U}_i \times SO((\xi_0)^{\perp},q_{W_1})^{\Delta}$ quasi-invariant pairing S of the form

$$S_j(F_1(\chi+s\otimes\chi',\varphi_1^{n_iu_im}),F_2(\gamma,\varphi^m)) = \Lambda_{\xi_0}(n_iu_i)S_j(F_1(\chi+s\otimes\chi',\varphi_1),F_2(\gamma,\varphi_2))\,.$$

We note that this is precisely the type of distribution discussed in §8 for the case $j = i+1$. Specifically we note that the distribution admits meromorphic continuation and satisfies the same functional equation properties as above when the data is analytic in (χ, χ', γ). In any case, we expect that these facts of meromorphic continuation remain true for all $j > i$. Specifically we note in the case where $j = \text{rank}(\mathbb{G})$ the required properties of meromorphic continuation are discussed in [S]. We discuss these issues in [R].

We now want to compare the various S_j for $j \geq i+1$.

We note that the choice of B'_j is dictated by j.

In fact, we assume that there exists an element $k_j \in K_{SO((\xi_0)^\perp, q_{W_1})}$ so that $k_j B'_{i+1} k_j^{-1} = B'_j$.

Then we note that there is an $SO((\xi_0)^\perp, q_{W_1})$ intertwining isomorphism between

$$\text{ind}_{B'_{i+1}}^{SO((\xi_0)^\perp, q_{W_1})}(\gamma') \rightsquigarrow \text{ind}_{B'_j}^{SO((\xi_0)^\perp, q_W}(\gamma')$$

given by the map

$$f_{\gamma'} \rightsquigarrow f_{\gamma'}^{k_j}(x) = f_{\gamma'}(k_j x)\,.$$

Specifically we note above that by examining each of the cases (a), (b), (c) and (d) above (in **Appendix 1 of** §**5**) we can choose such an element k_j as above.

Thus it follows, we can use S_j to define an $N^i\mathbb{U}_i \times SO((\xi_0)^\perp, q_{W_1})$ quasi-invariant pairing $S_j^{k_j}$ between

$$\text{ind}_{B_{\mathbb{G}}}^{\mathbb{G}}(\chi + s \otimes \chi') \otimes \text{ind}_{B'_{i+1}}^{SO(\xi_0^\perp, q_{W_1})}(\gamma)$$

given by

$$S_j^{k_j}(F_1(\chi + s \otimes \chi', \varphi_1), F(\gamma, \varphi_2)) = S_j(F_1(\chi + s \otimes \chi', \varphi_1), F(\gamma, \varphi_2)^{k_j})\,.$$

We now set $s = 0$ in the above formulae.

Now by the Principle of Uniqueness, stated in §**3**, we deduce that

$$S_j^{k_j} = c_j(\chi, \chi', \gamma) S_{i+1}$$

where $j > i+1$ and $c_j(\chi, \chi', \gamma)$ is some rational function in (χ, χ', γ). We note here that $S_{i+1} \neq 0$ by simple considerations coming from §**8**.

We now state the *Density Principle* alluded to in the beginning of this discussion.

For notational purposes below, we note if H'_1 is an analytic section below (coming from the Borel $B'_{\mathbb{G}}$), then H_1 is an analytic section (coming from $B_{\mathbb{G}}$) related to H'_1 by the recipe

$$H'_1 = \sum (F_1)'(\chi \otimes \chi')(\varphi_j) P_j(\chi, \chi') \leftrightarrow$$
$$H_1 = \sum F_1(\chi \otimes \chi')({}^{\bar{w}^{-1}}\varphi_j) P_j(\chi, \chi')$$

($P_j(\chi, \chi')$, polynomials in $q^{\pm \chi_i}$, $q^{\pm \chi'_\ell}$).

Theorem (Density Principle). *Let H_1' belong to the space $ind_{B'_{\mathbb{G}}}^{\mathbb{G}}(\chi \otimes \chi')$. Assume also H_1' is analytic in (χ, χ') as described in* §**5** *and $V_{i+1}(H_1'(\dots))$ is analytic in (χ, χ') and $supp(V_{i+1}(H_1'(\dots)) \subseteq \bigcup_{\xi} SO(W_1(R_j))\xi K_{SO((\xi_0)^{\perp}, q_{W_1})}$ (a finite union) which is independent of (χ, χ'). Moreover assume $V_{i+1}(H_1'(\dots))$ is $K_{SO((\xi_0)^{\perp}, q_{W_1})}$ invariant.. Then assume that*

$$S_{i+1}(H_1(\chi \otimes \chi', \ ), F(\gamma, \varphi_2)) \equiv 0$$

for all $F(\gamma, \varphi_2)$ which are $K_{SO((\xi_0)^{\perp}, q_{W_1})}$ invariant and (χ, χ', γ) lies in a nonempty open subset. Then

$$V_{i+1}(H_1'(\dots))(g) \equiv 0$$

for all $g \in SO((\xi_0)^{\perp}, q_{W_1})$.

Proof. The idea here is to use two steps. First use the *Principle of Uniqueness* above to show that $S_{\mathrm{rank}\mathbb{G}}(H_1(\chi + \chi', \ ), F(\gamma, \varphi_2)) \equiv 0$. Thus we are essentially in the Whittaker case studied in [S]. From this we deduce that $V_{\mathrm{rank}(\mathbb{G})}(H_1'(\dots))(g) \equiv 0$. The second point is to start from this vanishing to imply that $V_{i+1}(H_1'(\dots))(g) \equiv 0$.

We prove these two statements below.

We now show how $F_1' = \rho_{\Lambda_0}(F * \phi_0) - F^*$ satisfies the hypothesis of the above Theorem. Indeed, using the same arguments (see the **Proof of Theorem 5.1**), we see first that ρ_{Λ_0} can be chosen in a canonical way to be independent of (χ, χ'). Second we see also that $\mathrm{supp}(F_1')$ lies in $\mathbb{P}_j w_S \mathbb{U}_i SO((\xi_0)^{\perp}, q_{W_1})$ and does not depend on (χ, χ'). In fact, there is a compact subset A_{r_0} in $\mathbb{U}_i \times SO((\xi_0)^{\perp}, q_{W_1})$ so that $\mathrm{supp}(F_1') \subseteq P_j w_S A_{r_0}$ ($j = i + 1$ here).

Then we apply to F_1' above the Jacquet Whittaker integral (where convergent)

$$L_j(F_1'(\chi \otimes \chi'), \ )(x)$$
$$\int_{N^j_{+,-}} F_1'(\chi \otimes \chi_j')[w_{(i,j)}nx]\Psi_{i,j}(n)dn\,.$$

In such an instance we know that the above integral has an analytic continuation in (χ, χ') for x lying in a compact set. In any case, we note that for $x = w_S u b_1 k_1 g$ with $b_1 \in B$, $k_1 \in K_{SO(W_1(S) \cap W_1(R_j))}$, $u \in \mathbb{U}_i$ and $g \in SO((\xi_0)^{\perp}, q_{W_1})$

$$\begin{aligned} &L_j(F_1'(\chi \otimes \chi'), \ )(w_S u b_1 k_1 g) \\ &= \chi' \delta_B^{\frac{1}{2}}(b_1^{w_S}) L_j(F_1'(\chi \otimes \chi'), \ )(w_S b_1^{-1} u b_1 k_1 g) \end{aligned}$$

with $b_1^{-1} u b_1 \in \mathbb{U}_i$ (where $b_1 \in B \subseteq SO(W_1(S) \cap W_1(R_i)) \subseteq SO((\xi_0)^{\perp}, q_{W_1})$) and w_S preserves B via conjugation ($j = i + 1$ here).

We note that the equation persists via analytic continuation for all (χ, χ').

Then we note that for any (χ, χ'), if $L_j(F_1'(\chi \otimes \chi', \ ))(w_S u g) \neq 0$ we deduce that $F_1'(\chi \otimes \chi', \ )(p w_S u g) \neq 0$ for some $p \in \mathbb{P}_j$. This means that (viewing $g \in SO((\xi_0)^{\perp}, q_{W_1})$ as fixed)

$$\{u \in \mathbb{U}_i \mid F_1'(\chi \otimes \chi', \ )(p w_S u g) \neq 0 \text{ for some } p\}$$

has compact support $\mod N_\beta$ in $\mathbb{U}_i$. Thus the integral

$$V_j(F_1'(\chi\otimes\chi',\ )(g) = \int\limits_{N_\beta\backslash\mathbb{U}_i} L_j(F_1'(\chi\otimes\chi',\ )[w_S ug]\psi_{i,j,\xi_0}(u)du$$

is *absolutely convergent* for *all* (χ,χ').

But we also deduce from this integral that for all (χ,χ') as a function in g, $V_j(F_1'(\chi\otimes\chi',\ )(g)$ has support in $SO(W_1(S)\cap W_1(R_i))K_{r_0}$ where K_{r_0} is a compact subset in $SO((\xi_0)^\perp, q_{W_1})$ (K_{r_0} independent of (χ,χ')).

Thus we have that for $g=b_1k_1\cdot\Omega$ ($\Omega\in K_{r_0}$)

$$V_j(F_1'((\chi\otimes\chi'),\ ))(b_1k_1\Omega) = (\chi'\delta_B^{\frac{1}{2}})(b_1^{w_S})V_j(F_1'(\chi\otimes\chi',\ ))(k,\Omega)\,.$$

Following a similar argument as in the proof of **Theorem 5.1** we deduce that $V_j(F_1'(\chi\otimes\chi',\ ))(k,\Omega)$ is analytic in the (χ,χ') variable. Thus the function $(\chi,\chi')\to V_j(F_1'(\chi\otimes\chi',\ ))(g)$ (for fixed g) is *analytic*! ($j=i+1$ here).

We note here that $\langle\langle V_j(F_1')\mid T(w_0)\rangle\rangle$ then becomes a finite sum of terms of the form

$$\int\limits_{SO(W_1(S)\cap W_1(R_i))} \Big(\int F_1'(\chi+S\otimes\chi',\ )[w_S um\xi]\psi_{i,j,\xi_0}(u)du\Big) F(\gamma,\varphi_2)[w_0^H\eta_H m\xi]dm\,.$$

However, we note that when we set $s=0$, then the above function *need not be holomorphic* in $q^{\pm\chi}$, $q^{\pm\chi'}$ and $q^{\pm\gamma}$. But it is clearly rational in this data (see **Lemma 7.1** and **Lemma 8.1**).

We also recall here that if $F_1''=F*\phi_0-F^*$, then $V_j(F_1'')(g)=V_j(F_1')(g)$ for all $g\in SO((\xi_0)^\perp, q_{W_1})$ ($j=i+1$).

One comment we make here is that if we take the original $I_j(\tau,\sigma,s)$, then it is possible to have the full induced modules $\tau(\chi)$ and $\sigma(\chi')$ so that τ and σ are *submodules* in $\tau(\chi)$ and $\sigma(\chi')$. Thus the element $F_1'=\rho_{\Lambda_0}(F*\phi_0)-F^*$ is viewed as an element in $I_j(\tau,\sigma,s)$ and $I_j(\tau(\chi),\sigma(\chi'),s)=\mathrm{ind}_{B'_{\mathbb{G}}}^{\mathbb{G}}(\chi+s\otimes\chi')$ simultaneously. The point is that V_{i+1} is computed the same way as viewed in both spaces and admits exactly the same formula and the same formal properties as support, analyticity, etc. ....

Next we must show that F_1 (recall the relation between F_1' and F_1 satisfies the property

$$S_{i+1}(F_1, F(\gamma,\varphi_2))\equiv 0$$

for an *open set* of (χ,χ',γ).

To show the above fact we start with the integral

$$\langle\langle V_j(F'(\chi+s\otimes\chi',\varphi_1)\mid T(F(\gamma,\varphi_2)))\rangle\rangle$$

defined in **Appendix 1 to** §5 where the data is all spherical. In particular we deduce from **Theorem A** of **Appendix 1 of** §5 that for $j = i + 1$, the above pairing equals

$$T^{\mathbb{G}}_{(\chi+s)\otimes\chi')\otimes\gamma}(\varphi_1 \otimes \varphi_2)\,.$$

Since both $\varphi_1 = 1_{K_{\mathbb{G}}}$ and $\varphi_2 = 1_{K_{SO((\xi_0)^{\perp}, q_{W_1})}}$ we have determined in **Theorem 6.3** that the above functional equals (for a family of (χ, χ', γ, s) which lie in nonempty open subset)

$$\left(\frac{L_v(\tau_j(\chi)\otimes\Pi(\gamma), st, s+\frac{1}{2})}{L_v(\tau_j(\chi)\otimes\sigma(\chi'), st, s+1)L_v(\tau_j(\chi), {}^{\Lambda^2}_{\mathrm{sym}^2}, 2s+1)}\right)\langle g_\gamma | f_{\chi'}|_{T_{H'}}\rangle \zeta_{GL_j}(\chi\,|_{T_j})\,.$$

The point now is to compute the term $\langle g_\gamma | f_{\chi'}|_{T_{H'}}\rangle$ in terms of the inner product

$$\langle\langle V_j(F^*) \mid T(F(\gamma, \varphi_2))\rangle\rangle$$

defined above also. Using the formula for F^* given in §**5** above, we have equality with ($\tilde{\varphi}_2 = 1_{K_{SO((\xi_0)^{\perp}, q_{W_1})}}$ and $\tilde{\varphi}_1$ chosen suitably)

$$\int_{SO(W_1(R_j))} F(\gamma, \tilde{\varphi}_2)[w_0^H \eta_H m]\Big(\int_{N^j_{+,-}} F^*(\chi + s\otimes\chi', \tilde{\varphi}_1)[w_{ij} n w_S m] du\Big) dm\,.$$

Now we note that

$$F^*(\chi + s\otimes\chi'), \tilde{\varphi}_1)[w_{ij} n w_S b_\mu k_\mu w_S^{-1} w_S] = (\chi'\delta_{B_j}^{\frac{1}{2}})^{w_S}(b_\mu) F^*[\chi + s\otimes\chi', \tilde{\varphi}_1][w_{ij} n w_S]\,.$$

Thus the integral above has the form

$$\zeta_{GL_j}(\chi\,|_{T_j})\langle g_\gamma \mid f_{w_S(\chi')}\,|_{T_{H'}}\rangle\,.$$

We note that $\zeta_{GL_j}(\chi\,|_{T_j})$ is the value of the Whittaker functional for unramified spherical data at $\{e\}$ and $\langle g_\gamma \mid f_{w_s(\chi')}\,|_{T_{H'}}\rangle$ is defined as in §7 (again for unramified spherical data).

In any case, we note the slight twisting by w_S in the above formula. This only applies in the case where $\mathbb{G}$ corresponds to an even split form and $SO((\xi_0)^{\perp}, q_{W_1})$ corresponds to an odd dimensional quasi split form. In particular (with $j = i+1$), if we write $\chi' = \alpha_{j+1}\chi_{j+1} + \cdots + \alpha_n\chi_n$, then $w_S(\chi') = \alpha_{j+1}\chi_{j+1} + \cdots + \alpha_{n-1}\chi_{n-1} + (-\alpha_n)\chi_n$ (by the choice of w_S). In fact we see by direct calculation that

$$\langle g_\gamma \mid f_{w_S(\chi')}\rangle = \langle g_\gamma \mid f_{\chi'}\rangle\,.$$

Here we use the specific formula for the pairing $\langle\ |\ \rangle$ as developed in §6 and §7 that follow. We note here it is instructive for the reader to consider the pair $(SO(2,1), SO(1,1))$ to verify this simple point! The point is that both sides in the above formula are invariant under the map $\chi_n \to \pm\chi_n$.

Thus we have shown that

$$S_{i+1}(F_1(\chi + s \otimes \chi', \ ), F(\gamma, \varphi_2)) \equiv 0$$

for *all* (χ, χ', γ, s).

Now we return to the proof of **Theorem (Density Principle)**. Thus assume that $S_{i+1}(H(\cdots) \mid F(\gamma, \varphi_2)) = 0$ for all (χ, χ', γ) in an open set where $F(\gamma, \varphi_2)$ is the spherical vector. In any case, we deduce that

$$S_{\mathrm{rank}(\mathbb{G})}(H(\cdots), F(\gamma, \varphi_2)) \equiv 0$$

for a similar set of (χ, χ', γ). However, we note here that the vanishing above holds *identically* in (χ, χ', γ). Thus we have, in particular, (with S a maximal isotropic subspace or $j = \mathrm{rank}(\mathbb{G})$)

$$\begin{aligned} &\int_{N_{\{z_2,\ldots,z_\ell\}} U_{\{z_2,\ldots,z_\ell\}} SO(W_1(S)\cap W_1(R_i))\backslash SO((\xi_0)^\perp, q_{W_1})} \\ &\left(\int_{\Sigma_{ij}} F_1(\chi + \chi', \ )[w_S u m] \psi_{i,j,\xi_0}(u) du\right) \\ &\left(\int_{N_{\{z_2,\ldots,z_\ell\}} U_{\{z_2,\ldots,z_\ell\}} SO(W_1(S)\cap W_1(R_i))} F(\gamma, \varphi_2)[w^{\#}_{\ell-1} n u m_1 m] \Lambda_{\xi_0}(nu) dn du dm_1\right) dm \\ &\equiv 0 \end{aligned}$$

where the data is absolutely convergent (recall here that if $j = \dim W_1$ and i odd, we replace $z_\ell = z_{j-i}$ by $z^*_\ell = z^*_{j-i}$).

Then we consider first the integral

$$H_{\varphi_2}(\gamma, x) = \int_{N_{\{z_2,\ldots,z_\ell\}} U_{\{z_2,\ldots,z_\ell\}} SO(W_1(S)\cap W_1(R_i))} F(\gamma, \varphi_2)[w^{\#}_{\ell-1} n u m x] \Lambda_\xi(nu) dn du dm\,.$$

Thus as a function of $x \in SO((\xi_0)^\perp, q_{W_1})$ we have a Whittaker Jacquet integral. In fact

$$H_{\varphi_2} \in \mathrm{Ind}^{SO((\xi_0)^\perp, q_{W_1})}_{N_{\{z_2,\ldots,z_\ell\}} U_{\{z_2,\ldots,z_\ell\}} SO(W_1(S)\cap W_1(R_i))} (\overline{\psi}_{\ell-1} \otimes \overline{\psi}_{Z_\beta} \otimes 1)\,.$$

We note that we have above an $SO(\xi_0)^\perp, q_{W_1})$ intertwining map of $\mathrm{ind}_{B'}^{SO((\xi_0)^\perp, q_{W_1})}(\gamma)$ to the above Whittaker space. We note here that in this case the group $SO(W_1(S)\cap W_1(R_i))$ is *compact*. We note here, in fact, $SO(W_1(S)\cap W_1(R_i)) \subseteq K_{SO()(\xi_0)^\perp, q_{W_1})}$. We recall that the Casselman Shalika formula gives an explicit way to evaluate H_{φ_2} for $\varphi_2 = 1_{K_{SO((\xi_0)^\perp, q_{W_1})}}$. It is this specific formula we use to establish the following:

Theorem (Completeness of Spherical Whittaker models). *Suppose we are given a function* $H \in Ind_{N_{\{z_2,\ldots,z_\ell\}}U_{\{z_2,\ldots,z_\ell\}}SO(W_1(S)\cap W_1(R_i))}^{SO((\xi_0)^\perp,q_{W_1})}(\overline{\psi}_{\ell-1}\otimes\overline{\psi}_{Z_\beta}\otimes 1)$ *which is* $K_{SO((\xi_0)^\perp,q_{W_1})}$ *invariant and satisfies the following condition.*

There exists a function $\mathbb{H}$ *which is* $K_{SO((\xi_0)^\perp,q_{W_1})}$ *invariant and*

$$|H(g)H_{\varphi_2}(\gamma,g)\,|\leq \mathbb{H}(g)$$

(for all γ *where* $|Re(\gamma_i)|\leq T$ *for all* i *and* $T>0$*) with*

$$\mathbb{H}\in L^1_+(N_{\{z_2,\ldots,z_\ell\}}U_{\{z_2,\ldots,z_\ell\}}SO(W_1(S)\cap W_1(R_i))\backslash SO((\xi_0)^\perp,q_{W_1}))$$

where $L^1_+(\ )$ *corresponds to functions supported on a positive Weyl chamber in* $T^+_{\{z_2,\ldots,z_\ell\}}$ *(see below for the definition of this domain).*

Then if the family of integrals

$$\int_{N_{\{z_2,\ldots,z_\ell\}}U_{\{z_2,\ldots,z_\ell\}}SO(W_1(S)\cap W_1(R_i))\backslash SO((\xi_0)^\perp,q_{W_1})} H(g)H_{\varphi_2}(\gamma,g)dg\equiv 0$$

for all γ *with* $Re(\gamma_i)=0$ *(for all* i*). Then*

$$H\equiv 0\,.$$

Proof. We give the proof below. □

We now assume that $j=\mathrm{rank}(\mathbb{G})$.

Now we return to the proof of the Density Principle. From above we have concluded that

$$\int_{\Sigma_{ij}} F_1(\chi+\chi',\ )[w_S um]\psi_{i,j,\xi_0}(u)du\equiv 0$$

for *all* appropriate $\chi\otimes\chi'$ and *all* m! (where convergent).

Using the data in **Appendix 1 of §5** we have

$$\int_{N_-^i\cdot N_+^{j-i}\mathbb{U}_i} F_1(\chi+\chi'\ )[u_1u_2w_S um]\psi_{ij}(u_1u_2)\psi_{i,j,\xi_0}(u)du_1du_2du\equiv 0$$

(where convergent).

We consider then the family of integrals

$$L_{F^m}(g,\chi,\chi')=\int_{N_-^i\,\mathbb{U}_i} F(\chi+\chi',\ )[u_1gw_S um]\psi_{ij}(u_1)\psi_{i,j,\xi_0}(u)du_1du$$

with $g\in GL(\{z_1,\ldots,z_\ell\})$.

We discuss the issue of convergence of this integral below. But in any case, we have the following transformation formula

$$L_F(b_1 g, \chi, \chi') = \chi''(b_1)\delta^{\frac{1}{2}}_{B_{GL_{j-i}}}(b_1) L_F(g, \chi, \chi')$$

with $b_1 \in B_{j-i}$ Borel group of $GL_{j-i} = GL(\{z_1, \ldots, z_{j-i}\})$ (the flag here is $\{z_{j-i}, \ldots, z_1\} \supseteq \cdots \supseteq \{z_1\}$, i.e., $\ell = j - i$). In any case, what we have is that L_F lies in

$$\text{cont} - \text{ind}_{B_{j-i}}^{GL_{j-i}}(\chi'') .$$

Here cont– hyphenated means all the functions which have the transformation formula above and are *continuous.* In particular this means not the condition of smoothness. Another point is simply the space cont– is not necessarily $K_{GL_{j-i}}$ finite functions. We prove below that for $\chi \otimes \chi'$ suitably chosen L_F is a continuous function on GL_{j-i}.

Thus the condition of vanishing above in F_1 simply becomes

$$(***) \qquad \int_{N_+^{j-i}} L_{F_1^m}(u_2, \chi, \chi')\psi_{ij}(u_2)du_2 \equiv 0$$

for *all* $m \in SO((\xi_0)^\perp, q_{W_1})$. (Here $F_1^m(g) = F_1(gm)$.)

Thus the abstracted issue becomes the following. Assume we consider the space

$$\text{cont} - \text{ind}_{B_{j-i}}^{GL_{j-i}}(\chi'') .$$

Then GL_{j-i} acts on this space and we form the basic Whittaker Jacquet integral

$$S_{\phi_{\chi''}}(x) = \int_{N_+^{j-i}} \phi_{\chi''}[ux]\psi_{ij}(u)du$$

for the appropriate convergent domain in χ (discussed below).

Then the basic Proposition we require is the following.

Proposition. *Let $\phi_{\chi''}$ lie in cont– $\text{ind}_{B_{j-i}}^{GL_{j-i}}(\chi'')$. Assume that $L_{\phi_{\chi''}}$ is defined by an absolutely convergent integral. (The choice of χ'' is given below.) Then*

$$S_{\phi_{\chi''}}(x_1) \neq 0$$

for some $x_1 \in GL\{z_2, \ldots, z_{j-i}\}$).

Remark. We note that this fact is well known in case we consider the space of smooth or $K_{GL_{j-i}}$ finite functions. We repeat the same proof of this larger space of functions.

From the above Proposition applied to the function $L_{F^m}(g,\chi,\chi')$ we see that if $L_{F^m}(e,\chi,\chi') \neq 0$ for some $m \in SO((\xi_0)^\perp, q_{W_1})$, then there exists some $\xi \in GL(\{z_2,\dots,z_\ell\}) \subseteq SO((\xi_0)^\perp, q_{W_1})$ so that

$$\int_{N_+^{j-i}} L_{F_1^m}(u_2\xi,\chi,\chi')\psi_{ij}(u_2)du_2 \neq 0\,.$$

But this is the same as

$$\int_{N_+^{j-i}} L_{F_1^{w_S\xi w_S^{-1}m}}(u_2,\chi,\chi')\psi_{ij}(u_2)du_2 \not\equiv 0\,.$$

(Here $w_S\xi w_S^{-1}$ lies in $SO((\xi_0)^\perp, q_{W_1})$.)

This contradicts the above vanishing property (***). Thus assuming $L_{F_1^m}$ admits all the required continuity properties used above, we thus deduce that

$$L_{F^m}(e,\chi,\chi') \equiv 0$$

for all $m \in SO((\xi_0)^\perp, q_{W_1})$. Hence we have that

$$V_{i+1}(F_1'(\chi\otimes\chi',\dots)(m)) \equiv 0$$

for all $\chi+\chi'$ where we have convergence of V_{i+1}!

Thus to finish our above proof, we must show the compatibility of the convergence of the integrals defining $V_{i+1}, V_{\dim\mathbb{G}}$ and the absolute integrability condition required in the **"Completeness of Spherical Whittaker Models"**. This simply means to compare the convergence of L_{F^m}, the convergence of the Whittaker Jacquet models for GL_{j-i} and the absolute integrability condition alluded to in the last sentence.

We recall here the conditions for absolute convergence of

$$\int_{N_+^{j-i}(N^iU_i)} |F_1(\chi\otimes\chi',\ )|[u_1w_Sux]du_1du$$

given in **Appendix 1 to §5** (where we set $s=0$).

Indeed we require further absolute convergence above the set L_j to satisfy

(i) $Re(\chi_{\ell_1}) \pm Re(\chi_{\ell_2}) \gg 0 \qquad 1 < \ell_1 \leq \ell_2 < i$

(ii) $\chi_\ell \pm \chi_{\ell'} \gg 0 \qquad 1 \leq \ell \leq i,\ i+1 \leq \ell'$

(iii) $\chi_j \gg 0 \qquad 1 \leq j \leq i$ (if ε_j a root)

(iv) $\chi_{t_1} - \chi_{t_2} \gg 0 \qquad i+1 \leq t_1 < t_2 < j$.

In fact, we note that for the values $j=j_1$ and $j=j_2$, with $j_1<j_2$, then we note that $L_{j_1} \subset L_{j_2}$ and clearly $L_{\text{rank}(\mathbb{G})}$ is a *nonempty open set.*

Now since F_1'' is $K_{SO((\xi_0)^\perp, q_{W_1})}$ invariant the support properties of $V_j(F_1''(\dots))$ have been determined in **Appendix 1 to** §**5** more or less explicitly. Indeed, if we take $\{z_2, \dots, z_\ell\} \supseteq \cdots \supseteq \{z_\ell\}$, a flag in the space $(\xi_0)^\perp \subseteq W_1(R_i)$, then this flag determines a Borel subgroup of $SO((\xi_0)^\perp, q_{W_1})$ in cases (a) and (d) (of **Appendix 1 in** §**5**). We note that in case (b), the above space $\{z_2, \dots, z_\ell\}$ is not a maximal isotropic subspace. However, the subgroup $N_{\{z_2,\dots,z_\ell\}} U_{\{z_2,\dots,z_\ell\}} SO(W_1(S) \cap W_1(R_i))$ determines a specific Whittaker spherical model which gives the *identity* representaton on $SO(W_1(S) \cap W_1(R_i))$ (here $W_1(S) \cap W_1(R_i)$ is a 2 dimensional anisotropic subspace). Case (c) requires a bit more care. We explain this point more explicitly below. In any case, what we know in (c) is that there exists a flag $\{\hat{z}_1, z_2, \dots, z_\ell\} \supseteq \cdots \supseteq \{z_\ell\}$ ($\hat{z}_1$ is an isotropic vector in the space $\{z_1^-, Z_0\}$, see notation specifically in case (c) in **Appendix 1 to** §**5**) which determines a Borel subgroup of $SO((\xi_0)^\perp, q_{W_1})$.

Then we let $D(t_2, \dots, t_\ell)$ be an element in the torus $T_{\{z_2,\dots,z_\ell\}}$ adapted to the flag given by $\{z_2, \dots, z_\ell\} \supseteq \cdots \supseteq \{z_\ell\}$ (in cases (a), (b) and (d)). Then the required support property of the function $V_j(\dots)$ is the following:

$$|t_\ell| \leq |t_{\ell-1}| \leq \cdots \leq |t_2| \leq 1 \,.$$

In case (c) we let $D(t_1, \dots, t_\ell)$ be an element in the torus $T_{\{\hat{z}_1, z_2,\dots,z_\ell\}}$ adapted to the flag $\{\hat{z}_1, z_2, \dots, z_\ell\} \supseteq \cdots \supseteq \{z_\ell\}$. Then the support property of $V_j(\cdots)$ is

$$|t_\ell| \leq |t_{\ell-1}| \leq \cdots \leq |t_3| \leq |t_2| \leq |t_1| \,, \ |t_1| \leq |t_2|^{-1} \,.$$

Let us remark here again that cases (a), (d) and (c) are the "usual Whittaker" cases. Case (b) is not such a case (with the *compact group* $SO(W_1(S) \cap W_1(R_i))$ *replacing the last simple root subgroup of* $SO((\xi_0)^\perp, q_{W_1})$ here).

Then we deduce in cases (a), (b), and (d) the estimate

$$|V_j(F_1(\ ))(D(t_2, \dots, t_\ell))| \leq |\chi_{i+2}|(t_2) \dots |\chi_j^*|(t_j) \ \rho(t_2, \dots, t_j) \,.$$

We note in case (c), we have the estimate

$$\begin{aligned} |V_j(F_1(\dots))(D(\hat{t}_1, t_2, \dots, t_\ell))| &\leq |\chi_{i+2}|(t_2) \dots |\chi_j|(t_j) \rho(t_2, \dots, t_j) \\ &\quad \cdot q^{t_0 Re(\chi_{i+1})} \begin{cases} |\chi_{i+1}|(\hat{t}_1) & \text{if } |\hat{t}_1| < 1 \\ |\chi_{i+1}|(\hat{t}_1^{-1}) & \text{if } |\hat{t}_1| > 1 \end{cases} \end{aligned}$$

Here $|\chi|$ represents the character given by $Re(\chi)$ and χ_j^* represents either χ_j or $\chi_j^{w_S}$ ($j = \text{rank}(\mathbb{G})$) ($\chi_j^{w_S}$ only in case (a) where the pair $(i, \text{rank}(\mathbb{G}))$ has different parity). We note that ρ represents some fixed positive character in the variables $(t_2, \dots, t_j)$.

We note again that these estimates hold where $\chi = (\chi_1, \dots, \chi_{\text{rank}(\mathbb{G})})$ belongs to the set $L_{\text{rank}(\mathbb{G})}$. We compare these estimates with similar estimates in **Appendix 1 to** §**5** for $V_j(F'(\tau, \sigma, s)$ when $j < \text{rank}(\mathbb{G})$.

Now we finally consider the domain of convergence of the integral

$$\int_{N_{\{z_2,\ldots,z_\ell\}}U)_{\{z_2,\ldots,z_\ell\}}SO(W_1(S)\cap W_1(R_i))\backslash SO((\xi_0)^\perp,q_{W_1})} (H_{\varphi_2}(\gamma,x)\int_{\Sigma_{ij}} F_1(\chi+\chi')[w_S ux]\psi_{i,j,\xi_0}(u)du)dx\,.$$

We have given first the crude estimates on the inner integral above. Then we next give the estimates on $H_{\varphi_2}(\gamma,x)$. Here we use explicit formulae.

We note that H_{φ_2} is a Whittaker Jacquet integral. First we normalize H_{φ_2} by multiplying by the zeta function associated to $SO((\xi_0)^\perp,q_{W_1})$ define in §**6**, i.e.,

$$\zeta_{SO((\xi_0)^\perp,q_{W_1})}(\gamma,s)\,.$$

We let $s\equiv 1$. In fact, we know that $[\zeta_{SO((\xi_0)^\perp,q_{W_1})}(\gamma,1)]\neq 0$ if and only if $\Pi(\gamma)^0$ (= the unramified component of $\Pi(\gamma)$) admits a Whittaker model. Thus we let

$$H^*_{\varphi_2}=[\zeta_{SO((\xi_0)^\perp,q_{W_1})}(\gamma,1)]^{-1}H_{\varphi_2}\,.$$

Then $H^*_{\varphi_2}$ admits a very simple and precise formula in cases (a), (c) and (d).

First we note that the domain $D(t_2,\ldots,t_\ell)$ with $|t_\ell|\le|t_{\ell-1}\le\cdots\le|t_2|\le 1$ can be characterized simply as

$$D(\pi^{m_2},\ldots,\pi^{m_\ell})$$

where $m_\ell\ge m_{\ell-1}\ge\cdots\ge m_2\ge 0$ (m_i integers).

In case (c) we note that the domain $D(\hat{t}_1,t_2,\ldots,t_\ell)$ with $|t_\ell|\le\cdots\le|t_2|\le|t_1|$ and $|t_1|\le|t_2|^{-1}$ is simply

$$D(\pi^{m_1},\pi^{m_2},\ldots,\pi^{m_\ell})$$

where $m_\ell\ge\cdots\ge m_2\ge m_1\ge -m_2$ (m_i integers).

Then we consider cases (a), (d) and (c).

(a) The tuple $(m_\ell,\ldots,m_2)$ with $m_\ell\ge m_{\ell-1}\ge\cdots\ge m_2\ge 0$ determines an irreducible highest weight representation of $Sp_{\ell-1}(\mathbb{C})$ of the form $V_{(m_\ell,\ldots,m_2)}=$

$$(m_\ell-m_{\ell-1})\omega_1+(m_{\ell-1}-m_{\ell-2})\omega_2+\cdots+(m_3-m_2)\omega_{\ell-2}+m_2\omega_{\ell-1}$$

with ω_i the fundamental representations of $Sp_{\ell-1}(\mathbb{C})$.

Then

$$H^*_{\varphi_2}(D(\pi^{m_2},\ldots,\pi^{m_\ell}))=\delta_B^{\frac{1}{2}}(D(\pi^{m_2},\ldots,\pi^{m_\ell}))\,\mathrm{trace}_{V_{(m_\ell,\ldots,m_2)}}(A_\gamma)$$

with $A(\gamma)$ = the diagonal conjugacy class in $Sp_{\ell-1}(\mathbb{C})$ corresponding to the representation $\Pi(\gamma)^0$ and δ_B = the modular function for the Borel $B_{\{z_2,\dots,z_\ell\}}$ in $SO((\xi_0)^\perp, q_{W_1})$ ($\cong SO(\ell-1,\ell)$).

(c) The tuple $(m_\ell, m_{\ell-1}, \dots, m_2, m_1)$ with $m_\ell \geq m_{\ell-1} \geq \cdots \geq m_2 \geq m_1 \geq -m_2$ determines an irreducible highest weight representation of $SO(2\ell, \mathbb{C})$ of the form

$$V_{(m_\ell,\dots,m_2,m_1)} = (m_\ell - m_{\ell-1})\omega_1 + \cdots + (m_3 - m_2)\omega_{\ell-2} + (m_2 - m_1)\omega_{\ell-1} + (m_2 + m_1)\omega_\ell$$

with ω_i the fundamental representations of $SO(2\ell, \mathbb{C})$.

Then

$$H^*_{\varphi_2}(D(\pi^{m_1}, \pi^{m_2}, \dots, \pi^{m_\ell})) = \delta_B^{\frac{1}{2}}(D(\pi^{m_1}, \dots, \pi^{m_\ell})) \\ tr_{V_{(m_\ell,\dots,m_1)}}(A_\gamma)$$

where $A(\gamma)$ = the diagonal conjugacy class in $SO(2n, \mathbb{C})$ corresponding to the representation $\Pi(\gamma)^0$ and δ_B = the modular function for the Borel $B_{\{\hat{z}_1, z_2,\dots,z_\ell\}}$ in $SO((\xi_0)^\perp, q_{W_1})$.

Cases (b) and (b) have similar representations and will be given in [R].

But given any highest weight irreducible module of $Sp_t(\mathbb{C})$ or $SO(2t, \mathbb{C})$ and A_γ any diagonal conjugacy class we have the obvious estimate

$$|tr_{c_1\omega_1+\cdots+c_t\omega_t}(A_\gamma)| \leq tr_{c_1\omega_1+\cdots+c_t\omega_t}(A_{|\gamma|}) \\ \leq \prod |tr_{\omega_i}(A_{|\gamma|})|^{c_i}$$

Here $A_{|\gamma|}$ represents the diagonal matrix in Sp or $SO(\ )$ obtained by taking the absolute values of the diagonal entries.

Then for each ω_i we have the estimate:

$$|tr_{\omega_i}(A_{|\gamma|}) \leq \dim(\omega_i) q^{M_{\omega_i}(|Re(\gamma_1)|+|Re(\gamma_2)|+\cdots+|Re(\gamma_t)|)}.$$

Here M_{ω_1} is a positive integer which depends only on ω_i and $(\gamma_1, \dots, \gamma_t)$ defined by

$$A_\gamma = \begin{bmatrix} D_1 & 0 \\ 0 & D_1^{-1} \end{bmatrix}$$

with

$$D_1 = \begin{pmatrix} q^{-\gamma_1} & 0 \\ 0 & q^{-\gamma_t} \end{pmatrix}.$$

Thus finally we deduce that

$$|tr_{c_1\omega_1+\cdots+c_t\omega_t}(A_\gamma)| \leq \prod (\dim \omega_i)^{c_i} q^{c_i M_{\omega_i}(|Re(\gamma_1)|+\cdots+|Re(\gamma_t)|)}.$$

But clearly $(\dim \omega_i) \leq q^{M'_{\omega_i}}$ (M'_{ω_i} some positive integer). Hence we can find R large enough so that for $c_1\omega_1 + \cdots + c_t\omega_t$

$$|tr_{c_1\omega_1+\cdots+c_t\omega_t}(A_\gamma)| \leq q^{R(|Re(\gamma_1)|+\cdots+|Re(\gamma_t)|)(c_1+c_2+\cdots+c_t)}$$

(R is independent of the γ_i and the c_j).

Then we must compare these above estimate with the conditions defining the set $L_{\text{rank}(\mathbb{G})}$.

We note that in cases (a) and (c), the sum "$c_1 + c_2 + \cdots + c_t$"=

$$\begin{cases} m_\ell & \text{in case (a)} \\ m_\ell + m_2 & \text{in case (c)} \end{cases}$$

Thus we can put together the estimates used above to determine the domain of convergence of the integral

$$\int_{N_{\{z_2,\ldots,z_\ell\}}U_{\{z_2,\ldots,z_\ell\}}SO(W_1(S)\cap W_1(R_i))} H^*_{\varphi_2}(\gamma, x)(\int_{\Sigma_{ij}} F_1(\chi+\chi')[w_S ux]\psi_{i,j,\xi_0}(u)du)dx\,.$$

First we require that γ is constrained in the set

$$\{\gamma = (\gamma_1, \ldots, \gamma_\ell) \mid |Re(\gamma_i)| \leq T\}\,.$$

Then we choose $(\chi, \chi') \in L_{\text{rank}(\mathbb{G})}$ subject to the following extra condition.

In case (a), we require the convergence of

$$\sum_{m_\ell \geq m_{\ell-1} \geq \cdots \geq m_2 \geq 0} q^{-m_2(Re(\chi_{i+2})+A_2)-m_3(Re(\chi_{i+3})+A_3)\cdots \pm m_\ell(Re(\chi_\ell)+A_\ell)}$$

where $A_2, \ldots, A_\ell$ are certain real numbers which do not depend on the χ_i (only on the choice of T above). We note very specifically in the case (a) (where we have $(\chi')^{w_S} \neq \chi'$) that χ_ℓ "must" be chosen to satisfy $Re(\chi_\ell) \ll 0$. However, it is easy to see that such a choice is compatible for the construction of the set $L_{\text{rank}(\mathbb{G})}$.

Indeed, by iterating the series above all, we require for convergence is

$$Re(\chi_\ell) + A_\ell > 0, Re(\chi_{\ell-1}) + A_{\ell-1} > 0, \ldots, Re_{(\chi_{i+2})}) + A_2 > 0\,.$$

We note in the exceptional case where $\chi^{w_S} \neq \chi$, then we have $Re(\chi_\ell) + A_\ell < 0$ and the remaining $Re(\chi_t) + A_t > 0$.

In case (c) we require the convergence of

$$\sum_{m_\ell \geq m_{\ell-1} \geq \cdots \geq m_2 \geq m_1 \geq -m_2} q^{-|m_1|(Re(\hat{\chi}_{i+1})+A_1)-m_2(Re(\chi_{i+2})+A_2)-\cdots-m_\ell(Re(\chi_\ell)+A_\ell)}\,.$$

But repeating the same argument as above all we require for convergence of this series is

$$Re(\chi_\ell) + A_\ell > 0, Re(\chi_{\ell-1}) + A_{\ell-1} > 0, \ldots, Re(\hat{\chi}_{i+1}) + A_1 > 0\,.$$

Then we can use the "**Completeness of Spherical Whittaker Models Theorem**" for the function

$$H^*_{\varphi_2}(\gamma, g)\Big(\int_{\Sigma_{ij}} F_1(\chi + \chi')[w_S ug]\psi_{i,j,\xi_0}(u)du\Big).$$

We note that $T > 0$ so we can assume $Re(\gamma_i) = 0$ for all i.

Indeed, we majorize the above function by

$$\mathbb{H}_{\chi,\chi'}(g)$$

which is given by the formula

$$\begin{aligned}&\mathbb{H}_{\chi,\chi'}(D(\pi^{m_2},\dots,\pi^{m_\ell}))\delta_B^{-1}(D(\pi^{m_2},\dots,\pi^{m_\ell}))\\ &\quad = q^{-m_2(Re(\chi_{i+2})+A_2)\cdots \pm m_\ell(Re(\chi_\ell)+A_\ell)}\end{aligned}$$

in case (a). In case (c), we let $\mathbb{H}_{\chi',\chi}(g)$ be given by the formula

$$\begin{aligned}&\mathbb{H}_{\chi,\chi'}(D(\pi^{m_1},\pi^{m_2},\dots,\pi^{m_\ell}))\delta_B^{-1}(D(\pi^{m_1},\dots,\pi^{m_\ell}))\\ &\quad = (q^{-|m_1|(Re(\hat{\chi}_{i+1})+A_1)}q^{-m_2(Re(\chi_{i+2})+A_2)\cdots - m_\ell(Re(\chi_\ell)+A_\ell)}).\end{aligned}$$

In any case, we deduce that $\mathbb{H}_{\chi,\chi'}$ is integrable relative to the space

$$(N_{\{z_2,\dots,z_\ell\}}U_{\{z_2,\dots,z_\ell\}}SO(W_1(S)\cap W_1(R_i))\backslash SO((\xi_0)^\perp, q_{W_1})).$$

Thus the set of $(\chi,\chi') \in L_{\mathrm{rank}(\mathbb{G})}$ which also satisfy

(i) In the cases (a) and (d) $Re(\chi_t) + A_t > 0$ for $t = i+2,\dots,\ell$, where in the exceptional case $(\chi_\ell^{w_S} \neq \chi_\ell) Re(\chi_\ell) + A_\ell < 0$.

(ii) In case (c) $Re(\chi_t) + A_t > 0$ for $t = 1,\dots,\ell$ (including $\hat{\chi}_1$)

have the property that the integral

$$\int_{(N_{\{z_2,\dots,z_\ell\}}U_{\{z_2,\dots,z_\ell\}}SO(W_1(S)\cap W_1(R_i))\backslash SO((\xi_0)^\perp, q_{W_1}))} H^*_{\varphi_2}(\gamma)(g)\Big(\int_{\Sigma_{ij}} F_1(\chi+\chi')[w_S ug]\psi_{i,j,\xi_0}(u)du\Big)dg$$

is "absolutely convergent" and is such that we can replace $H^*_{\varphi_2}$ by its explicit formulae given above.

We note that the above set of (χ,χ') is an open nonempty set!

Finally we go back to the analysis of the continuity of the function $g \to L_{F^m}(g)$. Indeed we recall that L_{F^m} is given by the integral (where (χ,χ') are as above),

$$\int_{N_-^i \mathbb{U}_i} F(\chi \otimes \chi')[u_1 g w_S u m]\psi_{ij}(u_1)\psi_{i,j,\xi_0}(u)du_1 du.$$

Then by change of variable in the above integral ($w_S g w_S^{-1} = g^* \in SO(W_1(R_i))$) we have (for some integer t)

$$| \det g^* |^t \int_{N^i_- \mathbb{U}_i} F[\chi \otimes \chi'][u_1 w_S u g^* m] \psi_{ij}(u_1) \psi_{i,j,\xi_0}((g^*)^{-1} u (g^*)) du_1 du \, .$$

However, it is clear that $(g^*, u) \to \psi_{i,j,\xi_0}((g^*)^{-1} u g^*)$ is a continuous function. Noting that (χ, χ') is chosen so that the above integral is *absolutely convergent.* Thus by use of dominated convergence (and the fact that $g^* \to F(\chi \otimes \chi', \ )[u_1 w_S u g^* m]$ is a smooth function in g^* (i.e., locally constant).

One notes very specifically here that the condition that (χ, χ') satisfy

$$\chi_{\ell_1} - \chi_{\ell_2} \gg 0 \text{ (for } i+1 \leq \ell_1 < \ell_2 < \operatorname{rank}(\mathbb{G}() = j)$$

makes it possible that $(***)$ is thus convergent absolutely for the function L_{F^m}. Thus we can form the integral associated to $L_{F_1^m}$ (i.e., $(***)$) and thus we can apply **Proposition** to L_{F_1}!

Proof of "Completeness Theorem" of Spherical Whittaker Models. The idea of the proof is to use the explicit formula for $H_{\varphi_2}(\gamma, \ )$. We note when we form

$$H^*_{\varphi_2} = [\zeta_{SO((\xi_0)^\perp, q_{W_1}}(\gamma, 1)]^{-1} H_{\varphi_2}$$

we can in the Completeness Theorem replace H_{φ_2} by $H^*_{\varphi_2}$ provided $[\zeta_{SO((\xi_0)^\perp, q_{W_1})}(\gamma, 1)] \neq 0$ without changing the hypotheses.

The element γ corresponds to A_γ in either Sp, SO or O. In fact, A_γ defines an element of a maximal torus T in Sp or SO (cases (a) or (c) above).

We discuss case (a) here. The other cases will follow for similar considerations.

The first point is to establish that the integral (for any H)

$$\gamma \to (\int H(g) H^*_{\varphi_2}(\gamma, g) dg) = T(A_\gamma)$$

is analytic for γ where $\zeta_{SO((\xi_0)^\perp, q_{W_1})}(\gamma, 1) \neq 0$.

Indeed we can write

$$\begin{aligned} T(A_\gamma) &= \lim_\Lambda T_\Lambda(A_\gamma) \\ &= \sum_{(m_\ell, \ldots, m_2) \in \Lambda} H(D(\pi^{m_2}, \ldots, \pi^{m_\ell})) \delta_B^{\frac{1}{2}}(D(\pi^{m_2}, \ldots, \pi^\ell)) Tr_{V_{(m_\ell, \ldots, m_2)}}(A_\gamma) \end{aligned}$$

(where Λ is some sequence constructed from the tuples $(m_\ell, \ldots, m_2)$).

By the hypothesis in the **Theorem**, we deduce that each function T_Λ is uniformly bounded, i.e.

$$|T_\Lambda(A_\gamma)| \leq \int \mathbb{H}(g) dg < \infty \, .$$

Then we note each function $T_\Lambda(A_\gamma)$ is analytic in the open domain

$$\{\gamma \mid |Re\gamma_i\}| \leq T\,,\ \text{all } i,\ \text{and} \zeta_{SO((\xi_0)^\perp, q_{W_1})}(\gamma, 1) \neq 0\}\,.$$

We thus deduce that the family $\{T_\Lambda\}$ admits a convergent subsequence (on compact subsets) by use of Montel's Theorem for open domains. We let $T^* = \lim_{\Lambda'} T_{\Lambda'}$ where $\{\Lambda'\}$ determines the subsequence which converges.

Then we have that (For γ with $Re(\gamma_i)0$ for all i)

$$\lim_{\Lambda'} \sup |T_{\Lambda'}(A_\gamma) - T^*(A_\gamma)| \equiv 0\,.$$

But by hypothesis we have $\lim_{\Lambda'} T_{\Lambda'}(A_\gamma) \equiv 0$ for all such γ ($Re(\gamma_i) = 0$ for all i). Hence

$$\lim_{\Lambda'} \sup T_{\Lambda'}(A_\gamma) \equiv 0\,.$$

Simply if $k \in T^c$ = the compact torus given by the intersection of K^c (= the maximal compact subgroup of Sp or $SO(\ )$) with T. Thus

$$\lim_{\Lambda'} \sup_{k \in T^c} |T_{\Lambda'}(k)| \equiv 0\,.$$

This implies that

$$\lim_{\Lambda'} \sup |T_{\Lambda'}(k)|^2 \equiv 0\,.$$

Now the function T_Λ can also be viewed as a continuous function on K^c (the maximal compact subgroup of Sp or $SO(\ )$). Then T_Λ is $\mathrm{Ad}(K^c)$ invariant.

Hence we have

$$\int_K |T_\Lambda(k)|^2 dk \leq (\int_K dk) \sup_{k \in K} |T_\Lambda(k)|^2 = \sup_{t \in T^c} |T_\Lambda(t)|^2\,.$$

But

$$T_\Lambda(k) = \sum_{(m_\ell, \ldots, m_2)} (H(D(\pi^{m_2}, \ldots, \pi^{m_\ell}))\delta_B^{-\frac{1}{2}}(D(\pi^{m_2}, \ldots, \pi^{m_\ell})) \dim V_{(m_\ell, \ldots, m_2)})\Theta_{V_{(m_\ell, \ldots, m_2)}}(k)\,.$$

where $\Theta_{V_{(m_\ell, \ldots, m_2)}} = \frac{1}{\dim V_{(m_\ell, \ldots, m_2)}} tr_{V_{(m_\ell, \ldots, m_2)}}(k)$.

The functions Θ_{V_1} and Θ_{V_2} are orthogonal in $L^2(K, dk)$ if V_1 and V_2 are inequivalent K irreducibles.

Thus

$$\int_K |T_\Lambda(k)|^2 dk = \sum_{(m_\ell, \ldots, m_2) \in \Lambda} |H(D(\pi^{m_2}, \ldots, \pi^{m_\ell}))\delta_B^{-\frac{1}{2}}(D(\pi^{m_2}, \ldots, \pi^{m_\ell}) \dim V_{(m_\ell, \ldots, m_2)}|^2\,.$$

Hence we deduce that

$$H(D(\pi^{m_2}, \ldots, \pi^{m_\ell})) \equiv 0$$

for *all* $D(\pi^{m_2}, \ldots, \pi^{m_\ell})$.

Thus $H \equiv 0$!

We note a similar proof works in the case (c) above.

§6. Calculation of Local Factors

We know that the product

$$(\dim Hom_{\mathbb{G}}(X_{\sigma}^{\xi_0}, \Pi))(\dim Hom_{\mathbb{G}}(X_{\sigma^\nu}^{\xi_0}, \Pi^\nu)) \leq 1$$

(See **Appendix 2** to §3 for definitions, etc.)

Now we assume that $\mathbb{G}$, H and M are all quasi-split groups.

We let $B_{\mathbb{G}}, B_H$ and B_M be Borel subgroups of $\mathbb{G}, H$ and M respectively let $T_{\mathbb{G}}$, T_H and T_M be the corresponding maximal split tori.

We let χ and γ be characters on T_G and T_M.

We form the normalized induced modules:

$$\Pi(\chi) = \mathrm{ind}_{B_{\mathbb{G}}}^{\mathbb{G}}(\chi) \text{ and } \sigma(\gamma) = \mathrm{ind}_{B_M}^{M}(\gamma)$$

We know that for generic characters χ and γ, $\Pi(\chi)$ and $\sigma(\gamma)$ are spherical representations.

We note here the specific construction of $B_{\mathbb{G}}$, B_H, B_M and $T_{\mathbb{G}}$, T_H and T_M given in the beginning of §8. Also, we note the construction of the characters χ and γ on the groups $T_{\mathbb{G}}$ and T_M.

Comments in **Appendix 2** to §3 the space of $(N^i \times \mathbb{U}_i) \times M^\Delta$ intertwining map

$$Hom_{(N_i \times \mathbb{U}_i) \times M^\Delta}^{\mathbb{G} \times M}(\Pi(\chi) \otimes \sigma(\gamma), \Lambda_{\xi_0} \otimes 1)$$

is at most one dimensional.

Given any pair of Weyl group elements $(w, w') \in \mathbb{G} \times M$ (chosen relative to the Borel $B_{\mathbb{G}} \times B_M$) we let $I_w^{\mathbb{G}} \times I_{w'}^M$ be the usual intertwining operator defined from $\Pi(\chi) \otimes \sigma(\gamma) \rightsquigarrow \Pi((w, w')\chi) \otimes \sigma((w, w')\gamma)$ (See [C-S]). We let $\langle \mid \rangle$ be nonzero functional in the Hom space above. Then by uniqueness

$$\langle I_w^{\mathbb{G}}(f_\chi) | I_{w'}^M(g_\gamma) \rangle = \lambda_{w,w'}(\chi, \gamma) \langle f_\chi | g_\gamma \rangle$$

for any f_χ and $g_\gamma \in \Pi(\chi)$ and $\sigma(\gamma)$. We note that $\lambda_{w,w'}$ *depends* on the *choice* of the functional $\langle \ , \ \rangle$.

Remark 1. As an example of such a functional we consider the following case from §5. Namely we assume that $\Pi(\chi)$ has the form $I_{i+1}(\tau, \sigma', s)$ where τ and σ' are spherical representations of $GL(R_{i+1})$ and $SO(W_1(R_{i+1}))$. Moreover we let $\sigma(\gamma) = \Pi$.

Then we consider the pairing $\Pi(\chi)$ and $\sigma(\gamma)$ given in §5. That is for $F \in \Pi(\chi)$ and $w \in \sigma(\gamma)$ we consider the pairing

$$\langle\langle V_j(F) \mid T(w) \rangle\rangle \, .$$

This is the explicit pairing we denote by $\langle \mid \rangle_{(\mathbb{G},M)}$ in this section. We recall from §5 explicitly the formula for $\langle\langle V_j(F)|T(w)\rangle\rangle$. Indeed

$$\langle\langle V_j(F)|T(w)\rangle\rangle = \int \{V_j(F)(h)|T(w)(h)\}_\sigma dh$$
$$= \int \langle V_j(F)(h)|\pi(h)(w)\rangle_\sigma dh$$

Here the domain of integration is given as follows

$$N_{\{z_2,\cdots,z_\ell\}}U_{\{z_2,\cdots,z_\ell\}}SO(W_1(S)\cap W_1(R_i))\backslash SO((\xi_0)^\perp)q_{W_1})$$

We note in the formula above the inductive nature of the pairing $\langle\langle \mid \rangle\rangle$ with the (internal) pairing $\langle \mid \rangle_\sigma$.

We note that in **Theorem (A)** of **Appendix 1** to §5, we have computed $\langle\langle V_j(F) \mid T(w)\rangle\rangle$ for unramified data and determined the equality with the term $T^{\mathbb{G}}_{\chi\otimes\gamma}(\varphi_1\otimes\varphi_2)$ which is defined in §8 (for specific choice of $B_{\mathbb{G}}$, B_M etc. given in §8). Here φ_1 and φ_2 are the characteristic functions of the maximal compact subgroups of G and M. In this section we thus use $T^{\mathbb{G}}_{\chi\otimes\gamma}$ as defining the element $\langle \mid \rangle$ in

$$\mathrm{Hom}^{\mathbb{G}\times M}_{N_i\times\mathbb{U}_i\times M^\Delta}(\Pi(\chi)\otimes\sigma(\gamma),\Lambda_\xi\otimes 1)\,.$$

We recall here the group H' defined at the beginning of §**7** so that $H\supseteq M\supseteq H'$. Also we recall the Borel groups $B_{W_1(R_i)}$, B_H, B_M and $B_{H'}$ given in **Remark 6 of** §**7** (which in turn is defined in **Appendix 1** and **3** of §**5**, cases (a), (b), (c) and (d)). In any case, $T^H_{\chi\otimes\gamma}(1_{K_H}\otimes 1_{K_M})$ equals the integral

$$\int_{H'\backslash M} T_{\gamma\otimes\chi|T_{H'}}(1^{(1,m)}_{K_M}\otimes 1^{(1,m)}_{K_H})dm\,.$$

Here $T_{\gamma\otimes\chi|T_{H'}}$ is defined relative to the pair $(B_M, B_{H'})$.

The goal here is to determine the γ-factors $\lambda_{w,w'}(\chi,\gamma)$. Noting that the intertwining operators $I^{\mathbb{G}}_w$ and $I^M_{w'}$ satisfy the general factorization rule, it suffices to compute $\gamma_{w,w'}(\chi,\gamma)$ in the case when w and w' are the simple reflections corresponding to simple roots.

We note that the λ factor depends on the pair $(\mathbb{G}, M)$. In particular to show this extra dependence we write $\lambda_{w,w'} = \lambda^{(\mathbb{G},M)}_{w,w'}$.

The goal here is to first show the *inductive nature* of $\lambda^{(\mathbb{G},M)}_{w,w'}$.

We note that if the group GL_m is given then for a simple root α, w_α (the reflection in GL_m corresponding to the transposition $(i,i+1)$ the *Whittaker* Γ *factor* $\lambda^m_{w_{\alpha_i}}$ is defined as in [C-S]. We let $\mathbb{P}_\ell = GL(R_\ell)\times SO(W_1(R_\ell))\times\mathbb{U}_\ell$ be the parabolic subgroup of $\mathbb{G}$ defined above. Moreover, let the Borel subgroup be chosen so that

$$B_{\mathbb{G}} = B_\ell\times B_{W_1(R_\ell)}\rtimes\mathbb{U}_\ell$$

where B_ℓ and $B_{W_1(R_\ell)}$ are Borel subgroups of $GL(R_\ell)$ and $SO(W_1(R_\ell))$. Moreover let $T_{\mathbb{G}} = T_\ell\cdot T_{W_1(R_\ell)}$ be the corresponding decomposition of the maximal torus $T_{\mathbb{G}}$.

We note that the simple roots α of $SO(W_1)$ are such that either α is a simple root in $GL(R_{i+1})$ or α is a simple root in $H = SO(W_1(R_i))$!

Theorem 6.1. *(Determination of γ-factor)*

(1) Let w_{α_j} be a reflection corresponding to a simple root α_j in $GL(i+1)(j = 1,\dots,i)$. Then

$$\lambda_{w_\alpha,e}^{(\mathbb{G},M)}(\chi,\gamma) = \lambda_{w_\alpha}^{i+1}(\chi|T_{i+1}).$$

(2) Let w_α be a reflection corresponding to a simple root α_j in H. Then

$$\lambda_{w_\alpha,e}^{(\mathbb{G},M)}(\chi,\gamma) = \lambda_{w_\alpha,e}^{(H,M)}(\chi \mid T_H,\gamma).$$

Proof. This is proved as **Theorem 1** of §8. □

We establish certain consequences of **Theorem 1**.

We note that from **Theorem 5.1**, it is possible to determine certain analytic information about the

$$\langle f_\chi \mid g_\gamma \rangle$$

where f_χ and g_γ are spherical vectors in $\operatorname{ind}_{B_{\mathbb{G}}}^{\mathbb{G}}(\chi)$ and $\operatorname{ind}_{B_M}^{M}(\gamma)$ (normalized so that $f_\chi(e) = g_\gamma(e) = 1$). In fact we derive an exact formula for this specific inner product in this section (see **Theorem** 6.2).

Indeed, we consider $\chi = \chi_{i+1} + \boldsymbol{s} \otimes \chi'$ as given in §5. then we prove in **Theorem 5.1** the following type identity:

$$\langle f_{(\chi_{i+1}+\boldsymbol{s})\otimes\chi'}, g_\gamma \rangle = \left[\prod_{\ell=1}^{\ell=i+1} L(\chi_\ell, q^{-s}, \beta)\right]^{-1} R(\chi_{i+1}, \chi', \gamma, s)$$

where the factor $[\,]^{-1}$ is defined in §5 and $R(\chi_{i+1}, \chi', \gamma, s)$ is a polynomial in q^s and q^{-s} of the following type:

$$\sum_\lambda h_\lambda(\chi_{i+1}, \chi', \gamma) q^{-\lambda s}\,.$$

Here λ runs over a finite set of integers and h_λ is a rational function in $(\chi_{i+1}, \chi', \gamma)$. (See here **Remark** antecedent to **Theorem (A)** of **Appendix (I)** in §5).

Thus in the above identity when we set $\boldsymbol{s} = 0$ we have that

$$\langle f_\chi \mid g_\gamma \rangle = \frac{P^{(\mathbb{G},M)}(\chi,\gamma)}{Q^{(\mathbb{G},M)}(\chi,\gamma)}\,.$$

Here

$$Q^{(\mathbb{G},M)}(\chi,\gamma) = \prod_{\substack{1\le\ell\le i+1\\ 1\le\nu\le \operatorname{rank}(M)}} (1 - q^{-\chi_\ell - \gamma_\nu - \frac{1}{2}})(1 - q^{-\chi_\ell + \gamma_\nu + \frac{1}{2}})\,.$$

Here $\chi = (\chi_1, \chi_2, \dots)$ and $\gamma = (\gamma_1, \gamma_2, \dots)$ are the components of χ and γ given in the construction of §5 (characters on the tori $T_{\mathbb{G}}$ and T_M). Moreover, $P^{(\mathbb{G},M)}(\chi,\gamma)$ is a rational function in χ and γ.

We let $\zeta_{\mathbb{G}}(\chi, s)$ be the zeta polynomial associated to $\mathbb{G}$, that is

$$\zeta_{\mathbb{G}}(\chi, s) = \prod_{\substack{\alpha \in \mathbb{G} \\ \alpha > 0}} (1 - q^{-\langle \alpha^{\nu}, \chi \rangle} q_{\alpha}^{-s}).$$

Here α^{ν} denotes the dual root to α (given in [C-S]). We note

here that $q_{\alpha} = q$ for each root α where U_{α} the associated root group is isomorphic to the one dimensional affine line k. In the case where U_{α} is isomorphic to K (unramified quadratic extension

of k) then $q_{\alpha} = q^2$.

We note that if χ is a character on $T_{\mathbb{G}}$ then we have the simple identity:

$$\zeta_{\mathbb{G}}(\chi, s) = \zeta_{GL_i}(\chi | T_i, s) \zeta_H(\chi | T_H, s) \cdot \zeta_{\mathbb{U}_i}(\chi, s)$$

where

$$\zeta_{\mathbb{U}_i}(\chi, s) = \prod_{\substack{\alpha \in \mathbb{G} \\ \left\{ \begin{smallmatrix} \alpha \text{ a root} \\ \text{in } \mathbb{U}_i \end{smallmatrix} \right\}}} (1 - q^{-\langle \alpha^{\nu}, \chi \rangle} q_{\alpha}^{-s})$$

Remark 2. We note that in the case $\mathbb{G} = H$, then $\ell(\chi, \gamma)$ is given explicitly in **Corollary** to **Lemma 1.1**. Namely, we have the formula:

$$\langle f_{\chi} \mid g_{\gamma} \rangle_{(H,M)} = \frac{L_v(\sigma(\gamma), st, \chi_1 + \frac{1}{2})}{L_v(\pi(\chi \mid_{T_{H'}}), st, \chi_1 + 1)} \left(\frac{d_V(\chi_1 + \frac{1}{2})}{d_T(\chi_1)} \right) \langle g_{\gamma} | f_{\chi} |_{T_{H'}} \rangle_{\sigma'} .$$

We know that $\Pi(\chi \mid_{T_{H'}})$ denotes the representation of H' where the character $\chi \mid_{T_{H'}}$ denotes the restriction of the character $\chi \mid_{B_{W_1(R_i)}}$ to $B_{H'}$. Hence $\chi \mid_{T_{H'}}$ involves only the variables $\chi_2, \dots, \chi_{\nu}$. Moreover the pairing $\langle g_{\gamma} \mid f_{\chi} \mid_{T_{H'}} \rangle_{\sigma'}$ is given by the intetral

$$T_{\gamma \otimes \chi | T_{H'}} (1_{K_M} \otimes 1_{K_{H'}})$$

where $T_{\gamma \otimes \chi T_{H'}}$ is defined relative to the pair $(B_H, B_{H'})$. In face, $\langle g_{\gamma} \mid f_{\chi} \mid_{T_{H'}} \rangle_{\sigma'}$ has exactly the same form as the original $\langle f_{\chi} \mid g_{\gamma} \rangle_{(H,N)}$. Thus we exploit in the proof below the inductive nature of the calculation of $\langle f_{\chi} \mid g_{\gamma} \rangle_{(H,M)}$!

Then relative to the above data we have:

(1) $Q^{(H,M)}(\chi, \gamma) = L_v(\sigma(\gamma), st, \chi_1 + \frac{1}{2})^{-1} a(\chi)$ where

$$a(\chi) = \begin{cases} \zeta_K(\chi_1 + \frac{1}{2}) & \text{if the form defined by } M \text{ is even and nonsplit} \\ 1 & \text{otherwise} \end{cases}$$

(2) $P^{(H,M)}(\chi,\gamma) = \frac{a(\chi)}{L_v(\pi(\chi|_{H'}),st,\chi_1+1)}$

$$\left[\frac{d_V(\chi_1+\frac{1}{2})}{d_T(\chi_1)}\right]\langle g_\gamma|f_\chi|_{H'}\rangle = \langle g_\gamma|f_\chi|_{H'}\rangle$$

$$a(\chi)\Bigg(\prod_{2\leq\nu\leq rank(H)}(1-q^{-\chi_1-\chi_\nu-1})(1-q^{-\chi_1+\chi_\nu-1})\Bigg)$$

$$\begin{cases} 1 & \text{if } V \text{ is (even} \\ \zeta_\nu(2\chi_1+1) & \text{if } (V) \text{ is (odd)} \end{cases}$$

$$\begin{cases} \zeta_{K_\nu}(\chi_1+1)^{-1} & \text{if } H \text{ corresponds to an even form which is not split} \\ 1 & \text{otherwise} \end{cases}$$

Thus we can deduce directly from the above formula that

$$P^{(H,M)}(\chi,\gamma) = b_{H,M}(\chi)\frac{\zeta_H(\chi,1)}{\zeta_{H'}(\chi\,|_{T_{H'}},1)}\langle g_\gamma|f_\chi|_{H'}\rangle$$

where

$$b_{H,M}(\chi) = \begin{cases} \zeta(2\chi_1+1) & \text{if } H \text{ corresponds to an odd dimensional form and } M \text{ corresponds to an even nonsplit form} \\ 1 & \text{otherwise} \end{cases}$$

The problem here is to compute explicitly the $P^{(\mathbb{G},M)}(\chi,\gamma)$.

We note that some of the γ factors in the **Theorem** above can be computed. Indeed we know that

$$\lambda^{i+1}_{w_\alpha}(\chi|T_{i+1}) = \frac{1-q^{\chi_j-\chi_{j+1}-1}}{1-q^{\chi_{j+1}-\chi_j}}$$

We define $P^*(\chi,\gamma) = \zeta_{\mathbb{G}}(\chi,1)\langle g_\gamma|f_\chi|_{H'}\rangle[P(\chi,\gamma)]^{-1}$ (where $P(\chi,\gamma) = P^{(\mathbb{G},M)}(\chi,\gamma)$). Also we define

$$\overline{b_{H,M}}(\chi,\gamma) = \begin{cases} \zeta(2\chi_{i+1}+1) & \text{if } H \text{ corresponds to an odd form and } M \text{ corresponds to an even nonsplit form} \\ 1 & \text{otherwise} \end{cases}$$

Then a consequence of **Theorem 6.1** is the following invariance properties of $P^*(\chi,\gamma)$ on the χ variable.

Proposition 6.1. *(i) Let α be a simple root in GL_{i+1}. Then*

$$P^*(w_\alpha(\chi),\gamma) = P^*(\chi,\gamma)$$

(ii) Let α be a simple root in H. Then

$$\frac{P^*(w_\alpha(\chi),\gamma) = \zeta_{H'}(w_\alpha(\chi)\mid_{T_{H'}},1)\overline{b_{H,M}}(\chi)}{\zeta_{H'}(\chi\mid_{T_{H'}},1)\overline{b_{H,M}}(w_\alpha\chi)P^*(\chi,\gamma)}$$

Proof. We first note that (for a simple root α of $\mathbb{G}$) as above

$$I_{w_\alpha}(f_\chi) = c_{w_\alpha}(\chi)\ f_{w_{\alpha(\chi)}}$$

We note here the formula:

$$c_{w_\alpha}(\chi) = \frac{1-q^{-\langle\check{\alpha},\chi\rangle}q_\alpha^{-1}}{1-q^{-\langle\check{\alpha},\chi\rangle}}.$$

Thus we deduce (using the functional equation defining $\lambda_{w,e}$) that

$$\frac{P(w_\alpha\chi,\gamma)}{Q(w_\alpha\chi,\gamma)}c_{w_\alpha}(\chi) = \lambda_{w_\alpha,e}^{(\mathbb{G},M)}(\chi,\gamma)\frac{P(\chi,\gamma)}{Q(\chi,\gamma)}$$

Again for shorthand notation $Q = Q^{(\mathbb{G},M)}$ here.

Thus we deduce that (for any simple root α)

$$\frac{P^*(\chi,\gamma)}{P^*(w_\alpha\chi,\gamma)} = \left(\frac{\langle g_\gamma|f_\chi|_{H'}\rangle}{\langle g_\gamma|f_{w_\alpha(\chi)}|_{H'}\rangle}\right)\left(\frac{1-q^{-\langle\alpha^\nu,\chi\rangle}}{1-q^{+\langle\alpha^\nu,\chi\rangle}q_\alpha^{-1}}\right)\left(\lambda_{w_\alpha,e}^{(\mathbb{G},M)}(\chi,\gamma)\right)\frac{Q(w_\alpha\chi,\gamma)}{Q(\chi,\gamma)}$$

First we let α be a simple root in GL_{i+1}. Then it follows that $Q(w_\alpha\chi,\gamma) = Q(\chi,\gamma)$. Moreover, it is clear that $\langle g_\gamma|f_{w_\alpha(\chi)}|_{H'}\rangle = \langle g_\gamma|f_\chi|_{H'}\rangle$ for such a root α (in GL_{i+1}). Moreover we know from the **Theorem** above

$$\lambda_{w_\alpha}^{(\mathbb{G},M)}(\chi,\gamma) = \lambda_{w_\alpha}^{i+1}(\chi|T_{i+1}) = \left(\frac{1-q^{\langle\check{\alpha},\chi\rangle-1}}{1-q^{-\langle\check{\alpha},\chi\rangle}}\right)$$

Thus for such an α, $P^*(w_\alpha\chi,\gamma) = P^*(\chi,\gamma)$.

On the other hand let α be a simple root of $\mathbb{G}$ belonging to H. If α belongs to H', then again it follows that $Q(w_\alpha\chi,\gamma) = Q(\chi,\gamma)$. If α is the simple root which belongs to H but not H' then we can assume that w_α is given by the following: $w_\alpha(\chi_{i+1}) = \chi_{i+2}$, $w_\alpha(\chi_{i+2}) = \chi_{i+1}$ and $w_\alpha(\chi_j) = \chi_j$ for $j \neq i+1$ and $i+2$. Thus we deduce that

$$\frac{Q(w_\alpha\chi,\gamma)}{Q(\chi,\gamma)} = \frac{\prod_{1\le\nu\le rank(M)}(1-q^{-\chi_{i+2}-\gamma_\nu--1/2})(1-q^{-\chi_{i+2}+\gamma_\nu-1/2})}{\prod_{1\le\nu\le rank(M)}(1-q^{-\chi_{i+1}-\gamma_\nu-1/2})(1-q^{-\chi_{i+1}+\gamma_\nu-1/2})}$$

But we note that this last ratio can be interpreted as the ratio of Q factors relative to the pair (H, M). That is, we consider the induced module of H,

$$\mathrm{ind}_{B_H}^{H}(\chi|T_H).$$

Inductively we note that

$$\begin{aligned}&\mathrm{ind}_{B_H}^{H}(\chi|T_H) = \\ &\mathrm{ind}_{GL_1\times H'\times \mathbb{U}_i}^{H}(\chi_{i+1}\otimes \mathrm{ind}_{B_{H'}}^{H'}(\chi|T_{H'}))\end{aligned}$$

then the corresponding unramified functional associated to normalized spherical vectors $f_\chi$and g_γ (as above) in $\mathrm{ind}_{B_H}^{H}(\chi|T_H)$ and $\mathrm{ind}_{B_M}^{M}(\gamma)$ is given by:

$$\langle f_\chi|g_\gamma\rangle_{(H,M)} = \frac{P^{(H,M)}(\chi,\gamma)}{Q^{(H,M)}(\chi,\gamma)}$$

Here $Q^{(H,M)}$ and $P^{(H,M)}$ are determined in **Remark 2**. Thus we deduce first that (w_α as above)

$$\frac{Q(w_\alpha\chi,\gamma)}{Q(\chi,\gamma)} = \frac{Q^{(H,M)}(w_\alpha\chi|T_H,\gamma)}{Q^{(H,M)}(\chi|T_H,\gamma)}$$

Again by using **Theorem 1** and the above facts

$$\begin{aligned}\frac{P^*(\chi,\gamma)}{P^*(w_\alpha\chi,\gamma)} = &\left(\frac{\langle g_\gamma|f_\chi|_{H'}\rangle}{\langle g_\gamma|f_{w_\alpha(\chi)}|_{H'}\rangle}\right)\left(\frac{1-q^{-\langle\alpha^\nu,\chi\rangle}}{1-q^{\langle\alpha^\nu,\chi\rangle}q_\alpha^{-1}}\right)\\ &\cdot\lambda_{w_\alpha,e}^{(H,M)}(\chi|T_H,\gamma)\frac{Q^{(H,M)}(w_\alpha\chi|T_H,\gamma)}{Q^{(H,M)}(\chi|T_H,\gamma)}\end{aligned}$$

But we also have the same identity for $(P^{(H,M)})^*$. That is,

$$\begin{aligned}\frac{(P^{(H,M)})^*(\chi,\gamma)}{(P^{(H,M)})^*(w_\alpha\chi,\gamma)} = &\left(\frac{\langle g_\gamma|f_\chi|_{H'}\rangle}{\langle g_\gamma|f_{w_\alpha(\chi)}|_{H'}\rangle}\right)\left(\frac{1-q^{-\langle\check{\alpha},\chi\rangle}}{1-q^{\langle\check{\alpha},\chi\rangle}q_\alpha^{-1}}\right)\\ &\cdot\lambda_{w_\alpha,e}^{(H,M)}(\chi\,|_{T_H},\gamma)\frac{Q^{(H,M)}(w_\alpha\chi\,|_{T_H},\gamma)}{Q^{(H,M)}(\chi\,|_{T_H},\gamma)}.\end{aligned}$$

Thus we deduce (by direct calculation) that

$$\frac{P^*(\chi,\gamma)}{P^*(w_\alpha\chi,\gamma)} = \frac{(P^{(H,M)})^*(\chi,\gamma)}{(P^{(H,M)})^*(w_\alpha\chi,\gamma)} = \frac{\zeta_{H'}(\chi\,|_{T_{H'}},1)\overline{b_{H,M}(w_\alpha(\chi))}}{\zeta_{H'}(w_\alpha(\chi)\,|_{T_{H'}},1)\overline{b_{H,M}}(\chi)}.$$

This completes the proof. □

We now obtain an immediate Corollary to **Proposition 1**. Indeed, we first let

$$d_{H,M}(\chi) = \begin{cases}\prod_{\ell=1}^{i+1}\zeta(2\chi_\ell+1) & \text{if } H \text{ corresponds to an odd dimensional form and } M \text{ corresponds to an even nonsplit form}\\ 1 & \text{otherwise}\end{cases}$$

Then we note here that $d_{H,M}(w_\alpha(\chi)) = d_{H,M}(\chi)$ if w_α is a simple reflection coming from a simple root in GL_{i+1}. In particular, if we write $\chi = S_1 + S_2$ where $S_1 \in \mathrm{Span}\{\chi_1, \ldots, \chi_i\}$ and $S_2 \in \mathrm{Span}\{\chi_{i+1}, \ldots, \chi_r\}$, then (for w_α a reflection from a root α in H)

$$d_{H,M}(w_\alpha(S_1+S_2)) = d_{H,M}(S_1 + w_\alpha(S_2)) = \left(\prod_{\ell=1}^{\ell=i} \zeta(2\alpha_i + 1)\right) \overline{b_{H,M}}(w_\alpha(S))$$

(with $S_1 = \sum_{\ell=1}^{\ell=i} \alpha_\ell \chi_\ell$).

Corollary 1 to Proposition 6.1. *The function* $\chi \to \frac{P^*(\chi,\gamma)}{\zeta_{H'}(\chi|_{T_{H'}},1)} d_{H,M}(\chi)$ *is invariant under the full Weyl group* $W_{\mathbb{G}}$ *of* $\mathbb{G}$.

However, this **Corollary** does not determine $P^*(\chi, \gamma)$ completely!

Our goal here is to show, in fact, that

$$P^*(\chi, \gamma) = \zeta_{H'}(\chi\,|_{T_{H'}}, 1)(d_{H,M}(\chi))^{-1}.$$

We let $\mathbb{B}$ be the Iwahori subgroup of $\mathbb{G}$ adapted to the Borel subgroup $\mathbb{G}$ fixed vectors, $\mathrm{ind}_{B_{\mathbb{G}}}^{\mathbb{G}}(\chi)^{\mathbb{B}}$, is spanned by the set of functions $\{f_w(\chi)\}$ (which is *dual* to the set of intertwining functionals $h_\chi \to I_w(h_\chi)(e)$). We let w_0 be the long element of $W_{\mathbb{G}}$. The f_{w_0} is the element of $\mathrm{ind}_{B_{\mathbb{G}}}^{\mathbb{G}}(\chi)$, which is characteristic function of the open cell $B_{\mathbb{G}} w_0 N^+(\mathcal{O}_\nu)$. Then we know that

$$f(\chi) = \sum_{w \in W_{\mathbb{G}}} c_w(\chi) f_w(\chi)$$

where $c_w(\chi)$ are rational functions in χ determined by the formula

$$c_w(\chi) = \prod_{\{a>0|w(\alpha)<0\}} c_{w_\alpha}(\chi).$$

We let

$$\ell_0^{(\mathbb{G},M)}(\chi, \gamma) = \langle f_{w_0}(\chi) \mid g_\gamma \rangle_{(\mathbb{G},M)}.$$

Again for shorthand notation $\ell_0 = \ell_0^{(\mathbb{G},M)}$.

Then using the above data it is possible to develop a simple (yet uncomputable effectively) formula of $\ell(\chi, \gamma)$ in terms of the $\ell_0(\chi, \gamma)$.

Lemma 6.1.

$$\ell(\chi, \gamma) = \sum_{w \in W_{\mathbb{G}}} \frac{\lambda_{w,e}^{(\mathbb{G},M)}(w^{-1}\chi, \gamma)}{c_w(w^{-1}\chi)} c_{w_0}(w^{-1}\chi) \ell_0(w^{-1}\chi, \gamma)$$

Proof. (given in **Appendix** to §6). □

We let $\rho^\nu = \rho^\nu_{\mathbb{G}} = \frac{1}{2}(\sum_{\alpha^\nu>0} \alpha^\nu)$. We let $\Delta(\chi)$ be the function given by the formula:

$$\Delta(\chi) = \sum_{w\in W_{\mathbb{G}}} sgn(w) q^{\langle w\rho^\nu,\chi\rangle} = q^{\langle\rho^\nu,\chi\rangle} \prod_{\alpha>0}(1 - q^{-\langle\alpha^\nu,\chi\rangle})$$
$$= \prod_{\alpha>0}(q^{\langle\alpha^\nu/2,\chi\rangle} - q^{-\langle\alpha^\nu/2,\chi\rangle}).$$

Then we note $\Delta(w\cdot\chi) = sgn(w)\Delta(\chi)$ for $w \in W_{\mathbb{G}}$. Moreover

$$\Delta(\chi) = q^{\langle q^v,\chi\rangle}\zeta_{\mathbb{G}}(\chi,0).$$

Corollary to Lemma 6.1. *We have the equality*

$$\Delta(\chi) = \left[\frac{(P^{(\mathbb{G},M)})^*(\chi,\gamma)}{\zeta_H(\chi\,|_{T_{H'}},1)} d_{H,M}(\chi)\right]$$
$$\sum_{w\in W_{\mathbb{G}}} sgn(w) q^{\langle w\rho^\nu,\chi\rangle} \zeta_{H'}((w\chi)|T_{H'},1) Q(w\chi,\gamma)\ell_0(w\chi,\gamma)[\langle g_\gamma| f_{w(\chi)}|_{H'}\rangle d_{H,M}(w(\chi))]^{-1}$$

Proof. We note first that

$$\lambda^{(\mathbb{G},M)}_{w,e}(\chi,\gamma) = \frac{\ell(w\chi,\gamma)}{\ell(\chi,\gamma)}\cdot c_w(\chi) = \frac{P(w\chi,\gamma)}{P(\chi,\gamma)}\frac{Q(\chi,\gamma)}{Q(w\chi,\gamma)}\cdot c_w(\chi)$$

Then we substitute this expression for $\lambda^{(\mathbb{G},M)}_{w,e}(w^{-1}\chi,\gamma)$ in **Lemma 6.1** above to obtain

$$\ell(\chi,\gamma) = \sum_{w\in W_{\mathbb{G}}} \frac{\ell(\chi,\gamma)}{\ell(w^{-1}\chi,\gamma)}\frac{c_w(w^{-1}\chi)}{c_w(w^{-1}\chi)} c_{w_0}(w^{-1}\chi)\ell_0(w^{-1}\chi,\gamma)$$

Thus we deduce that

$$1 = \sum_{w\in W_{\mathbb{G}}} \frac{Q(w^{-1}\chi,\gamma)}{P(w^{-1}\chi,\gamma)} c_{w_0}(w^{-1}\chi)\ell_0(w^{-1}\chi,\gamma).$$

But we note from [C-S] that

$$c_{w_0}(\chi) = \zeta_{\mathbb{G}}(\chi,1) q^{\langle\rho^\nu,\chi\rangle}(\Delta(\chi))^{-1} = \zeta_{\mathbb{G}}(\chi,1)[\zeta_{\mathbb{G}}(\chi,0)]^{-1}$$

Then substituting in the above formula

$$1 = \sum_{w\in W_{\mathbb{G}}} \frac{Q(w^{-1}\chi,\gamma)}{P(w^{-1}\chi,\gamma)}\frac{\zeta_{\mathbb{G}}(w^{-1}\chi,1)}{\Delta(w^{-1}\chi)} q^{\langle\rho^\nu,w^{-1}\chi\rangle}\ell_0(w^{-1}\chi,\gamma)$$

Hence (using the invariance properties of $\Delta(\chi)$)

$$\Delta(\chi) = \sum_{w \in W_{\mathbb{G}}} sgn(w) q^{\langle \rho^\nu, w^{-1}\chi \rangle} Q(w^{-1}\chi, \gamma) P^*(w^{-1}\chi, \gamma) \ell_0(w^{-1}\chi, \gamma) \cdot [\langle g_\gamma | f_{w^{-1}(\chi)} |_{H'} \rangle d_{H,M}(w^{-1}(\chi))]^{-1}$$

Then we use the invariance properties of P^* given in **Corollary to Proposition1** to deduce the statement of the **Corollary**. □

Then we note that the factor $\ell_0^{(\mathbb{G},M)}$ is inductively determined.

Lemma 6.2. $\ell_0^{(\mathbb{G},M)}(\chi, \gamma) = \ell_0^{(H,M)}(\chi | T_H, \gamma)$

Proof. (See **Proposition 1** of §8)

Then to show that $P^{(\mathbb{G},M)})^*(\chi, \gamma) = \zeta_{H'}(\chi | T_{H'}, 1)[d_{H,M}(\chi)]^{-1}$ we use the following.

Lemma 6.3. *There is the identity*

$$\Delta(\chi) = \sum_{w \in W_{\mathbb{G}}} sgn(w) q^{\langle \rho^\nu, w\chi \rangle} \zeta_{H'}(w\chi | T_{H'}) Q(w\chi, \gamma) \ell_0(w\chi, \gamma) \cdot [\langle g_\gamma | f_{w(\chi)} |_{H'} \rangle d_{H,M}(w(\chi))]^{-1}$$

Proof. First we note that $\zeta_{H'}(\chi | T_{H'}), \ell_0(\chi, \gamma), Q(\chi, \gamma), \langle g_\gamma | f_\chi |_{H'} \rangle$ and $d_{H,M}(\chi)$ (as functions in χ variable) are W_{GL_i} invariant.

Secondly we note that $\rho^\nu_{\mathbb{G}} = \rho^\nu_{GL_t} + \rho^\nu_H + \rho^\nu_{\mathbb{U}}$. Thus the right hand side of the equation in this **Lemma** equals

$$\sum_{w \in W_{GL_i} \times W_H \backslash W_{\mathbb{G}}} \left(\sum_{w_1 \in W_{GL_i}} sgn(w_1) q^{\langle \rho^\nu_{GL_i}, w_1 w(\chi) \rangle} \right)$$

$$\left(\sum_{w_2 \in W_H} sgn(w_2) q^{\langle \rho^\nu_H, w_2 w(\chi) \rangle)} \zeta_{H'}(w_2 w(\chi) | T_{H'}) Q(w_2 w(\chi), \gamma) \ell_0(w_2 w(\chi), \gamma) \right.$$

$$\left. [\langle g_\gamma | f_{w_2 w(\chi)} |_{H'} \rangle d_{H,M}(w_2 w(\chi))]^{-1} \right) q^{\langle \rho^\nu_{\mathbb{U}_i}, w(\chi) \rangle} sgn(w)$$

Moreover we have

$$Q^{(\mathbb{G},M)}(\chi, \gamma) = Q_1^{(\mathbb{G},M)}(\chi, \gamma) Q_2^{(\mathbb{G},M)}(\chi, \gamma)$$

where

$$Q_1^{(\mathbb{G},M)}(\chi, \gamma) = \prod_{\substack{1 \le \ell \le i \\ 1 \le \nu \le rank(M)}} (1 - q^{-\chi_\ell - \gamma_\nu - 1/2})(1 - q^{-\chi_\ell + \gamma_\nu - 1/2})$$

and

$$Q_2^{(\mathbb{G},M)}(\chi,\gamma) = \prod_{1\le\nu\le rank(M)} (1-q^{-\chi_{i+1}-\gamma_\nu-1/2})(1-q^{-\chi_{i+1}+\gamma_\nu--1/2})$$

Then it follows that for $w_2 \in W_H$

$$Q^{(\mathbb{G},M)}(w_2\chi,\gamma) = Q_1(\chi,\gamma)Q_2(w_2\chi,\gamma).$$

On the other hand we can decompose as above $\chi = S_1 + S_2$ where $S_1 \in Sp\{\chi_1,\ldots,\chi_i\}$ and $S_2 \in Sp\{\chi_{i+1},\cdots,\chi_r\}$ $(r = \text{rank}(H))$. Here $\{\chi_1,\ldots,\chi_i\}$ spans the roots of GL_i and $\{\chi_{i+1},\ldots,\chi_r\}$ spans the root of H. In any case, we note the following simple identities:

(1) $q^{\langle\rho_H^\nu,S_1+S_2\rangle} = q^{\langle\rho_H^\nu,S_2\rangle}$, (2) $\zeta_{H'}((S_1+S_2)\mid_{T_{H'}},1) = \zeta_{H'}(S_2\mid_{T_{H'}},1)$, (3) $Q(S_1+S_2,\gamma) = Q_1(S_1,\gamma)Q_2(S_2,\gamma)$, (4) $\ell_0^{(\mathbb{G}\ ,M)}(S_1+S_2,\gamma) = \ell_0^{(H,M)}(S_2,\gamma)$. Thus we have that the sum (with $w(\chi) = S_1 + S_2$)

$$\sum_{w_2\in W_H} sgn(w_2)q^{\langle\rho_H^\nu,w_2w(\chi)\rangle}\zeta_{H'}(w_2w(\chi)\mid_{T_{H'}},1)$$
$$Q(w_2w(\chi),\gamma)\ell_0^{(\mathbb{G},M)}(w_2w(\chi),\gamma)[\langle g_\gamma \mid f_{w_2w(\chi)}\mid_{H'}\rangle d_{H,M}(w_2w(\chi))]^{-1}$$

equals the sum

$$\Bigg\{\sum_{w_2\in W_H} sgn(w_2)q^{\langle\rho_H^\nu,w_2(S_2)\rangle}\zeta_{H'}(w_2(S_2)\mid_{T_{H'}},1)$$
$$Q_2(w_2(S_2),\gamma)\ell_0^{(H,M)}(w_2S_2,\gamma)[\langle g_\gamma|f_{w_2(S_1+S_2)}|_{H'}\rangle\overline{b_{H,M}}(w_2(S_2))]^{-1}\Bigg\}$$
$$\cdot Q_1(\chi,\gamma)\left(\prod_{\ell=1}^{\ell=i}\zeta(2\alpha_\ell+1)\right).$$

However, we know that the specific **Lemma** we are now proving is true when $\mathbb{G} = H$. Then from **Remark 2** we have $(P^{(H,M)})^*(\chi,\gamma) = \zeta_{H'}(\chi\mid_{T_{H'}},1)(\bar{b}_{H,M}(\chi))^{-1}$. Also we

Make the simple observation that $Q_2^{(\mathbb{G},M)}(\chi,\gamma) = Q^{(H,M)}(\chi,\gamma)$. Thus we have

$$\Delta^H(\chi\mid_{T_{H'}}) = \sum_{w\in W_H} sgn(w)q^{\langle\rho_H^\nu,w\chi\rangle}\zeta_{H'}(w\chi\mid_{T_{H'}})$$
$$Q_2^{(\mathbb{G},M)}(w\chi,\gamma)\ell_0^{(H,M)}(w\chi\mid_{T_H},\gamma)[\langle g_\gamma|f_{w(\chi)}|_{T_{H'}}\rangle\bar{b}_{H,M}(w\chi)]^{-1}.$$

Thus the sum above (where $w(\chi) = S_1 + S_2$) equals

$$\Delta^H(w(\chi)\mid_{T_{H'}})Q_1(w(\chi),\gamma)\left(\frac{\overline{b_{H,M}}(w\chi)}{d_{H,M}(w\chi)}\right).$$

Thus continuing the above calculation (using **Lemma 2** and the explicit expressions for Δ^H and Δ^{GL_i}) we get

$$\sum_{w\in(W_{GL_i}\times W_H\backslash W_{\mathbb{G}})} sgn(w)\Delta^{GL_i}(w\chi\,|_{T_i})\Delta^H(w\chi\,|_{T_H})Q_1(w\chi,\gamma)\frac{\overline{b_{H,M}(w\chi)}}{d_{H,M}(w\chi)}q^{\langle q^{\nu}_{\mathbb{U}_i},w(\chi)\rangle}$$

$$= \sum_{w\in W_{\mathbb{G}}} sgn(w)q^{\langle\rho^{\nu}_{G},w\chi\rangle}Q_1(w\chi,\gamma)\left(\frac{\overline{b_{H,M}}(w\chi)}{d_{H,M}(w\chi)}\right).$$

But we note that the summation above is nothing but the projection of the element $q^{\langle\rho^{\nu}_{\mathbb{G}},\chi\rangle}Q_1^{(\mathbb{G},M)}(\chi,\gamma)\left[\frac{\overline{b_{H,M}(\chi)}}{d_{H,M}(\chi)}\right]$ onto the *skew-invariant* elements in the Laurent ring
$\mathbb{C}[q^{\langle\ ,\chi\rangle},q^{-\langle\ ,\chi\rangle}]$.

First we note that (where $n = \text{rank}\ (\mathbb{G})$)

$$\langle\rho^{\nu}_{\mathbb{G}},\chi\rangle = \begin{cases} n\chi_1+\cdots+1\cdot\chi_n & \text{if } \mathbb{G} \text{ is either odd dimensional or even dimensional nonsplit orthogonal group} \\ (n-1)\chi_1+\cdots+0\chi_n & \text{if } \mathbb{G} \text{ is an even dimensional } \textit{split} \text{ orthogonal group.} \end{cases}$$

Then using the explicit formula for $Q_1^{(\mathbb{G},M)}(\chi,\gamma)$ we deduce that the product

$$q^{\langle\rho^{\nu}_{\mathbb{G}},\chi\rangle}Q_1^{(\mathbb{G},M)}(\chi,\gamma)\left[\frac{\overline{b}_{H,M}(\chi)}{d_{H,M}(\chi)}\right]$$

is a linear combination of monomial terms having the form

$$q^{\alpha_1\chi_1+\alpha_2\chi_2+\cdots+\alpha_n\chi_n}$$

In the case where $\mathbb{G}$ is an even dimensional split orthogonal group and rank $(M) = r$ with $r \le n-2$ we note that $\alpha_i = n-i$ if $i \ge n-r$ and $\alpha_i \in [n-i,\cdots n-2r-i]$ for $i < n-r$. In any case it is easy to see that for $i < n-r, |\alpha_i| < n-i+1$. Then choose i maximal so that the α_i which occurs in $q^{\alpha_1\chi_1+\cdots+\alpha_n\chi_n}$ is *distinct* from $n-i$. Then there exists $j > i$ so that the coefficient $\alpha_j = \pm\alpha_i$. The if $\alpha_i \ne 0$ consider the Weyl group element $w \in W_{\mathbb{G}}$ defined by the rule:

$$\chi_i \rightsquigarrow (\pm)_i\chi_j, \chi_j \rightsquigarrow (\pm)_j\chi_i \text{ and fixes } \chi_k, k \ne i,j$$

Choose $(\pm)_i = (\pm)_j = 1$ if $\alpha_i > 0$ and $(\pm)_i = (\pm)_j = -1$ if $\alpha_i < 0$. Then $sgn(w) = -1$. If $\alpha_i = 0$ consider the element $w \in W_{\mathbb{G}}$ defined by the rule:

$$\chi_n \to \chi_i\,,\ \chi_i \to \chi_n \text{ and fixes } \chi_k\,,\ k \ne i,n\,.$$

Again $sgn(w) = -1$. Thus in both cases we have constructed an element $w \in W_{\mathbb{G}}$ so that w fixes

$$q^{\alpha_1\chi_1+\cdots+\alpha_n\chi_n=\langle\alpha^*,\chi\rangle}$$

and $sgn(w) = -1$. This clearly implies that

$$\sum_{w'\in W_{\mathbb{G}}} sgn(w')q^{\langle\alpha^*,w'\chi\rangle} \equiv 0.$$

for all vector $\alpha^* \neq \rho^{\nu}_{\mathbb{G}}$!

The same argument is used in the cases when $\mathbb{G}$ is odd or even quasi-split orthogonal group.

Thus we have the equality:

$$\Delta^{\mathbb{G}}(\chi) = \sum_{w\in W_{\mathbb{G}}} sgn(w)q^{\langle\rho^{\nu}_{\mathbb{G}},w\chi\rangle}Q_1^{(\mathbb{G},M)}(w\chi,\gamma)\left[\frac{\overline{b_{H,M}(w\chi)}}{d_{H,M}(w\chi)}\right].$$

This concludes the proof of **Lemma 3.** □

Remark 3. The key technical point of the proof of **Lemma 3** is the following interesting principle about L functions. Namely, we know by construction that $\tau_i(\chi) = \mathrm{ind}_{B_{R_i}}^{GL(R_i)}(\chi_1 \otimes \chi_2 \otimes \cdots \otimes \chi_i)$ is an induced module for $GL(R_i)$. Then the function

$$Q_1(\chi,\gamma)\left[\frac{\overline{b_{H,M}(\chi)}}{d_{H,M}(\chi)}\right] = [L_v(\tau_i(\chi)\otimes\sigma(\gamma), st, \frac{1}{2})]^{-1}.$$

Here $L_v(\tau_i \otimes \sigma(\gamma), st, s)$ is the standard L function for the pair $GL_i \times M$ (note GL_i not GL_{i+1}). Then the key identity in the proof of the **Lemma** above is that

$$\Delta^{G}(\chi) = \sum_{w\in W_{\mathbb{G}}} sgn(w)q^{\langle\rho^{\nu}_{\mathbb{G}},w\chi\rangle}(L_v(\tau_i(w\chi)\otimes\sigma(\gamma), st, 1/2)^{-1}.$$

The point here is that simply $q^{\langle\rho^{v}_{\mathbb{G}},\chi\rangle}L_v(\tau_i(\chi)\otimes\sigma(\gamma), st, 1/2)^{-1}$ has an expansion in terms of the monomials $q^{\langle\alpha^*,\chi\rangle} = q^{\alpha_1\chi_1+\alpha_2\chi_2+\cdots+\alpha_n\chi_n}$ in such a way that each term is fixed by some $w \in W_{\mathbb{G}}$ where $\mathrm{sgn}(w) = -1$ except when $\alpha^* = \rho^{v}_{\mathbb{G}}$. In other terms, the degree of $L_v(\tau_i(\chi) \otimes \sigma(\gamma), st, 1/2)^{-1}$ where expanded as a polynomial in $q^{-\chi_1}, q^{-\chi_2}, \ldots, q^{-\chi_n}$ has degree low enough so that the projection of $q^{\langle\rho^{v}_{\mathbb{G}},\chi\rangle}$ and $q^{\langle\rho^{v}_{\mathbb{G}},\chi\rangle}L_v(\tau_i(\chi) \otimes \sigma(\gamma), st, 1/2)^{-1}$ onto $\Delta^{\mathbb{G}}(\chi)$ coincide!

Thus we have established the following **Theorem**.

Theorem 6.2. *We have the identity:*

$$\ell^{(\mathbb{G},M)}(\chi,\gamma) = \left[\frac{\zeta_{GL_{i+1}}(\chi\mid_{T_{i+1}},1)\zeta_{\mathbb{U}_{i+1}}(\chi,1)d_{H,M}(\chi)}{Q^{(\mathbb{G},M)}(\chi,\gamma)}\right]\langle g_\gamma|f_\chi|_{H'}\rangle.$$

Proof. We use **Corollary 1 to Proposition 1**, **Corollary 1 to Lemma 1** and **Lemma 3** to deduce the identity. □

Remark 4. In the case that $\mathrm{ind}_{B_{\mathbb{G}}}^{\mathbb{G}}(\chi) \cong \mathrm{ind}_{GL_{i+1}\times H'\times \mathbb{U}_{i+1}}^{\mathbb{G}}(\tau_{i+1}\otimes |\ |^s \otimes \sigma' \otimes 1)$ and $\Pi = \mathrm{ind}_{B_M}^{M}(\gamma)$, then $\chi = (\chi_1|\ |^s \otimes \chi_2|\ |^s \otimes \cdots \otimes \chi_{i+1}|\ |^s) \otimes (\chi_{i+2} \otimes \cdots \otimes \chi_r)$. Thus we have that

$$\zeta_{GL_{i+1}}(\chi\ |_{T_{i+1}}) = \prod_{1 \le \ell < j \le i+1} (1 - q^{-\chi_\ell + \chi_j - 1}) .$$

This term represents the *evaluation* of the Jacquet Whittaker integral for the normalized spherical vector in τ_{i+1}.

On the other hand

$$\zeta_{\mathbb{U}_{i+1}}(\chi, 1) = \prod_{\alpha \in \mathbb{U}_i} (1 - q^{-\langle \check{\alpha}, \chi \rangle} q_\alpha^{-1}) = \frac{1}{L_v(\tau_{i+1} \otimes \sigma', st, s+1) L_v(\tau_{i+1}, \frac{\Lambda^2}{sym^2}, 2s+1)} .$$

Here $L_v(\tau_{i+1} \otimes \sigma', st, \lambda)$ is the standard L function of the tensor product $\tau_{i+1} \otimes \sigma'$. Here

$$L_v(\tau_{i+1}, \frac{\Lambda^2}{sym^2}, 2s+1) = \begin{cases} L_v(\tau_{i+1}, \Lambda^2, 2s+1) & \text{if } H \text{ corresponds to an even dimensional form} \\ L_v(\tau_{i+1}, sym^2, 2s+1) & \text{if } H \text{ corresponds to an odd dimensional form} \end{cases}$$

Here Λ^2 or sym^2 is either the exterior square or symmetric square representations of $GL_{i+1}(\mathbb{C})$.

Finally,

$$\frac{d_{H,M}(\chi)}{Q^{(\mathbb{G},M)}(\chi,\gamma)} = L_v(\tau_{i+1} \otimes \sigma(\gamma), st, s + \frac{1}{2}) .$$

Thus we can state the **Theorem** given in §4.

Theorem 6.3. *If we assume all the data is unramified, then the local integral given in §5 by the pairing*

$$\langle\langle V_j(F_1'(\cdots)|T(w_0)\rangle\rangle$$

(See **Remark** *1 above) equals*

$$\left(\frac{L_v(\tau_{i+1} \otimes \sigma(\gamma), st, s + \frac{1}{2})}{L_v(\tau_{i+1} \otimes \sigma', st, s+1) L_v(\tau_{i+1}, \frac{\Lambda^2}{sym^2}, 2s+1)} \langle g_\gamma | f_{\chi + \boldsymbol{s} \otimes \chi'} |_{H'} \rangle_\sigma \cdot \zeta_{GL_{i+1}}(\chi|_{T_{i+1}}) \right)$$

The factors $\langle g_\gamma | f_{\chi+\boldsymbol{s}\otimes\chi'}|_{H'}\rangle$ and $\zeta_{GL_{i+1}}(\chi|_{T_{i+1}})$ are independent of s.

Remark 5. **Theorem 6.2** is the generalization of **Corollary to Lemma 1.1** to the higher rank case with generalized Spherical-Whittaker models.

Appendix to §6.

We prove here **Lemma 6.1**. We let $I_{\mathbb{G}}$ be the Iwahori group of $\mathbb{G}$. In fact we prove a slightly more general result. We use freely here the notation in $[C-S]$. We consider the "extended" Iwahori algebra:

$$\mathfrak{A} = \mathcal{H}(\mathbb{G}//I_{\mathbb{G}}) \otimes \mathbb{C}[X_1, X_1^{-1}, \cdot, X_n, X_n^{-1}]$$

where $\mathbb{C}[X_1, X_1^{-1}, \cdots, X_n, X_n^{-1}] = \mathbb{C}[X_1, X_1^{-1}] \otimes \cdots \otimes \mathbb{C}[X_n, X_n^{-1}]$ the usual Laurent ring in the $X_1, \cdots, X_n$.

We let $\mathfrak{A}$ act on the space $\operatorname{ind}_{B_{\mathbb{G}}}^{\mathbb{G}}(\chi)$ in the usual way. That is, $f \otimes \varphi \in \mathfrak{A}$ acts by

$$\zeta_\chi * (f \otimes \varphi) \rightsquigarrow \varphi(\chi)(\zeta_\chi * f)$$

where $\zeta_\chi * f$ is the usual convolution action.

We determine elements Z_w in $\mathfrak{A}$ which have the property that

$$f_{w,\chi} * Z_{w'} = 0$$

if $w' \neq w$.

Recall there is a certain commutative semigroup contained in the Hecke algebra $\mathcal{H}(\mathbb{G}//I_{\mathbb{G}})$. Namely consider the family of functions given by

$$\Omega_\Lambda = \text{ characteristic function of } I_{\mathbb{G}} \Lambda I_{\mathbb{G}}$$

where $\Lambda \in T_{\mathbb{G}}^+$ (the set Λ^- in $[C,p]$). In any case we normalize $\Omega'_\Lambda = \frac{1}{vol(\mathbb{G}\Lambda\mathbb{G})} \Omega_\Lambda$. Then we have $\Omega'_{\Lambda_1} * \Omega'_{\Lambda_2} = \Omega'_{\Lambda_1 + \Lambda_2}$ (in the usual convolution structure on the group $\mathbb{G}$).

Moreover we have that (the eigenfunction property)

$$f_{w,\chi} * \Omega'_\Lambda = \delta_{B_{\mathbb{G}}}^{1/2}(\Lambda)(w\chi)(\Lambda) f_{w,\chi}.$$

The point is to construct Z_w using the elements Ω'_Λ above. Indeed we let

$$Z_w = \prod_{\{w' \neq w\}} (\Omega'_\Lambda - (w'\chi)(\Lambda)\delta_{B_{\mathbb{G}}}^{1/2}(\Lambda))$$

Then $Z_w \in \mathfrak{A}$. Moreover

$$f_{w,\chi} * Z_{w'} = \prod_{\{w'' \neq w'\}} (\delta_{B_{\mathbb{G}}}^{1/2}(\Lambda)[w\chi(\Lambda) - w''\chi(\Lambda)] f_{w,\chi} \cdot$$

Then it is clear that $f_{w,\chi} * Z_{w'} \equiv 0$ if $w' \neq w$; on the other hand if $w = w'$, then $f_{w,\chi} * Z_w = \left\{ \prod_{\{w'' \neq w\}} [w\chi(\Lambda) - w''\chi(\Lambda)] \delta_{B_{\mathbb{G}}}^{1/2}(\Lambda) \right\} \cdot f_{w,\chi}$. It is possible to choose Λ so that the term

$$\rho_w(\chi) = [\delta_{B_\mathbb{C}}(\Lambda)]^{\frac{1}{2} N_G} \prod_{\{w'' \neq w\}} [w\chi(\Lambda) - w''\chi(\Lambda)]$$

is nonzero ($N_{\mathbb{G}} =$ (order of $W_{\mathbb{G}}$) $- 1$ here).

Then using the formula of f_χ (the unique unramified vector in $\mathrm{ind}_{B_{\mathbb{G}}}^{\mathbb{G}}(\chi)$)

$$f_\chi * Z_{w_0} = \rho_{w_0}(\chi) c_{w_0}(\chi) f_{w_0,\chi}$$

($w_0 =$ the unique long element in $W_{\mathbb{G}}$).

Let ℓ be *any meromorphic* functional on the space $\mathrm{ind}_{B_{\mathbb{G}}}^{\mathbb{G}}(\chi)$ which satisfies the functional equations:

$$\ell(I_w(\xi_\chi)) = \gamma_w(\chi)\ell(\xi_\chi)$$

for $\xi_\chi \in \mathrm{ind}_{B_{\mathbb{G}}}^{\mathbb{G}}(\chi)$. Here ℓ depends rationally on χ in the sense that $\chi \rightsquigarrow \ell(\xi_\chi)$ is a rational function in $q^{\pm\chi}$. The factor γ_w is the associated gamma factor.

Thus we deduce from above that

$$\ell(I_w(f_\chi * Z_{w_0})) = \gamma_w(\chi)\rho_{w_0}(\chi)c_{w_0}(\chi)\ell(f_{w_0,\chi}).$$

On the other hand (using the fact that I_w is an intertwining operator)

$$\ell(I_w(f_\chi * Z_{w_0}) = c_w(\chi)\ell(f_{w\chi} * Z_{w_0}) = c_w(\chi)c_{w_0 w^{-1}}(w\chi)\rho_{w_0}(\chi)\ell(f_{w_0 w^{-1}, w\chi})$$

Thus we deduce that

$$\ell(f_{w_0 w^{-1}, w\chi}) = \left(\frac{\gamma_w(\chi)c_{w_0}(\chi)}{c_{w_0 w^{-1}}(w\chi)c_w(\chi)}\right)\ell(f_{w_0,\chi})$$

Then again using the expansion of f_χ in terms of the basis $\{f_{w,\chi}\}$ we deduce that

$$\ell(f_\chi) = \sum_{w \in W_{\mathbb{G}}} \frac{\gamma_w(w^{-1}\chi)c_{w_0}(w^{-1}\chi)}{c_w(w^{-1}\chi)}\ell(f_{w_0,w^{-1}\chi}).$$

this implies the statement of **Lemma 6.1**.

Remark. In fact we obtain a more general result than the formula above. In fact for any element of the Hecke algebra of the form Ω'_Λ

$$\ell(f_\chi * \Omega'_\Lambda) = \sum_{w \in W_{\mathbb{G}}} \frac{\gamma_w(w^{-1}\chi)c_{w_0}(w^{-1}\chi)}{c_w(w^{-1}\chi)}\ell(f_{w_0,w^{-1}\chi})\delta_{B_{\mathbb{G}}}^{1/2}(\Lambda)w_0 w^{-1}\chi(\Lambda)$$

This formula is a type of asymptotic formula determining the behavior of

$$\ell(f_\chi * \Omega'_\Lambda)$$

as Λ goes to infinity. This is the weak analogue of the formula for Whittaker functions given in [C-S].

We note that in the cases where ℓ is a matrix coefficient or a Whittaker functional it follows from the very specific form of ℓ (given by an integral) that $\ell(f_\chi \times \Omega'_\Lambda)$ represents the evaluation of the ℓ functional on a toral fundamental domain T_G^+.

§7. Determination of γ-factors (Spherical Case)

We let W_1 be a space with the nondegenerate form q_{W_1}. We decompose

$$W_1 = \mathcal{H}_1 \oplus \mathcal{H}_1^{\perp}$$

where $\mathcal{H}_1$ is a byperbolic plane and $\mathcal{H}_1^{\perp}$ = the q_{W_1} orthogonal complement to $\mathcal{H}$. Let $\xi_0 \in \mathcal{H}_1$ with $q_{W_1}(\xi_0) \neq 0$.

Then we let $H = SO(W_1)$, $M = SO((\xi_0)^{\perp}) = \{g \in H \mid g(\xi_0) = \xi_0\}$ and $H' = SO(\mathcal{H}_1^{\perp}) = \{g \in H \mid g \text{ fixes pointwise } \mathcal{H}_1\}$.

Then $H \supseteq M \supseteq H'$.

We let Π, σ, and ω be admissible irreducible representations of H, M and H' respectively.

We let $\langle \mid \rangle_\sigma \langle \mid \rangle_\omega$ resp.) be an element in

$$Hom_M(\Pi \otimes \sigma^{\vee}, 1)(Hom_{H'}(\sigma \otimes \omega^{\vee}, 1) \text{ resp.})$$

We concentrate here on the specific case when Π, σ and ω are of the form

$$ind_{B_H}^{H}(\chi), ind_{B_M}^{M}(\gamma) \text{ and } ind_{B_{H'}}^{H'}(\chi').$$

In this setup the pairing between Π and σ is dictated by analyzing the possible double cosets

$$B_H \times B_M \backslash H \times M / M^{\Delta}$$

and determine which of these sets can support a distribution $T : \mathcal{S}(H \times M) \to \mathbb{C}$ which has the following invariance property:

$$T(b_1, b_2) * \varphi * (m, m)) = (\chi \delta_{B_H}^{-\frac{1}{2}})(b_1)(\gamma \delta_{B_M}^{-\frac{1}{2}})(b_2) T(\varphi)$$

for all $(b_1, b_2) \in B_H \times B_M$ and $m \in M$. By the uniqueness properties stated in **Appendix 2** to§3 we deduce that (for generic values of (χ, γ)) that if such a T exists (for fixed value of (χ, γ)) then T is unique up to scalar multiple.

First we note that the set of double cosets $B_H \times B_M \backslash H \times M / M^{\Delta}$ is in bijective correspondence with the set of all double cosets of the form $B_H \backslash H / B_M$.

To give a more precise description of these double cosets we must describe more carefully the relation B_H and B_M inside the group H.

We note that $H \supseteq M$ are assumed to be *quasi-split* or *split*.

Specifically, this means that the forms (W_1, q_{W_1}) and $((\xi_0)^{\perp}, q_{W_1})$ correspond to

$$\mathcal{H} \oplus \mathcal{H}^* \oplus W_0$$

where $\mathcal{H}$, $\mathcal{H}^*$ maximal isotropic subspaces paired nonsingularly by q_{W_1} and W_0 an anisotropic form in 0, 1 or 2 variables. In the case W_1 is 2 dimensional, then (W_0, q_{W_1}) is a multiple of the norm form of an unramified quadratic extension K/k.

We let $B_H = T_H U_H$ and $B_M = T_M U_M$ be Borel subgroups.

Here T_H and T_M are the corresponding maximal tori and we assume that U_H and U_M the unipotent radicals of B_H and B_M satisfy $U_H \supseteq U_M$. But a more precise structural statement is available.

Lemma 7.1. *Let $H \supseteq M$ be a quasi-split pair except when $H = SO(r, r+1)$ and $M \cong SO((r-1, r-1) \oplus \Sigma_0)$ (with Σ_0 a 2 dimensional anisotropic). We let T_M^{spl} be the split part of the torus T_M. Then $T_M^{spl} \cong (k^\times)^r$ where $r =$rank (M). The quotient space U_H/U_M is a T_M^{spl} module, isomorphic to the left action of $(k^\times)^r$ on the vector space k^r.*

Proof. This is given in **Appendix** to §8. We note that the special case $SO(r, r+1) \geq SO((r-1, r-1) \oplus \Sigma_0)$ is also treated there). /**Q.E.D.**

Corollary to Lemma 7.1.

The group H admits a unique open dense orbit under the action of $B_H \times B_M$ (via $(b_1, b_2) : x \to b_1^{-1} x b_2$).

Proof. We note that relative to the action of $B_H \times B_H$ on H there is a unique open dense orbit in H. This set is given by

$$B_H w_0^H U_H$$

where w_0^H is the long Weyl group element in W_H. Then $T_M^{spl} \cdot U_M$ leaves this set stable through action on the right. Then we consider the map $B_H w_0^H U_H \underset{\text{proj}}{\longrightarrow} U_H \to U_H/U_M$. In particular it is easy to see that this map admits a local cross section. This implies that the inverse image of the set $(k^\times)^r$ in $k^r \cong U_H/U_M$ (which is the unique open orbit of T_M^{spl} in U_H/U_M) is an open dense set in $B_H w_0^H U_H$. Then there is a unique open dense orbit of $B_H \times B_M$ in H. For the case when $H \cong SO(r, r+1) \supseteq M \cong SO((r-1, r-1) \oplus \Sigma_0)$ the statement is proved in the **Appendix 1** to §7.)/**Q. E. D.**

Remark 1. The $B_H \times B_M$ orbit in H is a set of the form $B_H w_0^H Z$ where Z is open dense in U_H. We note that the orbit in fact $= B_H w_0^H \eta_H B_M$ where $\eta_H \in U_H$ so that the element $\eta_H U_M$ (in U_H/U_M) generates the unique open T_M^{spl} orbit in U_H/U_M) (see **Appendix** for modification of this statement in the exceptional case above). Thus it follows that the set $(B_H \times B_M)(w_0^H \eta_H, 1) M^\Delta = \{(b_1 w_0^H \eta_H m, b_2 m | (b_1, b_2) \in B_H \times B_M, m \in M\}$ is an open dense set in $H \times M$. On the other hand we know that the set $\{(m, m) \in M^\Delta | w_0^H \eta_H (m) (w_0 \eta_H)^{-1} \in B_H, m \in B_M\} = \{(m, m) \in M^\Delta | m \in (w_0^H \eta_H)^{-1} B_M (w_0 \eta_H) \cap B_H\}$. However it is straightforward to verify that this set reduces to the identity (in the exceptional case look in **Appendix** to §7). Thus it follows, that if $(h, m) \in H \times M$ satisfy $h = b_H w_0^H \eta_H m'$ and $m = b_M m'$ with $b_H \in B_H, B_H, b_m \in B_M$ and $m' \in M$ then the components b_H, b_M and m' are unique!

Then for a given $\varphi \in S(H \times M)$ we consider a functional of the following type. Here $d_\ell(b_1)$ and $d_\ell(b_2)$ are left invariant Haar measures on B_H and B_M.

$$T_{\chi \otimes \gamma}(\varphi) = \int\limits_{B_H \times B_M \times M} \varphi(b_1 w_0^H \eta_H m, b_2 m) \chi^{-1} \delta_{B_H}^{\frac{1}{2}}(b_1) \gamma^{-1} \delta_{B_M}^{\frac{1}{2}}(b_2) d_\ell b_1 d_\ell b_2 dm.$$

One notes very specifically that this integral is defined for the pair (B_H, B_M) of Borel subgroups of H and M satisfying the conditions in the above Remark; that is $B_H w_0^H \eta_H B_M$ is open in H.

Then "formally" this distribution satisfies the properties of quasi-invariance mentioned above.

Remark 2. For any pair (B'_H, B'_M) of Borel subgroups we can define an integral which defines a distribution similar to $T_{\chi\otimes\gamma}$ above. We note first that $B'_H \times B'_M$ has indeed a unique open dense orbit in H just as the specific pair $B_H \times B_M$ does, used above. Indeed there exist elements y_1 and y_2 in $\hat{H}$ and $\hat{M}$ so that $y_1 B_H y_1^{-1} = B'_H$ and $y_2 B_M y_2^{-1} = B'_M$ (here $\hat{H}$ and $\hat{M}$ are the associated orthogonal groups containing H and M). Now if both y_1 and y_2 lie in H and M and $B_H \xi B_M$ denotes the open dense orbit in H, we have that $y_1 B_H y_1^{-1} y_1 \xi y_2^{-1} y_2 B_M y_2^{-1} = y_1(B_H \xi B_M) y_2^{-1}$ and $y_1 \xi y_2^{-1}$ lies in H. Thus the set $B'_H(y_1\xi y_2^{-1})B'_M$ is the open dense $B'_H \times B'_M$ orbit in H. On the other hand the only case where y_1 may be in $\hat{H} - H$ is when H corresponds to an even split form. Then M corresponds to an odd dimensional form and if y_2 lies in M, then it is possible to find y_2^0 in $\hat{M} - M$ so that $y_2^0 B'_M y_2^{0^{-1}} = B'_M$. Thus, in particular, we can assume now y_2 lies in $\hat{M} - M$. Then $(y_1 B_H y_1^{-1})(y_1 \xi y_2^{-1})(y_2 B_M y_2^{-1}) = B'_H(y_1\xi y_2^{-1})B'_M$ (with $y_1\xi y_2^{-1} \in H$), which is the open dense $B'_H \times B'_M$ orbit in H. On the other hand, if M corresponds to an even split form and $y_2 \in \hat{M} - M$, then, as in the last argument, it is possible to assume that $y_1 \in \hat{H} - H$ also; then we can repeat the same argument so that $B'_H y_1 \xi y_2^{-1} B'_M$ is open dense in H.

Now let $\xi' \in H$ be any representative of the $B'_H \times B'_M$ open orbit in H. Then (ξ', e) is a representative for the $(B'_H \times B'_M) \times M^\Delta$ open orbit in $H \times M$. Then we define

$$_{\xi'}\tilde{T}_{\chi\otimes\gamma}(\varphi) = \int\limits_{B'_H \times B'_M \times M} \varphi(b'_1\xi' m, b'_2 m)\chi^{-1}\delta_{B'_H}^{\frac{1}{2}}(b'_1)\gamma^{-1}\delta_{B'_M}^{\frac{1}{2}}(b'_2)d_\ell(b'_1)d_\ell(b'_2)dm.$$

Then $'_\xi\tilde{T}_{\chi\otimes\gamma}$ and $T_{\chi\otimes\gamma}$ are related in the following precise way. We let $\gamma_1 \in \hat{H}$ and $\gamma_2 \in \hat{M}$ so that $\gamma_1 B_H \gamma_1^{-1} = B'_H, \gamma_2 B_M \gamma_2^{-1} = B'_M$ and the elements $\gamma_1^{-1}\xi'\gamma_2$ and $\gamma_1\gamma_2$ lie in H. Then we can make a change of variables so that

$$_{\xi'}\tilde{T}_{\chi\otimes\gamma}(\varphi) =$$
$$\int\limits_{B_H \times B_M \times M} \varphi(\gamma_1 b_1 \gamma_1^{-1}\xi' m, \gamma_2 b_2 \gamma_2^{-1} m)\chi_{\gamma_1}^{-1}\delta_{B_H}^{\frac{1}{2}}(b_1)\gamma_{\gamma_1}^{-1}\delta_{B_M}^{\frac{1}{2}}(b_2)d_\ell(b_1)d_\ell(b_2)dm$$

Here χ_{γ_1} and γ_{γ_2} are the characters on the groups B_H and B_M given by transport of structure formula, i.e. $\chi_{\gamma_1}(b_H) = \chi(\gamma_1^{+1} b_H \gamma_1^{-1})$ and $\gamma_{\gamma_2}(b_M) = \gamma(\gamma_2^{+1} b_M \gamma_2^{-1})$ We first make the change of variable $m \to \gamma_2^{-1} m$ above. Then we observe that $\gamma_1^{-1}\xi'\gamma_2$ lies in the $B_H \times B_M$ open orbit in H; thus $\gamma_1^{-1}\xi'\gamma_2 = (b_H)_0 w_0^H \eta_H (b_M)_0^{-1}$ for some choice of $((b_H)_0, (b_M)_0) \in B_H \times B_M$. If we let $\varphi^{(\gamma_1,\gamma_2^{-1}),(\gamma_2,\gamma_2^{-1})}(x,y) = \varphi(\gamma_1 x \gamma_2^{-1}, \gamma_2 y \gamma_2^{-1})$, there we have

$$'_\xi\tilde{T}_{\chi\otimes\gamma}(\varphi) = (\chi_{\gamma_1}\delta_{B_H}^{\frac{1}{2}})((b_0)_H)(\gamma_{\gamma_2}\delta_{B_M}^{\frac{1}{2}})^{-1}((b_0)_M)T_{\chi_{\gamma_1}\otimes\gamma_{\gamma_2}}(\varphi^{(\gamma_1,\gamma_2^{-1}),(\gamma_2,\gamma_2^{-1})})$$

The form of the "transported" formula for χ_{γ_1} can be made in fact more precise here. For instance we assume that $B_H(B'_H = \gamma_1 B_H \gamma_1^{-1}$ resp.) stabilizes the flag $\{U_1, \cdots, U_r\} \supseteq \cdots \supseteq \{U_r\}(\{\gamma_1(U_1), \cdots, \gamma_1(U_r)\} \supseteq \cdots \supseteq \{\gamma_1(U_r\}$ resp.). We note that the split torus $T_H(T'_H$ resp.) is given by the rule: $T_H(t_1, \cdots, t_r)$: $U_i \to t_{r-i+1}U_i, U_i^* \to t_{r-i+1}^{-1}U_i^*$ and $\mathbb{1}$ on $Sp\{U_1, \cdots, U_r, U_1^*, \cdots, U_r^*\}^\perp$ in W_1 (a similar formula on the basis $\{\gamma U_1, \cdots, \gamma U_i^*, \cdots\}$). Then $T'_H = \gamma T_H \gamma^{-1}$ coincides with the explicit identity: $T'_H(t_1, \cdot, t_r) = \gamma_1 T_H(t_1, \cdots, t_r)\gamma_1^{-1}$. This formula shows that the transported character $\chi_\gamma(T'_H(t_1, \cdots, t_r)) = \chi(\gamma_1^{-1}T'_H(t_1, \cdots, t_r)\gamma_1) = \chi(T_H(t_1, \cdots, t_r))$. Thus in particular we can speak unambiguously of $\chi|_{T_H}$ or $\chi|_{T_{H'}}$ when understood in the above context! In fact we omit χ_γ (write as χ) with the above privoso understood.

If we are given measures $d_\ell(b_1)$, $d_\ell(b_2)$ and dm on B_H, B_M and M^Δ it is possible to find Haar measure $dh \otimes dm'$ on $M \times M$ so that

$$dh' \otimes dm' = d_\ell(b_1) \otimes d_\ell(b_2) \otimes dm\,.$$

We note here that dm and dm' are not necessarily equal (only up to a scalar factor). This implies that the above functional can be expressed as an integral of the form

$$\int_{H \times M} \varphi(h, m')K_{\chi\otimes\gamma}(h, m')dhdm'$$

where $K_{\chi\otimes\gamma}$ is the kernel function given by the data:

$$K_{\chi\otimes\gamma}(h, m) = \begin{cases} \chi^{-1}\delta_{B_H}^{1/2}(b_1)\gamma^{-1}\delta_{B_M}^{1/2}(b_2) & \text{if } h = b_1 w_0 \eta m \text{ and } m = b_2 m \\ 0 & \text{otherwise} \end{cases}$$

We note here that we normalize the measures on B_H and B_M so that $K_H \cap B_H = B_H^0$ and $K_M \cap B_M = B_M^0$ has unit measure.

Remark 3. We let B_H and B_M be any pair of Borel subgroups of H and M. Then for $\varphi_1 \in S(H)$ and $\varphi_2 \in S(M)$ we consider the integrals

$$F_1(\chi, \varphi)(g) = \int_{B_H} \varphi_1(b_H g)\chi^{-1}\delta_{B_H}^{\frac{1}{2}}(b_H)d_\ell(b_H)$$

and

$$F_2(\gamma, \varphi_2)(g) = \int_M \varphi_2(b_M g)\gamma^{-1}\delta_{B_M}^{\frac{1}{2}}(b_M)d_\ell(b_M).$$

Then F_1 and F_2 define elements (for all (χ, γ)) in $\mathrm{ind}_{B_H}^H(\chi)$ and $\mathrm{ind}_{B_M}^M(\gamma)$. Moreover we have

$${}_\xi\tilde{T}_{\chi\otimes\gamma}(\varphi_1 \otimes \varphi_2) = \int_M F_1(\chi, \varphi_1)(\xi m)F_2(\gamma, \varphi_2)(m)dm$$

Then we note the Bruhat decomposition of $H \times M$ is given by

$$H \times M = \cup(B_H \times B_M)(w_i^H w_j^M)(U_H \times U_M)$$

where $w_i^H \in W_H$ and $w_j^M \in W_M$ (here W_H and W_M are the Weyl groups of H and M). In particular the open cell of $H \times M$ is given by

$$(B_H \times B_M)(w_0^H, w_0^M)(U_H \times U_M)$$

where w_0^H and w_0^M represent the longest Weyl group elements of W_H and W_M respectively.

The point here is that any element $g \in H \times M$ belonging to the open cell $(B_H \times B_M)(w_0^H, w_0^M)(U_H \times U_M$ can be expressed uniquely in terms of the Bruhat cell. Similarly any element $g \in H \times M$ belonging to the open set $(B_H \times B_M)(w_0^H \eta, 1)M^\Delta$ can be expressed uniquely in terms of this decomposition.

Then we summarize the properties of the distribution $T_{\chi \otimes \gamma}$.

Lemma 7.2. *There exists an open domain Ω whose closure is compact in $(\chi_1, \ldots, \chi_r, \gamma_1, \ldots, \gamma_t)$ $(t = r$ or $r-1)$ space so that $T_{|\chi| \otimes |\gamma|}(|\varphi_1 \otimes \varphi_2|)$ is absolutely convergent.*

Then the integral defining $T_{\chi \otimes \gamma}$ is an absolutely convergent integral. Moreover for fixed $\varphi \in S(H \times M)$ the function

$$(\chi, \gamma) \to T_{\chi \otimes \gamma}(\varphi)$$

is a rational function in the variables $q^{\pm \chi_i}$ and $q^{\pm \gamma_j}$.

More precisely, there exists a polynomial $\mathbb{P}$ in $q^{\pm \chi_i}$ and $q^{\pm \gamma_j}$ (all i and j) so that for all φ

$$(\chi, \gamma) \to \mathbb{P}(\chi, \gamma) T_{\chi \otimes \gamma}(\varphi)$$

is analytic.

Proof. We start with $\tilde{B}_H$ a Borel group of H. We note the equivalence of H modules

$$\mathrm{ind}_{\tilde{B}_H}^H(\chi) \cong \mathrm{ind}_{GL_1 \times H' \times \mathbb{U}_1}^H(|t_1|^{\chi_1} \otimes \sigma_{\chi'} \otimes \mathbb{1}_{\mathbb{U}_1}).$$

Here $\sigma_{\chi'} = \mathrm{ind}_{B_{H'}}^{H'}(\chi')$ with $\chi' = \chi|_{T_{H'}}$. Also $B_{H'} = H' \cap \tilde{B}_H$. Moreover we have that $T_H \supseteq T_{H'}$.

Thus given $\tilde{B}_H$ and $B_{H'}$ we can choose a Borel subgroup B_M of M in such a way that $B_M w_0^M \eta_M B_{H'}$ is the open dense orbit of $B_M \times B_{H'}$ in M given in the **Remark 1**. We emphasize here that the pair $(\tilde{B}_H, B_M)$ does not necessarily satisfy the hypotheses of **Remark 1**.

Then the H' invariant pairing between $\mathrm{ind}_{B_M}^M(\gamma)$ and $\mathrm{ind}_{B_{H'}}^{H'}(\chi')$ can be expressed in the following way: (dh' some Haar measure in H')

$$\langle F_1(\chi, \varphi_1) | F_2(\gamma, \varphi_2) \rangle = \int_{H'} F_2(\gamma, \varphi_2)(w_0^M \eta_M h') F_1(\chi, \varphi_1)(h') dh'$$

We then assume by induction that this integral is absolutely convergent (defined by the hypotheses of the **Lemma**). Here by absolute convergence we mean that the H' integral is *finite* where we replace $F_2(\gamma,\varphi_2)$ and $F_1(\chi,\varphi_1)$ by absolute values.

Then we have that

$$|\langle F_1(\chi,\varphi_1)*m|F_2(\gamma,\varphi_2)*m\rangle| \leq \int_{H'} |F_2(\gamma,\varphi_2)(w_0^M\eta_M h'm)||F_1(\chi,\varphi_1)(h'm)|dh'$$

We note that $|F_1(\chi,\varphi_1)| \leq F_1(|\chi|,|\varphi_1|)$ where $|\chi|$ is the quasicharacter obtained from χ by taking $Re(\chi)$. Moreover $|\varphi_1| \in S(H)$ and thus

$$F_1(|\chi|,|\varphi_1|) \in \mathrm{ind}_{B_H}^H(|\chi|)$$

In a similar way $|F_2(\gamma,\varphi_2)| \leq F_2(|\gamma|,|\varphi_2|)$ and $F_2(|\gamma|,|\varphi_2|) \in \mathrm{ind}_{B_M}^M(|\gamma|)$. We emphasize here that $F_1(|\chi|,|\varphi_1|)$ and $F_2(|\gamma|,|\varphi_2|)$ are nonnegative functions and we have

$$|\langle F_1(\chi,\varphi_1)*m|F_2(\gamma,\varphi_2)*m\rangle| \leq \langle F_1(|\chi|,|\varphi_1|)*m|F_2(|\gamma|,|\varphi_2|)*m\rangle.$$

We emphasize here that the first pairing in $|\langle\ |\ \rangle|$ is between $\mathrm{ind}_{B_M}^M(\gamma)$ and $\mathrm{ind}_{B_{H'}}^{H'}(\chi')$ and the second pairing (on the other side of the inequality) is between $\mathrm{ind}_{B_M}^M(|\gamma|)$ and $\mathrm{ind}_{B_{H'}}^{H'}(|\chi'|)$

The upshot of the above inequality is that we have verified that $(**)$ of §**1** is valid for the pair of $M\times H'$ modules

$$\mathrm{ind}_{B_M}^M(\gamma) \text{ and } \mathrm{ind}_{B_{H'}}^{H'}(\chi')$$

We note that $\tilde{\sigma} = \mathrm{ind}_{B_{H'}}^{H'}(|\chi'|)$ and $\tilde{\Pi} = \mathrm{ind}_{B_M}^M(|\gamma|)$ (using the notation in $(**)$ of §**1**).

Thus following §1 (see $(*)$ and the proof of **Lemma 1.1**) we deduce that the integral

$$\int_{H'\backslash M}\Big(\int_{H'} F_2(|\gamma|,|\varphi_2|)(w_0^M\eta_M h'm)F_1(|\chi|,|\varphi_1|)(h'm)dh'\Big)dm$$
$$= \int_M F_1(|\chi|,|\varphi_1|)(m)F_2(|\gamma|,|\varphi_2|)(w_0^M\eta_M m)dm$$

is absolutely convergent for $Re(\chi_1)$ sufficiently large. The measure dm is given as above (not the choice dm' here). Also we use the relevant quotient measure on $H'\backslash M$. However, how large $Re(\chi_1)$ must be is dependent on the data where the corresponding $\int_{H'}$ integral (with absolute values) is absolutely convergent. We can assume by induction that $(\int_{H'})$ (with absolute values) is absolutely convergent in some Ω' (open domain with compact closure) in the complex space spanned by $(\chi_2, \ldots, \chi_r)$, $(\gamma_1, \ldots, \gamma_t)$. Then by the arguments used in §**1** (with $\overline{\Omega}'$ compact) find χ_1 large enough so that an integral of the form

$$\int_M f_{|\chi_1|}(1, g)\langle v_1 \mid \sigma_{|\gamma|}(g)v_2\rangle dg$$

is absolutely convergent. Here $f_{|\chi_1|} = f_{|s|}$ in §1 and $\langle v_1, \sigma_{|\gamma|}(g)v_2\rangle$ is a matrix coefficient of $\sigma_{|\gamma|}$. We are using specifically here data from §1 so that the space of matrix coefficients of the form $\langle v_1 \mid \sigma_{|\gamma|}(g)v_2\rangle$ are majorized by $\Xi^t(g)$ (for some t uniform for $(\chi_2, \ldots, \chi_r)$, $(\gamma_1, \ldots, \gamma_t)$ in $\overline{\Omega}$, the compact set above). Thus we have shown that if we take the set $\{\chi_1 \mid Re(\chi_1) > R\} \times \Omega$ (R chosen to be dependent on $\overline{\Omega}$), then $T_{|\chi|\otimes|\gamma|}(|\varphi_1 \otimes \varphi_2|)$ is absolutely convergent.

The main problem now is to determine how the above integral can be tied to the expression for $T_{\chi\otimes\gamma}$. We note that there is a different type of integration (over M) in the integral above. However we see that in fact the above integral equals (up to a nonzero factor) to

$$T_{|\chi|\otimes|\gamma|}(|\varphi_1| \otimes |\varphi_2|).$$

The basic issue is concerned with the geometry of $B_H \times B_M$ acting on H.

Then we assert that relative to the pair $\tilde{B}_H \times B_M$ the orbit $\tilde{B}_H\xi B_M$ (with $\xi = (w_0^M \eta_M)^{-1}$) is open dense in H and thus determines the unique $B_H \times B_M$ open dense orbit in H. We note here that $\tilde{B}_H$ and B_M may not satisfy the hypothesis in **Remark 1** yet the set $\tilde{B}_H\xi B_M$ will be open dense in H. For this we let $R = \{(\xi\gamma\xi^{-1}, \gamma) \in \tilde{B}_H \times B_M\}$. Then we assert that $R = \{e\}$. If not then $\tilde{R} = \mathrm{proj}_M(R) \neq \{e\}$. However using the integral above we have

$$\int\limits_M F_1(|\chi|, |\varphi_1|)(\xi m)F_2(|\gamma|, |\varphi_2|)(m)dm =$$

$$\int\limits_{\tilde{R}\backslash M} \cdots (\int\limits_{\tilde{R}} |\chi|\delta_{B_H}^{1/2}(\xi\gamma\xi^{-1})|\gamma|\delta_{B_M}^{1/2}(\gamma)dr) \cdots$$

In any case the inner integral must be absolutely convergent. This implies that the group R is compact. However we know from **Appendix 1** of §**7** (see "**Proof of Certain Part of Lemma 7.2**" in **Appendix 1**) that such R can never be compact.

Thus $R = \{e\}$. Then, using **Remark 2** above, we can find $y \in \hat{H}$ and $x \in \hat{M}$ with $y^{-1}\underset{\sim}{B}_H y = \tilde{B}_H$ and $x^{-1}\underset{\sim}{B}_M x = B_M$ so that $\underset{\sim}{B}_H \times \underset{\sim}{B}_M$ satisfy the hypothesis of the **Remark 1**. Then the absolute convergence of

$$\int_M F_1(|\chi|, |\varphi_1|)(\xi m) F_2(|\gamma|, |\varphi_2|)(m) dm =$$

$$\int_{\tilde{B}_H \times B_M \times M} |\varphi_1(b_H \xi m)||\varphi_2(b_M m)||\chi|^{-1} \delta^{\frac{1}{2}}_{\tilde{B}_H}(b_H)$$

$$|\gamma|^{-1} \delta^{\frac{1}{2}}_{B_M}(b_M) d_\ell(b_H) d_\ell(b_M) dm$$

implies that the integral

$$\int_{\underset{\sim}{B}_H \times \underset{\sim}{B}_M \times M} |\varphi_1^{(y^{-1},x)}(\tilde{b}_H y \xi x^{-1} m)||\varphi_2^{(x^{-1},x)}(\tilde{b}_M m)||\chi_{y^{-1}}|^{-1}$$

$$\cdot \delta^{\frac{1}{2}}_{\underset{\sim}{B}_H}(\tilde{b}_H)|\gamma_{x^{-1}}|^{-1} \delta^{\frac{1}{2}}_{\underset{\sim}{B}_M}(\tilde{b}_M) d_\ell(\tilde{b}_H) d_\ell(\tilde{b}_M) dm$$

is also absolutely convergent. We note that $\chi_{y^{-1}}$ and $\gamma_{x^{-1}}$ can be identified to χ and γ following **Remark 2**. Then using **Remark 2** we deduce that $T_{|\chi| \otimes |\gamma|}(|\varphi_1^{(y^{-1},x)}| \otimes |\varphi_2^{(x^{-1},x)}|)$ is absolutely convergent (here $T_{|\chi|\otimes|\gamma|}$ is defined relative to the pair $\underset{\sim}{B}_H \times \underset{\sim}{B}_M$). In fact by another application of **Remark 2** we deduce $T_{|\chi|\otimes|\gamma|}$ is absolutely convergent defined for any pair of Borel subgroups $B_1 \times B_2$ satisfying the hypotheses of **Remark 1**. We emphasize here that (χ, γ) satisfies the hypotheses in the **Lemma**. Moreover we deduce the *exact identity*:

$$T_{\chi\otimes\gamma}(\varphi_1 \otimes \varphi_2) = \chi^{-1} \delta^{-\frac{1}{2}}_{\underset{\sim}{B}_H}((b_H)_0) \gamma^{+1} \delta^{+\frac{1}{2}}_{\underset{\sim}{B}_M}((b_M)_0)$$

$$\int_{(H' \backslash M)} \langle F_1(\chi, \varphi_1^{(y,x^{-1})})(m) | F_2(\gamma, \varphi_2^{(x,x^{-1})}) * m \rangle dm$$

The data in the identity is the following: (i) $T_{\chi\otimes\gamma}$ defined relative to the pair $(\underset{\sim}{B}_H, \underset{\sim}{B}_M)$ which satisfies the hypotheses of **Remark 1** (ii) F_1 and F_2 defined relative to a pair $(\tilde{B}_H, B_M)$ which are compatible in the following sense. First $\tilde{B}_H \supseteq B_{H'}$ with $\tilde{T}_H \supseteq T_{H'}$ and $(B_M, B_{H'})$ satisfy the conditions of **Remark 1** (iii) The pairing $\langle \, | \, \rangle$ is an H' invariant pairing which satisfies

$$\langle F_1(\chi, \varphi_1|_{H'}|) F_2(\gamma, \varphi_2) \rangle = T_{\gamma\otimes\chi'|_{T_{H'}}}(\varphi_2 \otimes \varphi_1|_{H'})$$

with $T_{\gamma\otimes\chi'|_{T_{H'}}}$ defined relative to $(B_M, B_{H'})$. (iv) $y_1^{-1} \underset{\sim}{\tilde{B}}_H y_1^{-1} = B_H, x_2^{-1} \underset{\sim}{B}_M x_2^{-1} = B_M$ and $y(w_0^M \eta_M)^{-1} x^{-1} = (b_0)_H w_0^H \eta_H (b_0)_M^{-1}$ with $(b_0)_H \in \underset{\sim}{B}_H$ and $(b_0)_M \in \underset{\sim}{B}_M$. Again we emphasize in the above identity the identification of χ_{ρ_1} and γ_{ρ_2} with χ and γ via **Remark 2**.

In the specific case above where y and x both lie in H and M respectively then the *exact identity* above can be replaced by

$$T_{\chi\otimes\gamma}(\varphi_1\otimes\varphi_2) = \chi\delta_{\underset{\sim}{B_H}}^{-\frac{1}{2}}((b_H)_0)\gamma^{-1}\delta_{\underset{\sim}{B_M}}^{\frac{1}{2}}((b_M)_0)$$

$$\int_{(H'\backslash M)} \langle F_1(\chi,\varphi^{(y,1)})(m) \mid F_2(\gamma)(\varphi^{(x,1)}) * m\rangle dm\,.$$

Using **Lemma 1.1** we deduce that

$$\int_{H'\backslash M} \langle F_1(\chi,\varphi_1)(m)|F_2(\gamma,\varphi_2) * m\rangle dm$$

is given as a finite linear combination of integrals of the form

$$(\int_M f_{\chi_1}(1,g)\langle v_1,\sigma_\gamma(g)v_2\rangle dg)$$

$$(\int_{H'} \tilde{F}_2(\gamma,\varphi_2)(w_0^M\eta_M h')\tilde{F}_1(\chi,\varphi_1)(h')dh')$$

for appropriate $\tilde{F}_2, \tilde{F}_1$ (here σ_γ is the representation of M associated to $\mathrm{ind}_{B_M}^M(\gamma)$). Here the first integral represents the convolution of $f_{\chi_1}(1,g)$ into the matrix coefficient σ_γ. The variable χ_1 in the formula above is the variable s in §1. Moreover the second integral represents (again) the H'-invariant pairing between $\mathrm{ind}_{B_{H'}}^{H'}(\chi')$ and $\mathrm{ind}_{B_M}^M(\gamma)$.

Moreover, we know that there exists by induction a polynomial $\mathbb{P}'$ in $q^{\pm\chi_j}$ and $q^{\pm\gamma_t}$ (all χ_j for $j \geq 2$) so that the function

$$(\chi',\gamma) \to \mathbb{P}'(\chi',\gamma)\int_{H'} \cdots$$

is analytic. On the other hand, by the standard arguments used in [PS-R], there exists a polynomial $\mathbb{P}''$ in $q^{\pm\chi_1}$ and $q^{\pm\gamma_j}$ (all j) so that

$$(\chi_1,\gamma) \to \mathbb{P}''(\chi_1,\gamma)\int_M f_{\chi_1}(1,g)\langle v_1,\sigma_\gamma(g)v_2\rangle dg$$

is analytic. This completes the proof of the last statement in **Lemma 7.2**.

The basic induction steps are (1) $H = SO(2,1), M = SO(1,1)$ and (2) $H = SO(K \oplus \langle 1,1\rangle), M = SO(2,1)$ (K = unramified quadratic extension). We prove these cases in **Appendix 2** to this section./**Q.E.D.**

The proof of the above Lemma carries the extra bonus in that there is an explicit formula for the distribution $T_{\chi\otimes\gamma}$.

Corollary 1 to Lemma 7.2.

Let $(\underset{\sim}{B}_H, \underset{\sim}{B}_M)$ and (B_H, B_M) be a pair of Borel subgroups of $H \times M$ which satisfy the following general properties:

(i) There exists T_H^{spl} and T_M^{spl} k split torii so that T_H^{spl} is a maximal k-split torus of $\underset{\sim}{B}_H$ and B_H and T_M^{spl} is a maximal k-split torus of $\underset{\sim}{B}_M$ and B_M where we have

$$T_H^{spl} \supseteq T_M^{spl}.$$

(ii) $\underset{\sim}{B}_H \supseteq B_{H'}$, $B_H \supseteq B_{H'}$ and $T_H^{spl} \supseteq T_{H'}^{spl}$

(iii) $(\underset{\sim}{B}_H, \underset{\sim}{B}_M)$ and $(B_M, B_{H'})$ satisfy the hypotheses given in **Remark 1**

Let $\varphi_1 \otimes \varphi_2 \in S(H \times M)$. Then $F_1(\chi, \varphi_1)$ and $F_2(\gamma, \varphi_2)$ belong to $ind_{\underset{\sim}{B}_H}^H(\chi)$ and $ind_{\underset{\sim}{B}_M}^M(\gamma)$.

Assume that there exists elements $\rho_1 \in N_{K_H}(T_H)$ and $\rho_2 \in N_{K_M}(T_M)$ so that $\rho_1^{-1}\underset{\sim}{B}_H\rho_1 = B_H$ and $\rho_2^{-1}\underset{\sim}{B}_M\rho_2 = B_M$. We let $(b_0)_H \in B_H$ and $(b_0)_M \in B_M$ be chosen so that $\rho_1(w_0^M\eta_M)^{-1}\rho_2^{-1} = (b_0)_H w_0^H \eta_H (b_0)_M^{-1}(w_0^M\eta_M$ and $w_0^H\eta_H$ are defined since (iii) above is valid).

Then we have the identity:

$$T_{\chi\otimes\gamma}(\varphi_1 \otimes \varphi_2) = \chi^{-1}\delta_{\underset{\sim}{B}_H}^{+\frac{1}{2}}((b_H)_0)$$
$$(\gamma^{+1}\delta_{\underset{\sim}{B}_M}^{-\frac{1}{2}})((b_M)_0)(\int\limits_{(H'\backslash M)} T_{\gamma\otimes\chi'|_{T_{H'}}}(\varphi_2^{(\rho_2,m)} \otimes \varphi_1^{(\rho_1,m)})dm)$$

Here $T_{\chi\otimes\gamma}$ is the pairing defined relative to $(\underset{\sim}{B}_H, \underset{\sim}{B}_M)$ and $T_{\gamma\otimes\chi'|_{T_{H'}}}$ is the pairing defined relative to $(B_M, B_{H'})$.

Proof. The Corollary is basically a restatement of parts of the proof of **Lemma 2 /Q.E.D.**

One point to be emphasized in the above identity is that $\chi\ |_{T_{H'}}$ in the formula above means the restriction of χ (defined on B_H via the transport of structure formalism given above) to $B_{H'}$. This is *not to be* confused with the restriction of χ (defined originally on $\underset{\sim}{B}_H$) to $B_{H'}$. We give an example below where this distinction is important!

Remark 5. It is possible to modify the statements in **Corollary 1** to **Lemma 7.2** to be valid for other cases. Namely we are given pairs $(\underset{\sim}{B}_H, B_M)$ and (B'_H, B_M) which satisfy

(i) $\underset{\sim}{B}_H$ and B'_H are K_H conjugate ($\underset{\sim}{B}_H = kB'_Hk^{-1}$ with $k \in K_H$)

(ii) $\underset{\sim}{B}_H \supseteq B_{H'}$, $B'_H \supseteq B_{H'}$ with $\underset{\sim}{T}_H \supseteq T_{H'}$ and $T'_H \supseteq T_{H'}$ $(\underset{\sim}{T}_H = kT_Hk^{-1})$

(iii) $(\underset{\sim}{B}_H, B_M)$ and $(B_M, B_{H'})$ satisfy the hypotheses of **Remark 1**.

Then a formula similar to the one is valid in the Corollary above.

Remark 6. We specifically apply the above **Remark** to the cases discussed in **Appendix 1** and **3** to §**5**. These are the cases (a), (b), (c) and (d) referred to there. In **Appendix 1** to §**5** the groups are given as follows: $\underset{\sim}{B}_H = \tilde{B}_{W_1(R_i)}$, $B'_H = B_{W_1(R_i)}$, $B_M = B_1$ and $B_{H'} = B_2$. However, to be consistent with the notation in **Appendix 3** to §**5**, now we let $\underset{\sim}{B}_H = B_H$, $B_1 = B_M$ and $B_2 = B_{H'}$.

In this specific circumstance we have that the $(b_H)_0$ and $(b_M)_0$ have toral parts which are, in fact, units (see the comments at the end of the discussion of cases (a), (b), (c), and (d) in **Appendix 3** to §5). Thus we have that the factor in front of the relation between $T_{\chi\otimes\gamma}$ and the integral involving $\int_{H'\backslash M}$ is equal to one. *This is, in fact, the exact same statement* as in **Theorem (A)** (for the case where $i = 0$). We note, however, the only difference between the discussion here and the discussion in **Theorem (A)** is the term w_S involved in $V_1(\cdots)$ ($j = 1$ in this case). However, what we do deduce is the following statement. We have that

$$\begin{aligned} T_{\chi\otimes\gamma}(\varphi_1 \otimes \varphi_2) &= \int_{H'\backslash M} \langle F_1(\chi, \varphi_1)(m) \mid F_2(\gamma, \varphi_2) * m\rangle dm \\ &= \int_{H'\backslash M} T_{\gamma\otimes\chi|_{T_{H'}}}(\varphi_1^{(\rho_1,m)} \otimes \varphi_2^{(\rho_2,m)})dm \end{aligned}$$

where φ_1 and φ_2 are the characteristic functions of the maximal compact subgroups K_H and K_M. Moreover, $T_{\chi\otimes\gamma}$ is define relative to the pair (B_H, B_M). Here F_1 is defined relative to $B_{W_1(R_i)}$, F_2 is defined relative to B_M and the pairing σ is defined as a $T_{\gamma\otimes\chi|_{T_{H'}}}$ functional (relative to the pair (B_M, B_{H_1})). *Thus we have shown that for unramified data the* 2 *M invariant bilinear forms when evaluated on spherical data coincide exactly.* Specifically, we note that via the theory of §1 and §2 we evaluate the form $\int_{H'\backslash M}\langle\cdots\rangle\, dm$ precisely in a inductive way where the answer depends on $\langle\ |\ \rangle_\sigma$. Thus we have the explicit value of $T_{\chi\otimes\gamma}(\varphi_1 \otimes \varphi_2)$ where φ_1 and φ_2 are the characteristic functions of the maximal compact subgroups K_H and K_M. We note here (apropos the comments after **Corollary 1** to **Lemma 7.2**) that *the character* $\chi\mid_{B_{H'}}$ *is the restriction of* $\chi\mid_{B_{W_1(R_i)}}$ *(and not the restriction of* $\chi\mid_{B_H}$*).* For instance, in example (a) in **Appendix 1** of §**5**, B_H ($B_{W_1(R_i)}$ and $B_{H'}$ resp.) stabilizes the flag $\{z_1, z_\nu, \ldots, z_2\} \supseteq \cdots \supseteq \{z_2\}$ ($\{z_\nu, \ldots, z_2, z_1\} \supseteq \cdots \supseteq \{z_1\}$ and $\{z_\nu, \ldots, z_2\} \supseteq \cdots \supseteq \{z_2\}$ resp.) Let χ be a character on B_H and χ (via transport of structure) be the corresponding character on $B_{W_1(R_i)}$. Let $\chi = (\chi_1, \ldots, \chi_\nu)$ denote the tuple defining χ in both cases. Then the restriction of χ (on $B_{W_1(R_i)}$) to $B_{H'}$ determines the tuple $(\chi_2, \ldots, \chi_\nu)$ and the restriction of χ (on B_H) to $B_{H'}$ determined the tuple $(\chi_1, \ldots, \chi_{\nu-1})$. We emphasize it is first case above that is important for considerations in this paper (see **Remark 2** in §**6**).

Corollary 2 to Lemma 7.2. *With the same notation in force as in* **Lemma 7.2**

$$T_{\chi\otimes\gamma}(\varphi_1\otimes\varphi_2) = c_H\cdot c_M$$
$$\int_{U_H}\left(\int_{U_M} F_1(\chi,\varphi_1)(w_0^H u_H u_M)F_2(\gamma,\varphi_2)(w_0^M u_M)du_M\right)$$
$$K_{\chi\otimes\gamma}(w_0^H u_H, w_0^M)du_H$$

(c_H, c_M defined below).

We note here that the measure dh (dm resp.) is disintegrated relative to the Bruhat decomposition of the open cell $B_H w_0^H U_H$ ($B_M w_0^M U_M$ resp.). Then $dh = c_H db_H \otimes du_H$ and $dm = c_M db_M \otimes du_M$. In fact, we have here normalized the db_H, du_H, db_M, du_M to have mass 1 on $K_H \cap B_H$, $K_H \cap U_H$, $K_M \cap B_M$ and $K_M \cap U_M$.

Remark 7. This integral representation of $T_{\chi\otimes\gamma}$ can be interpreted in the following way. The inner integral represents the convolution of $F_1(\chi,\varphi_1)(w_0^H u_H)$ and $F_2(\gamma,\varphi_2)(w_0^M u_M)$ relative to the subgroup U_M. The outer integral represents the Kernel function $K_{\chi\otimes\gamma}(w_0^H u_H, w_0^M)$ convoluted into the previous step.

It is also possible to represent $T_{\chi\otimes\gamma}$ in yet another manner. Indeed we recall that the "convolution" of the function φ_2 in $S(M)$ into φ_1 in $S(H)$ is given as follows:

$$(\varphi_1 *_M \varphi_2)(h) = \int_M \varphi_1(hm)\varphi_2(m)dm$$

Since $\varphi_2 \in S(M)$ it follows that $\varphi_1 *_M \varphi_2$ lies in $S(H)$. Then we have the next Corollary.

Corollary 3 to Lemma 7.2.

$$T_{\chi\otimes\gamma}(\varphi_1\otimes\varphi_2) = \int_{B_H\times B_M} (\varphi_1 *_M \varphi_2)(b_H w_0^H \eta_H b_M^{-1})$$
$$(\chi^{-1}\delta_{B_H}^{\frac{1}{2}})(b_H)(\gamma^{-1}\delta_{B_M}^{\frac{1}{2}})(b_M)d_\ell(b_H)d_\ell(b_M) = \int_{B_M} F_1(\varphi_1 *_M \varphi_2,\chi)$$
$$\cdot\,(w_0^H \eta_H b_M)\gamma\delta_{B_M}^{\frac{1}{2}}(b_M)d_\ell(b_M)$$

We recall that if I_w^H and $I_{w'}^M$ are intertwining operators from $\mathrm{ind}_{B_H}^H(\chi)$ to $\mathrm{ind}_{B_H}^H(w^{-1}\chi)$ and from $\mathrm{ind}_{B_M}^M$ to $\mathrm{ind}_{B_M}^M((w')^{-1}(\gamma))$, then we have the functional equation:

$$\langle I_w^H(F_1(\chi,\varphi_1)|I_{w'}^M(F_2(\gamma,\varphi_2))\rangle =$$
$$\gamma_{w,w'}(\chi,\gamma)\langle F_1(\chi,\varphi_1)|F_2(\gamma,\varphi_2)\rangle$$

Here $\gamma_{w,w'}(\chi,\gamma)$ (the so called γ factor) is a rational function in the variables $q^{\pm\chi_i}$ and $q^{\pm\gamma_i}$.

The basic problem is to have an effective procedure to determine these factors.

As another application of the above formulae for the distribution $T_{\chi\otimes\gamma}$ we can determine more exact information in the case when $F_1(\chi,\varphi_1)$ and $F_2(\gamma,\varphi_2)$ correspond to the unique spherical vectors in $ind_{B_H}^H(\chi)$ and $\mathrm{ind}_{B_M}^M(\gamma)$ (normalized in such a way that the value at the identity element equals 1). In fact we consider the following quantity:

$$\ell_{\chi\otimes\gamma}(\varphi_1\otimes\varphi_2)=T_{\chi\otimes\gamma}(\varphi_1\otimes\varphi_2)T_{\chi^{-1}\otimes\gamma^{-1}}(\varphi_1\otimes\varphi_2).$$

Here φ_1 and φ_2 are chosen so that $F_1(\chi,\varphi_1)$, $F_1(\chi^{-1},\varphi_1)$ $F(\gamma,\varphi_2)$ and $F(\gamma^{-1},\varphi_2)$ are normalized spherical vectors. In particular we assume that φ_1 and φ_2 are K_H and K_M bi-invariant functions on H and M.

We note that $\mathrm{ind}_{B_H}^H(\chi^{-1})\cong(\mathrm{ind}_{B_H}^H(\chi))^\nu$ as H modules $(\mathrm{ind}_{B_M}^M(\gamma^{-1})\cong(\mathrm{ind}_{B_M}^M(\gamma)^\nu$ as M Modules).

We know that the *adjoint L functions* of $\Pi(\chi)$ and $\sigma(\gamma)$ can be defined through the formulae:

$$(1)\quad L(\Pi(\chi),Ad,s)=\prod_{\alpha\in\Delta(H)}(1-q^{-\langle\alpha^\nu,\chi\rangle-s})^{-1}[(1-q^{-s})]^{-rank(H)}$$

$$(2)\quad L(\sigma(\gamma),Ad,s)=\prod_{\beta\in\Delta(M)}(1-q^{-\langle\beta^\nu,\gamma\rangle-s})^{-1}[1-q^{-s}]^{-rank(M)}.$$

Here $\Delta(H)$ ($\Delta(M)$ resp.) is the set of all roots of H relative to the pair (B_H,T_H) (roots of M relative to the pair (B_M,T_M)).

On the other hand we also can construct the tensor product L functions of the pair of representations of (H,M) given by $(\Pi(\chi),\sigma(\gamma))$. Indeed the L group of $H\times M$ is given by either $SO(2n,\mathbb{C})^\sim\times Sp_{n-1}(\mathbb{C})$ or $Sp_n(\mathbb{C})\times SO(2n,\mathbb{C})^\sim$ (where $SO(2n,\mathbb{C})^\sim$ equals either $SO(2n,\mathbb{C})$ in the case H is the split $SO(n,n)$ or the semidirect product $SO(2n,\mathbb{C})\ltimes Z_2$ (where Z_2 acts on $SO(2n,\mathbb{C})$ through the unique outer automorphism of $SO(2n,\mathbb{C})$) in the case $H\cong SO((n-1,n-1)\oplus\sum_0)$ where $\sum_0$ is an anisotropic form in 2 variables corresponding to a multiple of a norm form of an unramified quadratic extension). Then we form the tensor product L function associated to the standard module of ${}^L(H\times M)$ given above. We denote this L function by $L(\Pi(\chi)\otimes\sigma(\gamma),s)$.

The basic structure *Theorem* about $\ell_{\chi\otimes\gamma}$ is the following calculation.

Theorem 7.1. *Let the residual characteristic of $k\neq 2$. Then*

$$\ell_{\chi\otimes\gamma}(\varphi_1\otimes\varphi_2)=c(\chi,\gamma)\frac{L(\Pi(\chi)\otimes\sigma(\gamma),1/2)}{L(\Pi(\chi),Ad,1)L(\sigma(\gamma),Ad,1)}$$

where $c(\chi,\gamma)$ is a rational function in $q^{\pm\chi_i}$ and $q^{\pm\gamma_j}$ with no zeroes or poles.

Remark 8. We note that this Theorem should be a first approximation to the condition $Hom_M(\Pi(\chi)^0,\sigma(\gamma)^0)$ to be *nonzero*. See [R]. More precisely we denote

$\Pi(\chi)^0$ and $\sigma(\gamma)^0$ as the unique H and M spherical components (or unramified representations) in $\Pi(\chi)$ or $\sigma(\gamma)$. Indeed the nonvanishing of $\ell_{\chi\otimes\gamma}(\varphi_1\otimes\varphi_2)$ should be a first approximation to $\mathrm{Hom}_M(\Pi(\chi)^0, \sigma(\gamma)^0) \neq 0$.

Proof of Theorem. We first observe that **Remark 5** and **6** apply here. Indeed the fact that the pair (H, M) consist of quasi-split groups insures that $2s$ will be unit (as required in **Remark 6**).

We note by **Corollary 1** to **Lemma 2** that

$$T_{\chi\otimes\gamma}(\varphi_1\otimes\varphi_2) = \int\limits_{H\backslash M} \langle F_1(\chi,\varphi_1)(m) \mid F_2(\gamma,\varphi_2) * m\rangle_\sigma dm.$$

Here $\langle\,|\,\rangle_\sigma$ represents a H' invariant pairing between $\mathrm{ind}_{B_M}^M(\gamma)$ and $(\mathrm{ind}_{B_{H'}}^{H'}(\chi|T_{H'}))$. This specific pairing is given again by an integral of the form $T_{\gamma\otimes\chi|_{H'}}$. Now we use **Corollary** to **Lemma** 1.1. Here $\sigma = \pi(\chi|T_{H'})$, $\Pi = \sigma(\gamma)$ and $s = \chi_1$. Then the integral

$$\int\limits_{H'\backslash M} \langle F_1(\chi,\varphi_1)(m) \mid F_2(\gamma,\varphi_2) * m\rangle_\sigma dm$$
$$= \frac{L(\sigma(\gamma), st, \chi_1 + 1/2)}{L(\Pi(\chi|T_{H'}), st, \chi_1 + 1)} \frac{d_V(\chi_1 + 1/2)}{d_T(\chi_1)} \langle F_1(\chi,\varphi_1)(e) \mid F_2(\gamma,\varphi_2)\rangle_\sigma.$$

Thus we deduce that

$$\begin{aligned}\ell_{\chi\otimes\gamma}(\varphi_1\otimes\varphi_2) =& \frac{L(\sigma(\gamma), st, \chi_1 + 1/2) L(\sigma(\gamma^{-1}), st, -\chi_1 + 1/2)}{L(\Pi(\chi|T_{H'}), st, \chi_1 + 1) L(\Pi(\chi^{-1}|T_{H'}), st, -\chi_1 + 1)} \\ & \left[\frac{d_V(\chi_1 + 1/2) d_V(-\chi_1 + 1/2)}{d_T(\chi_1) d_T(-\chi_1)}\right] \langle F_1(\chi,\varphi_1)(e) \mid F_2(\gamma,\varphi_2)\rangle_\sigma \\ & \cdot \langle F_1(\chi^{-1},\varphi_1)(e) \mid F_2(\gamma^{-1},\varphi_2)\rangle_\sigma.\end{aligned}$$

Then we note by direct calculation that

$$\frac{d_V(\chi_1 + 1/2) d_V(-\chi_1 + 1/2)}{d_T(\chi_1) d_T(-\chi_1)} = \begin{cases} \dfrac{1}{\zeta(2\chi_1 + 1)\zeta(-2\chi_1 + 1)} & \text{if } \dim T \textit{ even} \\ 1 & \text{if } \dim T \textit{ odd} \end{cases}$$

Then by use of the induction hypothesis the term

$$\langle F_1(\chi,\varphi_1)(e) \mid F_2(\gamma,\varphi_2)\rangle \langle F_1(\chi^{-1},\varphi_1)(e) \mid F_2(\gamma^{-1},\varphi_2)\rangle$$

equals the term

$$c'(\gamma, (\chi \mid T_{H'})) \frac{L(\sigma(\gamma) \otimes \pi(\chi|T_{H'}), 1/2)}{L(\sigma(\gamma), Ad, 1) L(\pi(\chi|T_{H'}), Ad, 1)}$$

where c' is a rational function in $q^{\pm\gamma_i}$ and $q^{\pm\chi_j}$ with no zeros or poles.

Then we note that the term

$$L(\pi(\chi), Ad, 1) = L(\pi(\chi|T_{H'}), \chi_1 + 1)L(\pi(\chi^{-1})T_{H'}), -\chi_1 + 1)$$
$$\cdot \left\{ \begin{array}{ll} \zeta_v(2\chi_1 + 1)\zeta_v(-2\chi_1 + 1) & \text{if } \dim T \text{ even} \\ 1 & \text{if } \dim T \text{ odd} \end{array} \right\} \cdot L(\pi(\chi|_{T_{H'}}), Ad, 1)$$

(here equality is up to a nonzero factor independent of χ)

Thus the proof follows. We note that the beginning inductive step is when $\dim T$ is 3 and $\dim V$ is 2. In fact $\ell_{\chi\otimes\gamma}$ in this case equal (as follows from **Corollary** to **Lemma 1.1**)

$$\frac{L(\sigma(\gamma), st, \chi_1 + 1/2)L(\sigma(\gamma), st, -\chi_1 + 1/2)}{\zeta_v(2\chi_1 + 1)\zeta_v(-2\chi_1 + 1)}.$$

However we note that $L(\pi(\chi), Ad, 1) = \zeta(2\chi_1 + 1)\zeta(-2\chi_1)\zeta(1)$. On the other hand $L(\sigma(\gamma), Ad, 1) = \zeta(1)$ also. Moreover we note that

$$L(\sigma(\gamma), st, \chi_1 + 1/2)L(\sigma(\gamma), st, -\chi_1 + 1/2) = L(\pi(\chi) \otimes \sigma(\gamma), 1/2)!$$

QED

Appendix 1. (Proof of Lemma 1).

We let $W_1 = L_1 \oplus S \oplus L_1^*$ where L_1, L_1^* span isotropic subspace which are nonsingularly paired by q_{W_1}. Moreover let S be an nondegenerate subspace of dimension either 1, 2 or 3.

(1) We let S be 2 dimensional space. In such a case either $H \cong SO(r+1, r+1)$ or $\cong SO((r,r) \oplus \Sigma_0)$ where Σ_0 is 2 dimensional anisotropic. Here M is the stabilizer of a vector (of a non zero length relative to q_{W_1}) in S.

Let $B_H = T_H \cdot U_H$ be the Borel group of H.

Then we choose a basis of W_1 where $\langle e_1, \cdots, e_r\rangle = L, \langle e_r^*, \cdots, e_1^*\rangle = L^*$ and $S = \langle e_{r+1}, e_{r+1}^*\rangle$ or $\langle x, y\rangle$. Here we have $q_{W_1}(e_i^*, e_j) = \delta_{ij}$ (in the case $S = \langle x, y\rangle$, then S is anisotropic).

Then we let

$$U_H = V_H \times Y_H$$

where

$$Y_H = \begin{bmatrix} I_r & * & * \\ 0 & \boxed{I_2} & * \\ 0 & 0 & I_r \end{bmatrix} \text{ and } V_H = \begin{bmatrix} \begin{smallmatrix} 1 & & * \\ & \ddots & \\ 0 & & 1 \end{smallmatrix} & 0 & 0 \\ 0 & \boxed{I_2} & 0 \\ 0 & 0 & \begin{smallmatrix} 1 & & * \\ & \ddots & \\ 0 & & 1 \end{smallmatrix} \end{bmatrix}$$

Moreover

$$T_H = \begin{bmatrix} x_1 & & & & & & \\ & \ddots & & 0 & & 0 & \\ & & x_r & & & & \\ & 0 & & \boxed{A} & & 0 & \\ & & & & x_r^{-1} & & \\ & 0 & & 0 & & \ddots & \\ & & & & & & x_1^{-1} \end{bmatrix}$$

where $A = \begin{pmatrix} x_{r+1} & 0 \\ 0 & x_{r+1}^{-1} \end{pmatrix}$ or $A \in SO(2)$ (which is anisotropic).

Then it is possible to choose B_M to have the following structure. First $B_M = T_M \cdot U_M$.

Then $T_M = \{t \in T_H | A = I_2\}$.

Then $U_M = V_H \times Y_H^M$ where

$$Y_H^M = \left\{ \begin{bmatrix} I_r & X & Y \\ 0 & I_2 & Z \\ 0 & 0 & I_r \end{bmatrix} \in Y_H | (\alpha) Z = 0 \right\}$$

Here α represents a 1×2 matrix (which represents the vector in S which M fixes pointwise).

Then we deduce easily that (as a T_M module)

$$(Y_H^M \backslash Y_H) \cong \oplus k(r - \text{times})$$

(2) We let S be one dimensional. In such a case $H \cong SO(r, r+1)$ and $M =$ the stabilizer of a nonzero vector in S.

We choose the basis of W_1 as in (1) where S is now anisotropic (one dimensional space).

Then $U_H = V_H \times Y_H$ where again

$$V_H = \begin{bmatrix} 1 & & * & & & & \\ & \ddots & & 0 & & 0 & \\ 0 & & 1 & & & & \\ & 0 & & \boxed{1} & & 0 & \\ & & & & 1 & & * \\ & 0 & & & & \ddots & \\ & & & & 0 & & 1 \end{bmatrix} \text{ and } Y_H = \begin{bmatrix} I_r & * & * \\ 0 & \boxed{1} & * \\ 0 & 0 & I_r \end{bmatrix}$$

Moreover

$$T_H = \begin{bmatrix} x_1 & & & & & & \\ & \ddots & & & & 0 & \\ & & x_r & & & & \\ & & & \boxed{1} & & & \\ & & & & x_r^{-1} & & \\ & 0 & & & & \ddots & \\ & & & & & & x_1^{-1} \end{bmatrix}$$

Then it is possible to choose B_M so that $B_M = T_M U_M$.

First $T_H = T_M$.

Then $U_M = V_H \times Y_H^M$ where

$$Y_H^M = \left\{ \begin{bmatrix} I_r & \begin{matrix} 0 \\ \vdots \\ 0 \end{matrix} & Z \\ & \boxed{1} & 0\cdots0 \\ 0 & & I_r \end{bmatrix} \in Y_H \right\}$$

Again it is straightforward to verify that (as a T_M module)

$$Y_H^M \backslash Y_M = \oplus k(r - \text{times})$$

Thus in cases (1) and (2) we have established that U_H/U_M is isomorphic to Y_H/Y_M. Moreover Y_H/Y_M is as a T_M^{spl} module isomorphic to the left action of $(k^\times)^r$ on the vector space k^r (see **Lemma 1**).

The case where $H = SO(r, r+1) \geq M = SO((r-1, r-1) \oplus \sum_0)(\sum_0$ 2 dimensional anisotropic) is considered as (3) below.

(3) We let S be a 3 dimensional space (so that Witt rank $(S) = 1$). We choose a basis $\{e_{r+1}, x, e^{*_{r+1}}\}$ of S where $\{e_{r+1}, e^*_{r+1}\}$ span a nondegenerate hyperbolic plane. Here $H = SO(r \cdot H_2 \oplus S)$ and $M = SO(r \cdot H_2 \oplus S_0)$ where S_0 a 2 dimensional anisotropic subspace of S.

Then $U_H = V_H \times Y_H$ where

$$V_H = \begin{bmatrix} \begin{matrix} 1 & & * \\ & \ddots & \\ 0 & & 1 \end{matrix} & & 0 \\ & \boxed{\begin{matrix} 1 & & * \\ & 1 & \\ 0 & & 1 \end{matrix}} & \\ 0 & & \begin{matrix} 1 & & * \\ & \ddots & \\ 0 & & 1 \end{matrix} \end{bmatrix} \quad \text{and } Y_H = \begin{bmatrix} I_r & * & * \\ 0 & \boxed{I_3} & * \\ 0 & 0 & I_r \end{bmatrix}$$

Moreover

$$T_H = \begin{bmatrix} \begin{matrix} x_1 & & \\ & \ddots & \\ & & x_r \end{matrix} & & 0 \\ 0 & \boxed{\begin{matrix} x_{r+1} & & \\ & 1 & \\ & & x_{r+1}^{-1} \end{matrix}} & 0 \\ 0 & 0 & \begin{matrix} x_r^{-1} & & \\ & \ddots & \\ & & x_1^{-1} \end{matrix} \end{bmatrix}$$

Then it is possible to choose $B_M = T_M U_M$.

First $T_M = T'_H \cdot T''_M$ where $T'_H = \{t \in T_H | x_{r+1} = 1\}$ and

$$T''_M = \left\{ \begin{bmatrix} I_r & 0 & 0 \\ 0 & \boxed{\gamma} & 0 \\ 0 & 0 & I_r \end{bmatrix} \mid \gamma \in SO(2) \right\}$$

Here $\gamma \in SO(2) = SO(S_0) \subseteq SO(3) \cong SO(S)$ above.

On the other hand $U_M = V'_H \times Y_H^M$ where

$$V'_H = \left\{ \begin{bmatrix} \begin{matrix} 1 & & * \\ & \ddots & \\ 0 & & 1 \end{matrix} & & 0 \\ & \boxed{I_3} & \\ 0 & & \begin{matrix} 1 & & * \\ & \ddots & \\ 0 & & 1 \end{matrix} \end{bmatrix} \in V_H \right\} \text{ and}$$

$$Y_H^M = \left\{ \begin{bmatrix} I_r & X & Y \\ 0 & \boxed{I_3} & Z \\ 0 & 0 & I_r \end{bmatrix} \in Y_H \mid (\alpha) Z = 0 \right\}$$

Here α represents a 1×3 matrix (which represents the vector in S which M fixes).

Then following the same reasoning as above (as T'_H modules)

$$(Y_H^M \backslash Y_H) = \oplus k \ (r - \text{times}).$$

Unlike (1) and (2) we note that in (3) $U_M \backslash U_M$ is not isomorphic to $Y_H^M \backslash Y_H$. There is an extra piece in this case coming from $(V'_H \backslash V_H)$.

We modify the Bruhat decomposition of H in this case in the following way. We consider the decomposition of H into double cosets relative to $B_H \times P_H$ where P_H is the parabolic group of H given by $T'_H \cdot V'_H \cdot Y_H \cdot SO(S)$. Again the unique open cell in H is given by

$$B_H w_0^H (SO(S)) V'_H \cdot Y_H$$

On the other hand we know that the group $PGL_2(k)$ is isomorphic to the group $SO(S)$. In particular we consider the adjoint representation of GL_2. This defines a linear representation $Ad : GL_2 \to SO(S)$ which gives an explicit realization of $PGL_2 \cong SO(S)$.

On the other hand we know from the Iwasawa decomposition in $PGL_2(k)$ that

$$PGL_2(k) = B^-_{PGL_2} \cdot (K^x / k^x)$$

Here K is an unramified quadratic extension of k and $K^x/k^x \cong \{x \in K^x | N_{K/k}(x) = 1\} \cong SO(2)$ (by Hilbert's Theorem 90 for norm one elements). Moreover $B^- = \{\left(\begin{smallmatrix} * & 0 \\ * & * \end{smallmatrix}\right) \in GL_2\}$ and $B^-_{PGL_2} \cong k^x \backslash B^-$ = the projection of B^- into PGL_2. Thus from the comments above we deduce that

$$B_H w_0^H T''_M V'_H \cdot Y_H$$

is open dense in H.

Thus from the comments above

$$B_H w_0^H \eta_H B_M$$

is an open dense set in H (where η is the element of Y_H so that ηY_H^M corresponds to a representative of the unique open orbit of T'_H in $Y_H^M \backslash Y_H$).

Thus we have proved in case (3) above the Corollary to **Lemma 1.**

Proof of Certain Parts of Lemma 7.2.

We determine the general representatives of $B_H \times B_M$ orbits in G. First we know that for G there is a Bruhat decomposition

$$G = \bigcup_{w \in W_G} B_H w U_H$$

with $W_G =$ the Weyl group of G.

Then the above calculations ((1), (2) and (3) above) show that there exist elements $\tilde{\eta}$ in U_H/U_M so that

$$B_H w U_H = \bigcup_{\tilde{\eta}} B_H w \tilde{\eta} U_M$$

where $\tilde{\eta}$ runs through a certain set or representatives of Y_H/Y_H^M.

In any case we can determine the structure of the $\text{Stabilizer}_{B_H \times B_M}(w\tilde{\eta})$ in the following way. We recall that in cases (1) and (2) above the groups $B_H = T_H U_H$ and $B_M = T_M U_M$ have the property that $T_H \supseteq T_M$ and $U_H \supseteq U_M$. In fact Y_H^M is a normal subgroup of Y_H!

In case (3) above the groups $B_H = T_H U_H$ and $B_M = T_M U_M$ satisfy $T_H \supseteq T_M^{spl}$ and $U_H \supseteq U_M$. Again Y_H^M is normal in Y_H.

Then we define $B'_M = \begin{cases} B_M & \text{in cases (1) and (2)} \\ T_M^{spl} U_M & \text{in case (3)} \end{cases}$

In the cases above there is a normal subgroup $\overline{Y}_H$ in Y_H which is given as follows:

$$\overline{Y}_H = \left\{ \begin{bmatrix} I & 0 & * \\ 0 & I & 0 \\ 0 & 0 & I \end{bmatrix} \middle| \, * \text{ skew symmetric} \right\}$$

We note that there exists an exact sequence of groups

$$1 \to \overline{Y}_H \to Y_H \to \overline{Y}_H \backslash Y_H \to 1$$

where $\overline{Y}_H \backslash Y_H$ is Abelian!

Thus we first determine $\text{Stabilizer}_{B_H \times B'_M}(w\tilde{\eta})$.

Indeed if (t_1u_1, t_2u_2) lies in $\text{Stabilizer}_{B_H \times B'_M}(w\tilde{\eta})$ then

$$u_1^{-1}t_1^{-1}w\tilde{\eta}t_2u_2 = w\tilde{\eta}$$

By the Bruhat decomposition this implies that $t_1^{-1}wt_2w^{-1} = e$ and $u_1^{-1}wt_2^{-1}\tilde{\eta}t_2u_2 = w\tilde{\eta}$. Thus $u_1^{-1}w(t_2^{-1}\tilde{\eta}t_2u_2\tilde{\eta}^{-1}) = w$.

Hence $t_2^{-1}\tilde{\eta}t_2u_2\tilde{\eta}^{-1} \in U_w^H = \{u \in U_H | wuw^{-1} \in U_H\}$. This implies that $u_2 \in t_2^{-1}\tilde{\eta}^{-1}t_2U_w^H\tilde{\eta}$. In other words $\text{Stab}_{B_H x B'_M}(w\tilde{\eta}) = \{(t_1u_1, t_2u_2) \in (B_H x B'_M) | (1) t_1 = wt_2w^{-1} (2) u_1 = w(t_2^{-1}\tilde{\eta}t_2u_2\tilde{\eta}^{-1})w^{-1}$ when $u_2 \in t_2^{-1}\tilde{\eta}^{-1}t_2U_w^H\tilde{\eta} \cap U_M\}$.

Thus we note that Lie algebra U_w^H can be decomposed into root spaces in Lie algebra U_H. We have a decomposition Lie algebra $U_w^H =$

$$\text{Lie algebra } (U_w^H) \cap \text{Lie algebra } (V_H)) \oplus (\text{Lie algebra } (U_w^H) \cap \text{Lie algebra } (Y_H)).$$

Thus $u \in U_w^H$ has the form

$$u = \exp(X) = \exp(X_1)\exp(X_2)$$

where $X_1 \in$ Lie algebra $(U_w^H) \cap$ Lie algebra (V_H) and $X_2 \in$ Lie algebra $(U_w^H) \cap$ Lie algebra (Y_H).

On the other hand we also have a decomposition Lie algebra $(Y_H) =$ Lie algebra $(\overline{Y}_H) \oplus \Sigma_H$ where Σ_H is a sum of root subspaces (that occur in $\overline{Y}_H \backslash Y_H$, which is Abelian).

Thus $u \in Y_H$ has the form

$$u = \exp(Z_1)\exp(Z_2)$$

with $Z_1 \in \overline{Y}_H$ and $Z_2 \in \Sigma_H$. Hence we can write $\exp(X_2) = \exp(X_2^0)\exp(X_2^{00})$ with $X_2^0 \in \Sigma_H$ and $X_2^{00} \in$ Lie algebra of $\overline{Y}_H$ when X_2^0 and X_2^{00} also lie in the Lie algebra of U_w^H.

Thus the condition (with $X \in$ Lie algebra U_w^H)

$$t_2^{-1}\tilde{\eta}^{-1}t_2 \exp(X)\tilde{\eta} = (t_2^{-1}\tilde{\eta}^{-1}t_2\tilde{\eta}) [\tilde{\eta}^{-1}\exp(X_1)\exp(X_2^0)\tilde{\eta}][\tilde{\eta}^{-1}\exp(X_2^{00})\tilde{\eta}] \in U_M$$

is equivalent to

$$[t_2^{-1}\tilde{\eta}^{-1}t_2\tilde{\eta}][\tilde{\eta}^{-1}\exp(X_1)\exp(X_2^0)\tilde{\eta}] = t_2^{-1}\tilde{\eta}t_2\exp(X_1)\exp(X_2^0)\tilde{\eta} \in U_M$$

(since $\tilde{\eta}^{-1}\exp(X_2^{00})\tilde{\eta} \in U_M$, normality of $\overline{Y}_H$ in Y_H).

In particular we note that the group

$$\{(u_1, u_2) \mid u_1 = w\tilde{\eta}u_2\tilde{\eta}^{-1}w \ , \ u_2 \in \tilde{\eta}^{-1}(U_w^H \cap \overline{Y}_H)\tilde{\eta} \cap Y_H^M\}$$

lies in $\mathrm{Stab}_{B_H X B'_M}$. We note here that $\overline{Y}_H \subseteq Y_H^M$ and hence the condition $\widetilde{\eta}^{-1}(U_w^H \cap \overline{Y}_H)\widetilde{\eta} \cap Y_H^M$ is the same as $\widetilde{\eta}^{-1}(U_w^H \cap \overline{Y}_H)\widetilde{\eta}$.

Now assume $X_2^0 \neq 0$. Thus the Lie algebra $(U_w^H) \cap \Sigma_H \neq 0$.

We also let $\Sigma_{(M,H)}$ be a T_M^{spl} stable complement to Lie algebra Y_H^M in Lie algebra (Y_H). We note that T_M^{spl} acts on $\Sigma_{(M,H)}$ via the left regular action of $T_M^{spl} \cong (k^x)^r$ on the vector space $\oplus k(r \text{ times})$. We let $\{Z_i\}$ be a basis $\oplus k(r \text{ times})$. Then we let $\widetilde{\eta} = \exp\,(\Sigma \alpha_i Z_i)$ for some set of numbers α_i.

We assume that Lie algebra $(U_w^H) \cap \sum_H \neq (0)$. Then let W_α be a one dimensional root subspace in Lie algebra $(U_w^H) \cap \sum_H$. Then relative to the decomposition of Lie algebra (Y_H) into $\Sigma_{(M,H)}$ and Lie algebra (Y_H^M) we consider the projection of (W_α) to $\Sigma_{(M,H)}$. Either $(W_\alpha) \subseteq Y_H^M$ or there exists Z_i so that

$$s_0 W_\alpha = \alpha_i Z_i + \Omega_i(s_0).$$

with $s_0 \neq 0$ and $\Omega_i(s_0) \in$ Lie algebra (Y_H^M). In the latter case then the element $\widetilde{\eta}\,\exp\,(\Omega_i(s_0)) = \exp\,(\alpha_i Z_i + \Omega_i(s_0) + \sum_{j\neq i} \alpha_j Z_j) = \exp\,(s_0 W_\alpha)\,\exp\,(\sum_{j \neq i} \alpha_j Z_j)$ (mod Y_H^M). Here mod Y_H^M means relative to the left multiplication. In any case

$$w\widetilde{\eta}\,\exp\,(\Omega_i(s_0)) = (w\,\exp\,(s_0 W_\alpha)w^{-1}w\,\exp\,(\sum_{j\neq i} \alpha_j Z_j)\ \mathrm{mod}\ (Y_H^M)$$

This implies that the $B_H x B_H$ orbit through $w\widetilde{\eta}$ contains the point

$$w\,\exp\,(\sum_{j \neq i} \alpha_j Z_j).$$

The key point here is that the new element $\widetilde{\eta}' = \exp\,(\sum_{j\neq i} \alpha_j Z_j)$ has a fewer number of nonzero Z_j in its expansion.

In the case that $(W_\alpha) \subseteq Y_H^M$ then we note that

$$\{(u_1, u_2) | u_1 = w\widetilde{\eta} u_2 \widetilde{\eta}^{-1} w^{-1}, u_2 \in (\widetilde{\eta}^{-1} U_w^H \cap \exp\,(tw_\alpha)\widetilde{\eta}) \cap Y_H^M\}$$

is a *nontrivial* subgroup of $\mathrm{Stabilizer}_{B_H x B'_M}$.

Then we assume that $X_1 \neq 0$ but $X_2^0 = X_2^{00} = 0$. Then we decompose

$$V_H = V_H^{'} \cdot V_H^{''}$$

where $V_H^{''}$ equals 1 in cases (1) and (2) above and $V_H^{''}$ = unipotent radical of Borel subgroup in $SO(2,1)$ in case (3). Then we write $\exp\,(X_1) = \exp\,(\check{X}_1)\,\exp\,(\hat{X}_1)$ relative to the decomposition above. In particular we note $t_2^{-1}\widetilde{\eta}^{-1} t_2\,\exp\,(X_1)\widetilde{\eta} =$

$$\exp\,(\check{X}_1)\,\exp\,(\hat{X}_1)[\exp\,(-X_1)t_2^{-1}\widetilde{\eta}^{-1}t_2\,\exp\,(X_1)\widetilde{\eta}] \in U_M.$$

Now since exp $(\check{X}_i) \in U_M$ this implies that $\hat{X}_1 = 0$ (since $V_H^{''} \cap U_M = (e)$) and

$$\exp(-\check{X}_i)t_2^{-1}\widetilde{\eta}^{-1}t_2 \exp(\check{X}_i)\widetilde{\eta} \in (U_M \cap Y_H) = Y_H^M .$$

This implies that if $\widetilde{\eta} = \exp(Z)$ ($Z = \sum \alpha_i Z_i$ as above) then

$$Z - \text{ Ad } (\exp(-\check{X}_i)t_2^{-1})(Z) \in \text{ Lie algebra } (Y_H^M).$$

We then note that the space $\Sigma_{(M,H)} \subseteq \sum_H$ can be chosen so that $\sum_H = \Sigma_{(M,H)} \oplus$ Lie algebra $(Y_M^H) \cap \Sigma_H$ where each summand is stable under the adjoint action of $V_H^{'}$. Then we suppose that $U_{w^\bullet}^H \cap V_H^{'} \neq (e)$. In particular we can assume that $U_w^H \cap V_H^{'}$ contains a *one parameter group* of the form

$$x_{ij}(t) = \{I + te_{ij} \mid t \in k\} .$$

Then we deduce that

$$Z - \text{ Ad } (x_{ij}(-t) \cdot t_2^{-1})(Z) = 0$$

or

$$Z = \text{ Ad } (x_{ij}(-t)t_2^{-1})(Z).$$

Thus we note that $t_2^{-1}\widetilde{\eta}^{-1}t_2 \exp(x_1)\widetilde{\eta} \in U_M$ if and only if $Z = \text{Ad}(\exp(-x)t_2^{-1})(Z)$ with $\exp(Z) = \widetilde{\eta}$ given above.

For this discussion we assume all the $\alpha_i \neq 0$ and that in fact $\alpha_i = 1$. Thus when determining explicitly the above equation we deduce that

$$\begin{gathered} T_1 - 1 = 0 \\ (T_2 + a_{12}T_1) - 1 = 0 \\ \vdots \\ (T_i + a_{i-1,i}T_{i-1} + \cdots + a_{1i}T_i) - 1 = 0 \\ \vdots \\ (T_r + a_{r-1,r}T_{r-1} + \cdots + a_{1r}T_1) - 1 = 0 \end{gathered}$$

Here $t_2^{-1} = \begin{pmatrix} T_r & & 0 \\ & \ddots & \\ 0 & & T_1 \end{pmatrix}$ and the matrix

$$x_{ij}(-t) = \begin{bmatrix} 1 & a_{r-1,r} & \cdots & a_{1r} \\ 0 & 1 & & a_{1r-1} \\ & & \ddots & \vdots \\ & 0 & & 1 \end{bmatrix}$$

Now it is clear there is a smallest value of i so that $T_\ell = 1 (\ell < i)$ and $T_i = 1 - s$.

This implies that an element of the form

$$t_2 \cdot u_2 = \begin{bmatrix} * & & & & 0 \\ & \ddots & & & \\ & & 1-s & & \\ & & & \ddots & \\ 0 & & & & 1 \end{bmatrix} \cdot u_2(s)$$

(with $*$ determined by the equations above) satisfies the criterion that

$$u_2(s) = t_2^{-1}\widetilde{\eta}^{-1}t_2 x_{ij}(s)\widetilde{\eta}$$

lies in U_M.

We require here that $s \neq 1$.

Thus we can deduce the following structural description about the group

$$\mathrm{Stab}_{B_H x B'_M}(w\widetilde{\eta}).$$

Proposition. *Let* $\widetilde{\eta} = \exp(\Sigma \alpha_i Z_i)$. *Then*

(i) if some $\alpha_i = 0$, $Stab_{B_H x B'_M}(w\widetilde{\eta})$ *contains*

$$\{(wt_2w^{-1}, t_2) \in T_H \times T_M^{sp\ell} | t_2 = (x_1, \ldots, x_r) \text{ where } x_i = 1 \text{ if } \alpha_i \neq 0\} \cong \prod (k^x)$$

(the number of k^x *equals the number of* i *so that* $\alpha_i = 0$.*)*

(ii) Let $U_w^H \cap \overline{Y}_H \neq (e)$ *then* $\{(u_1, u_2) \in U_H \times U_M | u_1 = w\widetilde{\eta}u_2\widetilde{\eta}^{-1}w^{-1},\ u_2 \in \widetilde{\eta}^{-1}U_w^H \cap \overline{Y}_H\widetilde{\eta}\} \subseteq Stab_{B_H x B'_M}(w\widetilde{\eta})$.

(iii) Let Lie algebra $(U_w^H) \cap \Sigma_H \neq 0$ *and some root subgroup* X_α *of Lie algebra* $(U_w^H) \cap \Sigma_H$ *contained in Lie algebra* $(Y_H^M) \cap \Sigma_H$ *properly then*

$$\begin{aligned} &(u_1, u_2) \in U_H \times U_M | u_1 = w\widetilde{\eta}u_2\widetilde{\eta}^{-1}w^{-1}, \\ &\quad u_2 \in \widetilde{\eta}^{-1}(U_w^H \cap \exp(X_\alpha)\widetilde{\eta}\} \subseteq Stab_{B_H x B'_M}(w\widetilde{\eta}). \end{aligned}$$

(iv) Let $U_w^H \cap V'_H \neq (e)$. *In fact we let* $x_{ij}(t)$ *be the one parameter group (given above) contained in* $U_w^H \cap V'_H$. *Then the set of elements*

$$t_2 \cdot u_2 = \begin{bmatrix} 1 & & & & 0 \\ & \ddots & & & \\ & & 1-s & & \\ & & & j\ \ddots & \\ 0 & & & & * \end{bmatrix} u_2(s)$$

with $u_2(s) = t_2^{-1}\widetilde{\eta}^{-1}t_2 x_{ij}(s)\widetilde{\eta}$ *satisfies* $u_2(s) \in U_M$ *(for* $s \neq 1$*). In particular the set of elements* $\{(t_1u_1, t_2u_2) | t_1 = wt_2w^{-1}, u_1 = w(t_2^{-1}\widetilde{\eta}t_2u_2\widetilde{\eta}^{-1})w^{-1}$ *where* $u_2 = u_2(s)$ *and* t_2 *as above* $\} \subseteq Stab_{B_HxB'_M}(w\widetilde{\eta})$.

Proof. We recapitulate the steps given above. In essence case (i) is clear. Steps (ii), (iii) and (iv) also follow from discussion above. /**QED**

The importance of the above **Proposition** is that we have determined the structure of the $\text{Stab}_{B_HxB'_M}(w\widetilde{\eta})$ to some extent. In fact if $\alpha_i \neq 0$ for some i then by case (i) above we have a nontrivial torus $\prod(k^x)$ contained in $\text{Stab}_{B_HxB'_M}(w\widetilde{\eta})$. Now if $\alpha_i \neq 0$ for all i then we consider the various cases when U_w^H intersects $\overline{Y}_H$, Σ_H and V'_H. We note in case (3) above (where $V''_H \neq (e)$) then the Bruhat decomposition of G is taken relative to B_H and $P_H = SO(2,1)$ (see case (3) above). Thus we do not need to consider (in case (3)) those w which satisfy $U_w^H \cap V''_H \neq (e)$. In any case we have shown in cases (ii) and (iii) above that $\text{Stab}_{B_HxB'_M}(w\widetilde{\eta})$ contains (at least) a *one dimensional* unipotent group (isomorphic to k). We note here in the case where Lie algebra $U_w^H \cap \Sigma_H \neq 0$ and (iii) is not satisfied then the arguments above imply we can choose the $\widetilde{\eta}$ (for this specific w) so that at least one $\alpha_i = 0$. Thus we have completed the proof of "Certain Parts of **Lemma** 7.2." Indeed we have shown that if $w\widetilde{\eta}$ does not lie in the "big" orbit then the group $\text{Stab}_{B_HxB'_M}(w\widetilde{\eta})$ is not *compact*! Indeed in case (iv) above we have constructed a subset in $\text{Stab}_{B_HxB'_M}(w\widetilde{\eta})$ which is noncompact (i.e. the set varies with $s \in k^x$ and hence noncompact). The subtle point here is that the set $\{(t_1u_1, t_2u_2) | t_1 = wt_2w^{-1}, u_1 = w(t_{i2}^{-1}\widetilde{\eta}t_2u_2\widetilde{\eta}^{-1})w^{-1}$ where $u_2 = u_2(s)$ and t_2 as above $\}$ is not a group but is a noncompact set.

Appendix 2. Proof of Convergence of low dimensional Integrals.

The basic integral defining $T_{\chi\otimes\gamma}(\varphi_1 \otimes \varphi_2)$ can be expressed as

$$\int_M F_1(\chi, \varphi_1)(w_0^H \eta_H m) F_2(\gamma, \varphi_2)(m) dm.$$

Using the Iwasawa decomposition of $M = B_M \cdot K_M$ we can write the above integral as

$$\int_{B_M \times K_M} F_1(\chi, \varphi_1)(w_0^H \eta_H b_M k_M)(\gamma \delta_{B_M}^{1/2})(b_M) F_2(\gamma, \varphi_2)(k_M) d_\ell(b)m) dk_M$$

The convergence of this integral is then determined by an integral of the form

$$\int_{T_M \times U_M} F_1(|\chi|, |\varphi_1|)(w_0^H \eta_H t_M u_M)(|\gamma| \delta_{B_M}^{1/2})(t_M) d(t_M) d(u_M).$$

We now examine these integrals when (i) $H = SO(2,1), M = SO(1,1)$ and (ii) $H = SO(K \oplus \langle 1,1 \rangle)$ and $M = SO(2,1)$ (K/k unramified ext. of degree 2 over k). In case (i) the integral becomes

$$\int_{k^\times} F_1(|\chi|, |\varphi_1|)(w_0^H u_H(x)) |x|^{\chi_1 - \gamma_1 + 1/2} d^\times(x)$$

Here $U_H(x) =$ the one parameter unipotent group in H. The integral thus can be decomposed into a sum of two terms

$$\int_{|x|\leq 1} + \int_{|x|>1}$$

The first integral is dominated by a term of the form

$$(\sup_{|x|\leq 1} F_1(|\chi|, |\varphi_1|)(w_0^H u_H(x)))(\int_{|x|\leq 1} |x|^{\chi_1-\gamma_1+1/2} d^\times(x))$$

Thus convergence is assured for the 2nd term if $\chi_1 - \gamma_1 + 1/2 > 0$. The 2nd integral above (over $|x| > 1$) requires that we express $w_0^H u_H(x)$ in Iwasawa form. However a direct calculation shows

$$w_0^H u_H(x) \in t_M(x^{-2})U_H \cdot K_H.$$

Here $t_M(*)$ denotes the one parameter multiplicative group defining t_M (see first **Appendix** in this section). In any case the second integral is dominated by

$$\sup_{k\in K_H} |F_1(|\chi||\varphi_1|)(k)|(\int_{|\chi|>1} |x|^{-\chi_1-\gamma_1--1/2} d^\times(x))$$

Thus convergence is given when $\chi_1 + \gamma_1 + 1/2 > 0$.

Thus we have in case (i) that convergence is given where $\chi_1 + 1/2 > |\gamma_1|$.

Next we consider case (ii). In such a case $U_H \cong K$ (here K/k is a quadratic extension of k)). In fact

$$\eta_M t_M u_M = t_M(t_M^{-1}\eta_M t_M)u_M = t_M u_H(t^{-1}, s)$$

(where we have chosen a basis (ξ_0, ξ_1) of K/k so $u_H(t^{-1}, s) \cong t^{-1}\xi_0 + s\xi_1$ with $u_H(0, s) = u_M(s) =$ the one parameter group defining u_M).

Thus the relevant integral in case (ii) becomes

$$\int_{k^\times \times k} F_1(|\chi|, |\varphi_1|)[w_0^H u_H(t, s)]|t|^{\chi_1-\gamma_1+1/2} d^\times(t)ds$$

Then we separate the integral into 2 parts relative to $k^\times$, i.e.

$$= \int_{(|t|\leq 1)\times k} + \int_{(|t|>1)\times k}$$

The first term is dominated by terms of the form

$$(\int F_1|\chi|, |\varphi_1|)[w_0 u_H(0, s)]ds)(\int_{|t|\leq 1} |t|^{\chi_1-\gamma_1+1/2} d^\times(t))$$

Thus convergence is assured for the first term if $Re(\chi_1) > 0$ and for the second term if $Re(\chi_1) - Re(\gamma_1) + 1/2 > 0$.

The we again decompose

$$\int_{(|t|>1)\times k} = \int_{(|t|>1)\times(|s|\leq 1)} + \int_{(|t|>1)\times(|s|>1)}$$

Then by a direct calculation

$$w_0^H U_H(t,s) \in t_H(\varphi(t,s)^{-2})U_H K_H$$

where $|\varphi(t,s)| = \max(|t|,|s|)$. Then

$$\int_{(|t|>1)\times(|s|\leq 1)} \leq \sup_{k\in H_H} F_1(|\chi|,|\varphi_1|)(k)(\int_{|s|\leq 1} ds)(\int_{|t|>1} |t|^{-\chi_1-\gamma_1-3/2} d^\times(t))$$

Thus we have convergence for this term provided $Re(\chi_1+\gamma_1)+3/2 > 0$.

Then we analyze the integral

$$\int_{(|t|>1)\times(|s|>1)} \cdots d^\times(t)ds = \int_{\{s||s|>1\}} \left\{ \int_{\{|t|\geq|s|\}} d^\times(t) \right\} ds + \int_{\{s||s|>1\}} \left\{ \int_{\{|t|<|s|\}} d^\times(t) \right\} ds\,.$$

The first term is majorized by an integral of the form

$$(\int_{\{s||s|>1\}} (\int_{\{|t|\geq|s|\}} |t|^{-\chi_1-\gamma_1-3/2} d^\times(t)))ds) \cdot |\sup_{k\in K_H} F_1(|\chi|,|\varphi_1|)(k)|.$$

Then the integral above is proportional (up to a nonzero number independent of χ)

$$\int_{|t|>1} |t|^{-\chi_1-\gamma_1-3/2} d^\times(t))(\int_{|s|>1} |s|^{-\chi_1-\gamma_1-1/2} d^\times)$$

Thus we have convergence here if $Re(\chi_1)+Re(\gamma_1)+3/2 > 0$ and $Re(\chi_1)+Re(\gamma_1)+1/2 > 0$. Here the second estimate suffices!

Similarly the term

$$\int_{\{s||s|>1\}} (\int_{\{t||t|<|s|\}} \cdots d^\times(t))ds$$

is dominated by

$$\int_{|s|>1} \Big(\int_{\{t|1\leq|t|<|s|\}} |t|^{-\chi_1-\gamma_1-3/2} d^\times(t)\Big) ds$$

The inner integral equals

$$\sum_{r=0}^{r=\log|s|} q^{(\chi_1+\gamma_1+3/2)r}$$

This term is dominated by

$$\ln|s||s|^{-\chi_1-\gamma_1-3/2}$$

Then the convergence of

$$\int_{|s|>1} \ln|s||s|^{-\chi_1-\gamma_1-1/2} d^\times(s)$$

is governed by the condition $Re(\chi_1) + Re(\gamma_1) + 1/2 > 0$!

Thus we have shown in cases (i) and (ii) above that the basic integral defining $T_{\chi\otimes\gamma}$ is absolutely convergent when $Re(\chi_1) + 1/2 > |Re(\gamma_1)|$ and $Re(\chi_1) > 0$!

Appendix 3. Conjugation by Compact Elements.

We consider a 3 dimensional space $V = Sp\{v, \xi, v^*\}$ where the form $q_V(v, v) = q_V(v^*, v^*) = 0, q_V(v, v^*) = 1$ and $q_V(\xi, \xi) = 2s$ with $2s$ a unit.

Then we choose another $\mathcal{O}$ basis of X given by $X = v - sv^* + \xi, Y = - -\frac{1}{4s}(v - sv^* - \xi), Z = v + sv^*$. Then $q_V(X, X) = q_V(Y, Y) = 0, q_V(X, Y) = 1$ and $q_V(Z, Z) = 2s$. Then let τ be the linear map defined by $\tau(v) = Y, \tau(v^*) = X$ ad $\tau(\xi) = Z$. Then $\tau \in$ Orthogonal group of V and we compute $det(\tau) = 1$. Moreover the $\mathcal{O}$ module spanned by v, ξ and v^* is a maximal lattice relative to the form $q_V(\cong \langle 1\rangle \oplus \langle -1\rangle \oplus \langle 2s\rangle)$ with $2s$ a unit. Thus $\tau \in$ the maximal compact subgroup of $SO(V)(\cong$ the subgroup of $SO(V)$ stabilizing the $\mathcal{O}$ module above).

§8. Determination of γ factors (Spherical-Whittaker case)

We let $W_1 = R_i \oplus W_1(R_i) \oplus R_i^*$ be the orthogonal decomposition relative to q_{W_i}. Here R_i and R_i^* are isotropic subspaces nonsingularly paired by q_{W_i}.

We let $R_i = Sp\{v_1, \cdots, v_i\}$ and $R_i^* = Sp\{v_1^*, \cdots, v_i^*\}$ with $q_{W_1}(v_i, v_j^*) = \delta_{ij}$.

We let $GL(R_i)$ = the general linear group of the space R_i. Using the flag $\{v_1, \cdots, v_i\} \supseteq \{v_2, \cdots, v_i\} \supseteq \cdots \supseteq \{v_i\}$ we let $B_{GL(R_i)}$ the corresponding Borel group or stabilizer of the above flag. In particular let N^i be the unipotent radical of $B_{GL(R_i)}$.

We let $\mathbb{G}$ be $SO(W_1)$. We let $\mathbb{P}_i$ be the parabolic of $\mathbb{G}$ which stabilizes a i dimensional isotropic subspace R_i. Then $\mathbb{P}_i$ can be presented as

$$GL(R_i) \times SO(W_1(R_i)) \times \mathbb{U}_i.$$

In particular we let

$$H = SO(W_1(R_i)).$$

Moreover let ξ be a vector in $W_1(R_i)$ of nonzero length. Then we let

$$M = SO((\xi_0)^{\perp})$$

the orthogonal group of the space $(\xi_0)^{\perp}$ in $W_1(R_i)$.

We recall from §3 that $\mathbb{U}_i \cong \{N((\xi, s)) | \xi = (\xi_1, \cdots, \xi_i)$ with $\xi_\ell \in W_1(R_i)$ and $s \in Skew_i(k)\}$. Then the group structure on $\mathbb{U}_i$ is given in §3. Moreover we let the following character on ψ_{ξ_0} on $\mathbb{U}_i$ be defined as follows:

$$N((\xi_1, \cdots, \xi_i), s) \overset{\psi_{\xi_0}}{\rightsquigarrow} \psi(q_{W_1}(\xi_1, \xi_0)).$$

Then we know that N^i fixes this character pointwise and thus we can extend ψ_ξ to $N^i\mathbb{U}_i$ via the formula:

$$\Lambda_{\xi_0}(nu) = \psi_i(n)\psi_{\xi_0}(u)$$

where ψ_i is the unitary character on N^i which is trivial on each nonsimple root subgroup of N^i and nontrivial in each simple root subgroup.

Recall that the general orthogonal Whittaker model is given by

$$\mathrm{ind}_{M \times N^i \times \mathbb{U}_i}^{\mathbb{G}}(\sigma \otimes \Lambda_{\xi_0})$$

As before we must determine

$$Hom_{\mathbb{G}}(\Pi, \mathrm{ind}_{M \times N^i \times \mathbb{U}_i}^{\mathbb{G}}(\sigma \otimes \Lambda_{\xi_0}))$$

where $\Pi = \mathrm{ind}_{B_{\mathbb{G}}}^{\mathbb{G}}(\chi)$ and $\sigma = \mathrm{ind}_{B_M}^{M}(\gamma)$ where $B_{\mathbb{G}}$ and B_M are Borel subgroups of $\mathbb{G}$ and M.

Then for this we must determine the double cosets

$$B_{\mathbb{G}} \times B_M \backslash \mathbb{G} \times M / M^{\Delta} \cdot (N^i \mathbb{U}_i \times 1)$$

which support distributions $T : S(\mathbb{G} \times M) \to \mathbb{C}$ which have the following invariance property:

$$T((b_1, b_2) * \varphi * (m, m) \cdot (n \cdot u, 1)) = \chi \delta_{B_{\mathbb{G}}}^{-\frac{1}{2}}(b_1) \gamma \delta_{B_M}^{-\frac{1}{2}}(b_2) \Lambda_{\xi_0}(nu) T(\varphi)$$

for all $(b_1, b_2) \in B_{\mathbb{G}} \times B_M, m \in M$ and $n \cdot u \in N^i \mathbb{U}_i$ ($M^{\Delta} = \{(m, m) \subset \mathbb{G} \times M | m \in M\}$). By the uniqueness properties in §6 we deduce that T is unique for generic values of (χ, γ).

First we let $\varphi_1 \in S(\mathbb{G})$. We project φ_1 to $\mathrm{ind}_{B_{\mathbb{G}}}^{\mathbb{G}}(\chi)$ by the map

$$\varphi_1 \rightsquigarrow F_1(\chi, \varphi_1)(g) = \int_{B_{\mathbb{G}}} \varphi_1(b_{\mathbb{G}} g) \chi^{-1} \delta_{B_{\mathbb{G}}}^{\frac{1}{2}}(b_{\mathbb{G}}) d_\ell b_{\mathbb{G}}$$

where $d_\ell b_{\mathbb{G}}$ is a left invariant measure.

At this point we define a specific Borel subgroup $B_{\mathbb{G}}$ of $\mathbb{G}$. First in the space $W_1(R_i)$ we consider an orthogonal decomposition

$$W_1(R_i) = Sp\{Z_1, \cdots, Z_\nu\} \oplus W_{anis} \oplus Sp\{Z_1^*, \cdots, Z_\nu^*\}$$

where $q_{W_1}(Z_i, Z_j^*) = \delta_j$ and W_{anis} is anisotropic relative to q_{W_1}. Then we consider a flag $\mathfrak{A}$ in W_1 consisting of the sequence of subspaces $\{Z_1, \cdots, Z_\nu, v_1, \cdots, v_i\} \supseteq \{Z_2, \cdots, Z_\nu, v_1, \cdots, v_i\} \supseteq \cdots \supseteq \{v_1, \cdots, v_i\} \supseteq \cdots \supseteq \{v_i\}$. We let $B_{\mathbb{G}}$ be the Borel subgroup of $\mathbb{G}$ corresponding to the flag $\mathfrak{A}$; then $B_{\mathbb{G}} = B_{GL(R_i)} \cdot B_H \mathbb{U}_i$ where B_H is the Borel group in H of the flag $\{Z_1, \cdots, Z_\nu\} \supseteq \cdots \supseteq \{Z_\nu\}$.

We note here that the basis $\{Z_1, \cdots, Z_\nu\}$ has no *a priori* relation to the choice of basis used for $W_1(R_i)$ in the earlier part of this paper. We note that the maximal k-split torus $T_{GL(R_i)}$ and $T_H^{sp\ell}$ are given as follows. First $T_{GL(R_i)} = \{T(t_i, \cdots, t_1) | t_\ell \in k^\times\}$ where the action $T(t_i, \cdots, t_1)$ is defined by the formulae: $T(t_i, \cdots, t_1)(v_\ell) = t_{\ell-i+1} v_\ell$ and $T(t_i, \cdots, t_1)(v_\ell^*) = t_{\ell-i+1}^{-1}(v_\ell^*)(\ell = 1, \cdots, i)$. Secondly $T_H = \{T(s_\nu, \cdots, s_1) | s_i \in k^\times\}$ where the action $T(s_\nu, \cdots, s_1)$ is defined by the formulae: $T(s_\nu, \cdots, s_1)(Z_i) = s_{\nu-i+1} Z_i$ and $T(s_1, \cdots, s_\nu)(Z_i^*) = s_{\nu-i+1}^{-1} Z_i^*$

Next we note the simple root subgroups in $B_{\mathbb{G}}$ are given in the following way.

If α is a simple root in $GL(R_i$ of the form $\varepsilon_\ell - \varepsilon_{\ell+1}$ (for $\ell = 1, \cdots, i-1$) then we consider the one parameter group $\eta_\alpha(t)$ defined by: $\eta_\alpha(t)(v_{i-\ell}) = v_{i-\ell} + t v_{i-\ell+1}, \eta_\alpha(t)(v_t) = v_t$ if $t \neq i - \ell$. Then we have the relation:

$$T(t_i, \cdots, t_1) \eta_\alpha(s) T(t_i, \cdots, t_1) = \eta_\alpha \left(\frac{t_\ell}{t_{\ell+1}} s \right)$$

Then we note that the simple root $\alpha = \varepsilon_i - \varepsilon_{i+1}$ has a one parameter subgroup given by $N_i\left((tZ_\nu^*, 0, \cdots, 0), 0\right)$ Indeed we have that

$$T(s_\nu, \cdots, s_1) T(t_i, \cdots, t_1) N_i\left((tZ_\nu^*, 0, \cdots, 0), 0\right) \left[T(s_\nu, \cdots, s_1) T(t_i, \cdots, t_1)\right]^{-1}$$

$$= N\left((\frac{t_i}{s_1} tZ_\nu^*, 0, \cdots, 0), 0 \right)$$

(i.e. transforms according to the character $\frac{t_i}{s_1}$.)

The remaining simple roots α determine root subgroups in H (relative to T_H). We do not elaborate the one parameter root subgroups in this case which are similar to the $GL(R_i)$ case.

We emphasize here that when we take an unramified character $\chi = \chi_1 \otimes \cdots \otimes \chi_i \otimes \cdots \otimes \chi_{i+\nu}$ on $T_{GL(R_i)} x T_H^{sp\ell}$ we mean precisely

$$\chi\left(T(t_i, \cdots, t_1)T(s_\nu, \cdots, s_1)\right) = \prod \chi_i(t_i) \prod \chi_{i+\ell}(s_i).$$

Now we choose the basis $\{Z_1, \ldots, Z_\nu\}$ very specifically. Namely, we use the construction of B_H given in cases (a), (b), (c) and (d) in **Appendix 1** and **Appendix (3)** to §**5**. For instance, in case (a) (of **Appendix 1** of §**5**) the basis $Z_1 = z_1, Z_2 = z_\nu, \ldots, Z_\nu = Z_2$ (here $\xi_0 = z_1 + \gamma z_1^*$ for γ a unit). Thus specifically, the flag $\{Z_1, \ldots, Z_\nu\} \supseteq \cdots \supseteq \{Z_\nu\}$ coincides with the flag $\{z_1, z_\nu, \ldots, z_2\} \supseteq \cdots \supseteq \{z_2\}$. Again we emphasize this is how B_H is determined. Then B_{H_1} and B_M are defined in similar fashion from the data given.

We recall from §3 that W_i is a Weyl element in $\mathbb{G}$ which satisfies $w_i(R_i) = R_i^*$. In fact we choose w_i more precisely. First we choose w_0^G and w_0^H the long Weyl elements in $\mathbb{G}$ and H in the specific way given in **Appendix 3 of** §**5** (for each of the cases (a), (b), (c) and (d)). We note that in any case $w_i = w_0^{\mathbb{G}} w_0^H = t' w_0^H w_0^{\mathbb{G}}$ with $t' \in T_H \cap K_H$. Moreover, $w_i N^i w_i^{-1} = N_-^i =$ the opposed unipotent group in $GL(R_i)$ to N^i and $w_i \mathbb{U}_i w_i^{-1} = \mathbb{U}_i^- =$ the opposed unipotent radical associated to $\mathbb{P}_i^-$ (stabilizer of R_i^*).

For notational purposes, now we let $\xi = \xi_0$ for the rest of this section.

Then we can form the integral (for $g \in GL(R_i) \times H$)

$$F_1^*(\chi, \varphi_1, \xi)(g) = \int_{N^i \times \mathbb{U}_i} F_1(\chi, \varphi_1)(g w_i n u) \Lambda_\xi(nu) dn du$$

$$= \int_{B_{\mathbb{G}} \times N^i \times \mathbb{U}_i} \varphi(b_{\mathbb{G}} g w_i n u) \chi^{-1}(b_G) \delta_{B_{\mathbb{G}}}^{+\frac{1}{2}}(b_{\mathbb{G}}) \Lambda_\xi(nu) d_\ell(b_{\mathbb{G}}) dn du$$

This integral has a domain of absolute convergence given in the following way. Indeed $|F_1^*(\chi\varphi_1, \xi)|$ can be majorized by an integral of the form

$$\int_{N^i \times \mathbb{U}_i} F_1(|\chi|, |\varphi_1|)(g w_i n u) dn du =$$

$$\int_{B_{\mathbb{G}} \times N^i \times \mathbb{U}_i} |\varphi(b_{\mathbb{G}} g w_i n u)| |\chi|^{-1} \delta_{B_{\mathbb{G}}}^{+\frac{1}{2}}(b_{\mathbb{G}}) d_\ell(b_{\mathbb{G}}) du du$$

However this integral represents an intertwining integral on $\mathbb{G}$ relative to the roots in $N^i \times \mathbb{U}_i$. In fact we have absolute convergence for all χ so that

$$(*) \qquad Re(\chi, \check{\alpha}) > 0$$

(for all roots α in $N^i \times \mathbb{U}_i$). Thus $F_1^*(\chi, \varphi_1, \xi)$ is defined for χ in the above domain.

Remark 1. It is also possible to write $F_1^*(\chi, \varphi_1, \xi)$ as an integral of the form

$$\int_{B_{\mathbb{G}} \times N^i \times \mathbb{U}_i} \varphi(b_{\mathbb{G}} w_i n u g^{w_i}) \chi^{-1}(b_{\mathbb{G}}) \delta_{B_{\mathbb{G}}}^{\frac{1}{2}}(b_{\mathbb{G}}) \Lambda_\xi(nu) d_\ell(b_{\mathbb{G}}) du du$$

where $g^{w_i} = w_i g w_i^{-1}$ with $g \in M$.

We note that ξ_0 and ξ_1 are chosen as in **Appendix 1** and **3** to §**5** (cases (a), (b), (c) and (d)) (see **Appendix 3** to §**5**).

Moreover, $w_i(\xi_0) = \pm \xi_0$ and $w_i(\xi_1) = \pm \xi_1$.

We deduce that $w_i M w_i^{-1} = M (M = Stab_H(\xi))$. Moreover we know that $w_i B_H w_i^{-1} = B_H$ and $w_i B_M w_i^{-1} = B_M$. We note that if $\chi|_{T_H} = \sum_{i=1}^{i=\nu} c_i \chi_i$ then either $w_i(\chi)|_{T_H} = \chi|_{T_H}$ or $\sum_{i=1}^{\nu-2} c_i \chi_i + c_{\nu-1}\chi_{\nu-1} - c_\nu \chi_\nu$ (in fact this only happens when $W_{anis} = \{0\}$ with the pair $(i+\nu, \nu)$ both (even, odd) or (odd, even)). Thus we have that as sets $w_i(B_H w_0^H \eta_H B_M) w_i = B_H w_0^H \eta_H B_M$. In any case the element $w_0^H \eta_H$ determines an open orbit for the action of $B_H \times B_M$ in H through left and right action. Indeed then the open orbit of $B_H \times B_M \times \{M^\sim = \{(m^{w_i}, m) | m \in M\}\}$ in $H \times M$ (through the action $(b_1, b_2, (m^{w_i}, m)) : (x, y) \rightsquigarrow (b_1^{-1} \times m^{w_i}, b_2^{-1} y m))$ admits as a representative an element of the form $(w_0^H \eta_H, e)$.

Then we define

$$\begin{aligned}
&T_{\chi \otimes \gamma}^{\mathbb{G}}(\varphi_1 \otimes \varphi_2) \\
&= \int_M F_1^*(\chi, \varphi_1, \xi)(w_0^H \eta_H (m^{w_i}) F_2(\gamma, \varphi_2)(m) dm \\
&= \int_{B_{\mathbb{G}} \times B_M \times (N^i \times \mathbb{U}_i) \times M} \varphi_1(b_{\mathbb{G}} w_0^H \eta_H m^{w_i} w_i n u) \varphi_2(b_M m) \\
&\chi^{-1} \delta_{B_{\mathbb{G}}}^{\frac{1}{2}}(b_{\mathbb{G}})(\gamma^{-1} \delta_{B_M}^{\frac{1}{2}})(b_M) \Lambda_\xi(nu) d_\ell(b_{\mathbb{G}}) d_\ell(b_M) dn du dm \\
&= \int_{B_{\mathbb{G}} \times B_M \times (N^i \times \mathbb{U}_i) \times M} \varphi_1(b_{\mathbb{G}} w_0^H \eta_H w_i n u m) \varphi_2(b_M m) \\
&\chi^{-1} \delta_{B_{\mathbb{G}}}^{\frac{1}{2}}(b_{\mathbb{G}})(\gamma^{-1} \delta_{B_M}^{\frac{1}{2}})(b_M) \Lambda_\xi(nu) d_\ell(b_{\mathbb{G}}) d_\ell(b_M) dn du dm
\end{aligned}$$

Then we deduce that at least formally the distribution

$$\varphi_1 \otimes \varphi_2 \rightsquigarrow T_{\chi \otimes \gamma}^{\mathbb{G}}(\varphi_1 \otimes \varphi_2)$$

satisfies the invariance property of T cited above.

We note that the open cell in $\mathbb{G}$ relative to the parabolic $\mathbb{P}_i$ is given by

$$\mathbb{P}_i w_i \mathbb{U}_i$$

In the group $\mathbb{P}_i$ we consider the open dense set given by

$$(B_{GL(R_i)} \cdot N_-^i) \cdot H \cdot \mathbb{U}_i$$

Then the set

$$(B_{GL(R_i)} \cdot N_-^i) \cdot H \cdot \mathbb{U}_i w_i \mathbb{U}_i$$

is open dense in $\mathbb{G}$. On the other hand we consider the set in H given by

$$B_H w_0^H \eta_H M.$$

Then we have the equality of sets:

$$\begin{aligned}
&(B_{GL(R_i)} \cdot N_-^i)(B_H w_0^H \eta_H M) \cdot \mathbb{U}_i w_i \mathbb{U}_i = \\
&(B_{GL(R_i)} \cdot B_H \cdot \mathbb{U}_i)(w_0^H \eta_H) \cdot M \cdot N_-^i w_i \mathbb{U}_i = \\
&B_{\mathbb{G}}(w_0^H \eta_H) M \cdot w_i N^i \cdot \mathbb{U}_i .
\end{aligned}$$

Then the decomposition of an element $(g, m) \in \mathbb{G} \times M$ according to $B_{\mathbb{G}} w_0^H \eta_H M w_i \cdot \mathbb{U}_i N^i$ and $B_M \cdot M$ is unique. Here we mean that if $g = b_G w_0^H \eta_H (m')^{w_i} w_i n u$ and $m = b_M m'$ then the components $b_{\mathbb{G}} \in B_{\mathbb{G}}, b_M \in B_M, m' \in M, n \in N^i, u \in \mathbb{U}_i$ are uniquely determined! (See comments in §7, i.e. **Remark 1**).

Again here the measures are normalized so that $d_\ell(b_{\mathbb{G}})$, $d_\ell(b_M)$, dn, du_i have mass 1 on the unit groups $K_{\mathbb{G}} \cap B_{\mathbb{G}}$, $K_M \cap B_M$, $K_{\mathbb{G}} \cap N^i$ and $K_{\mathbb{G}} \cap N^i$ and $K_{\mathbb{G}} \cap \mathbb{U}_i$.

At this point we want to contrast $K^\xi_{\chi \otimes \gamma}$ to $K_{\chi|_{T_H} \otimes \gamma}$ given in §7.

For this we define $K^{\text{twist}}_{\chi|_T \otimes \gamma}$ as follows:

$$K^{\text{twist}}_{\chi|_{T_H} \otimes \gamma}(h, m') = \begin{cases} \chi^{-1} \delta^{\frac{1}{2}}_{B_H}(b_H) \gamma^{-\frac{1}{2}} \delta^{\frac{1}{2}}_{B_M}(b_M) & \text{if } h = b_H w_0^H \eta_H m^{w_i} \\ & \quad m' = b_M m \\ 0 & \text{otherwise} \end{cases}$$

Then we note first that $\delta^{\frac{1}{2}}_{B_{\mathbb{G}}} \mid_{T_H} = \delta^{\frac{1}{2}}_{B_H} \mid_{T_H}$. Then the precise relation between K^ξ and K^{twist} is given as:

$$K^\xi_{\chi \otimes \gamma}(h w_i, m) = K^{\text{twist}}_{\chi|_T \otimes \gamma}(h, m) .$$

We note that $T^{\text{twist}}_{\chi \otimes \gamma}$ can be defined in a similar manner as $T_{\chi \otimes \gamma}$. Indeed

$$\begin{aligned}
&T^{\text{twist}}_{\chi \otimes \gamma}(\varphi_1 \otimes \varphi_2) \\
&= \int_{B_H \times B_M \times M} \varphi_1(b_H w_0^H \eta_H m^{w_i}) \varphi_2(b_M m) \chi^{-1} \delta^{\frac{1}{2}}_{B_H}(b_H) \gamma^{-1} \delta^{\frac{1}{2}}_{B_M}(b_M) d_\ell(b_H) d_\ell(b_M)\, dm .
\end{aligned}$$

In any case, it is possible as in §7 to find a Haar measure $dh' \otimes dm'$ on $H \times M$ so that

$$T^{\text{twist}}_{\chi\otimes\gamma}(\varphi_1 \otimes \varphi_2) = \int_{H\times M} \varphi_1(h')\varphi_2(m')K^{\text{twist}}_{\chi|_{T_H}\otimes\gamma}(h', m')dh'dm' .$$

At this point we choose a constant c_H^{twist} so that

$$dh' = c_H^{\text{twist}} db_H \otimes du_H$$

relative to the Bruhat decomposition $dh' \simeq db_G \otimes du_H (H' \sim B_H w_0^H U_H)$.

Now we also normalize the $db_{\mathbb{G}} = du_i \otimes db_{GL_i} \otimes db_H$. Then we can write

$$\begin{aligned} &db_{\mathbb{G}} \otimes db_M \otimes dm \otimes dn \otimes du \\ &\quad = du_i \otimes db_{GL_i} \otimes (db_H \otimes db_M \otimes dm) \otimes dn \otimes du \\ &\quad = (du_i \otimes db_{GL_i} \otimes dh' \otimes dn \otimes du_i) \otimes dm' \\ &\quad = c_H^{\text{twist}} du_i \otimes db_{GL_i} \otimes db_H \otimes du_H \otimes dn \otimes du \otimes dm' . \end{aligned}$$

On the other hand we note that

$$\begin{aligned} &\int\limits_{\mathbb{U}_i \times B_{GL_i} \times B_H \times U_H \times N^i \times \mathbb{U}_i} F(u_i b_{GL_i} b_H w_0^H u_H w_i n u) du_i db_{GL_i} db_H du_H dndu \\ &= \int F(u_i b_{GL_i} b_H t' w_0^{\mathbb{G}} (w_i^{-1} u_H w_i) nu) du_i db_{GL_i} db_h du_H dndu_i . \end{aligned}$$

With a change of variables $b_{GL_i} b_H \to b_{GL_i} b_H t'$ $(t' \in T_H \cap K_H)$ and $u_H \to w_i^{-1} u_H w_i$ we deduce the last integral then equals

$$\begin{aligned} &\int\limits_{\mathbb{U}_i \times B_{GL_i} \times B_H \times U_H \times N^i \times \mathbb{U}_i} F(u_i b_{GL_i} b_H w_0^{\mathbb{G}} u_H nu) du_i db_{GL_i} db_H du_H dndu \\ &= \int\limits_{B_{\mathbb{G}} \times U_H \times N^i \times \mathbb{U}_i} F(b_{\mathbb{G}} w_0^{\mathbb{G}} u_H nu) db_{\mathbb{G}} du_H dndu . \end{aligned}$$

We then choose a decomposition of measure

$$\begin{aligned} dg \otimes dm' &= du_i \otimes db_{GL_i} \otimes dh' \otimes dn \otimes du \otimes dm' \\ &= c_H^{\text{twist}} db_{\mathbb{G}} \otimes du_H \otimes dn \otimes du_i . \end{aligned}$$

This implies that the distribution

$$T^{\mathbb{G}}_{\chi\otimes\gamma}(\varphi_1 \otimes \varphi_2) = \int_{\mathbb{G}\times M} \varphi_1(g)\varphi_2(m')K^{\xi}_{\chi\otimes\gamma}(g, m)dgdm'$$

where $K^{\xi}_{\chi\otimes\gamma}(\,\cdot\,,\,\cdot\,)$ is the kernel function given by the data:

$$K^{\xi}_{\chi\otimes\gamma}(g, m) = \begin{cases} \chi^{-1}\delta_{B_{\mathbb{G}}}^{+\frac{1}{2}}(b_{\mathbb{G}})\gamma^{-1}\delta_{B_M}^{+\frac{1}{2}}(b_M)\Lambda_\xi(nu) & \text{if } g = b_{\mathbb{G}} w_0^H \eta_H(m')^{w_i} w_i nu, \\ & m = b_M m' \\ 0 & \text{otherwise} \end{cases}$$

Then we summarize the basic properties of $T^{\mathbb{G}}_{\chi\otimes\gamma}$.

Lemma 8.1. *Let χ satisfy the condition $(*)$. Moreover let $w_i(\chi)|_{T_H}$ and $\gamma|_{T_M}$ satisfy the conditions of* **Lemma 7.2** *of §7 relative to (T_H, T_M) in (H, M) (i.e., $(w_i(\chi)\ |_{T_H}, \gamma\ |_{T_M})$ lies in the domain Ω given in* **Lemma 7.2***). Then the integral $T^{\mathbb{G}}_{\chi\otimes\gamma}$ is absolutely convergent. For fixed $\varphi_1\otimes\varphi_2 \in S(\mathbb{G}\times M)$ the function $(\chi,\gamma) \longrightarrow T^{\mathbb{G}}_{\chi\otimes\gamma}(\varphi\otimes\varphi_2)$ is a rational function in $q^{\pm\chi_i}$ and $q^{\pm\gamma_j}$.*

Proof. We first note the inequality:

$$\int\limits_M |F_1^*(\chi,\varphi_1,\xi)(\tilde{w}_0^H\tilde{\eta}_H m^{w_i})||F_2(\gamma,\varphi_2)(m)|dm \leq$$

$$\int\limits_{B_M\times M}\left(\int\limits_{B_{\mathbb{G}}N^i\mathbb{U}_i} |\varphi_1(b_{\mathbb{G}}\tilde{w}_0^H\tilde{\eta}_H m^{w_i}nu)||\chi|^{-1}\delta^{\frac{1}{2}}_{B_{\mathbb{G}}}(b_{\mathbb{G}})d_\ell(b_{\mathbb{G}})dndu\right)$$

$$|\varphi_2(b_M m)||\gamma|^{-1}\delta^{\frac{1}{2}}_{B_M}(b_M)d_\ell(b_M)dm$$

Indeed then $h \rightsquigarrow \int\limits_{B_{\mathbb{G}}N^i\mathbb{U}_i} |\varphi_1(b_{\mathbb{G}}h^{w_i}nu)||\chi|^{-1}\delta^{\frac{1}{2}}_{B_{\mathbb{G}}}(b_{\mathbb{G}}d_\ell(b_{\mathbb{G}})dndu$ lies in the space $\mathrm{ind}^H_{B_H}(|w_i(\chi)|_{T_H}|)$ (when χ satisfies $(*)$). On the other hand if $w_i(\chi)|_{T_H}$ and $\gamma|_{T_M}$ satisfy the hypotheses of **Lemma 7.2** we deduce that

$$\int\limits_M F_2(|\varphi_2|,|\gamma|)(m)\left(\int\limits_{B_{\mathbb{G}}\times N^i\mathbb{U}_i} |\varphi_1(b_{\mathbb{G}}\tilde{w}_0^H\tilde{\eta}_H m^{w_i}nu)|\right.$$

$$\left.|\chi|^{-1}\delta^{\frac{1}{2}}_{B_{\mathbb{G}}}(b_{\mathbb{G}})d_\ell(b_{\mathbb{G}})dndu\right)dm$$

is finite.

Thus we have proven the convergence of $T_{\chi\otimes\gamma}$. Indeed we can choose $\chi_1,\dots,\chi_i$ so that (i) $Re(\chi_{\ell_1}-\chi_{\ell_2}) \gg 0$ $(1 \leq \ell_1 < \ell_2 \leq i)$ (ii) $Re(\chi_\ell)\pm Re(\chi_t) \gg 0$ and $Re(\chi_\ell) \gg 0$ $(1 \leq \ell \leq i,\ i+1 \leq t \leq m)$ and (iii) the set $(w_i(\chi_{i+1},\dots,\chi_m),\gamma_1,\dots,\gamma_t)$lies in an open region Ω (with compact closure) given in **Lemma 7.2**.

We note then that since

$$Hom_{\mathbb{G}}(\Pi, Ind^{\mathbb{G}}_{M\times N^i\times\mathbb{U}_i}(\sigma\otimes\Lambda_\xi))$$

has dimension at most 1 (for generic $\Pi = \mathrm{ind}^{\mathbb{G}}_{B_{\mathbb{G}}}(\chi)$ and $\sigma = \mathrm{ind}^M_{B_M}(\gamma)$. Thus it follows that the basic integral defining $T^{\mathbb{G}}_{\chi\otimes\gamma}(\varphi_1\otimes\varphi_2)$ admits

meromorphic continuation (see [G-PS-]). Hence $T^{\mathbb{G}}_{\chi\otimes\gamma}(\varphi_1\otimes\varphi_2)$ is a rational function in $q^{\pm\chi_i}$ and $q^{\pm\gamma_j}$./ **Q.E.D.**

As above we have the same Corollary about the structure of $T^{\mathbb{G}}_{\chi\otimes\gamma}$.

Corollary to Lemma 8.1. *The distribution*

$$\begin{aligned}
&T^{\mathbb{G}}_{\chi\otimes\gamma}(\varphi_1\otimes\varphi_2)\\
&= \int_{\mathbb{U}_i\times B_{GL_i}\times H\times N^i\times\mathbb{U}_i\times M} \varphi_1(u_i b_{GL_i} h' w_i n u)\varphi_2(m')\\
&\quad K^{\xi}_{\chi\otimes\gamma}(b_{GL_i} u_i h' w_i n u, m)\, db_{GL_i}\, du_i\, dh'\, dm'\, dn^i\, du_i\\
&= c_H^{\mathrm{twist}} \int_{\mathbb{U}_i\times B_{GL_i}\times B_H\times U_H\times N\times\mathbb{U}_i\times M} \varphi_1(u_i b_{GL_i} b_H w_0^{\mathbb{G}} u_H n u)\varphi_2(m')\\
&\quad K^{\xi}_{\chi\otimes\gamma}(u_i b_{GL_i} b_H w_0^{\mathbb{G}} u_H n u, m')\, du_i\, db_{GL_i}\, db_H\, du_H\, dn\, du\, dm'\\
&= c_H^{\mathrm{twist}} \int_{\mathbb{U}_i\times B_{GL_i}\times B_H\times U_H\times N^i\times\mathbb{U}_i\times M} \varphi_1(u_i b_{GL_i} b_H w_0^{\mathbb{G}} u_H n u)\varphi_2(m')\\
&\quad K^{\mathrm{twist}}_{\chi|_{T_H}\otimes\gamma}(w_0^H(w_i u_H w_i^{-1}), m')\chi^{-1}\delta^{\frac{1}{2}}_{B_{GL_i}\times B_H}(b_{GL_i} b_H)\\
&\quad \Lambda_\xi(nu)\, du_i\, db_{GL_i}\, db_H\, du_H\, dn\, du\, dm'\,.
\end{aligned}$$

Proof. The first equality follows from the definition of $T^{\mathbb{G}}_{\chi\otimes\gamma}$ and the choice $dh'\otimes dm' = db_H\otimes db_M\otimes dm$. The second equality follows from the substitution for $h' = b_H w_0^H u_H$ and the use of $h'w_i = b_H w_0^H w_i(w_i^{-1}u_H w_i) = b_H t'(w_0^H)^2 w_0^{\mathbb{G}}(w_i^{-1}u_H w_i)$. Then we know that $(w_0^H)^2 = 1$ (by its specific choice in **Appendix 3 of** §**5**). Also $t'\in T_H\cap K_H$. Also we can make a chance of variable $u_H\to w_i^{-1}u_H w_i$ (where the module has absolute value $=1$). Finally the third equality is obtained from the second by replacing $w_0^{\mathbb{G}} = t_1 w_0^H w_i u_H$ with $t_1\in T_H\cap K_H$. Then $w_0^H w_i u_H = w_0^H w_i u_H w_i^{-1} w_i$. We also then use the equality

$$K^{\xi}_{\chi\otimes\gamma}(w_0^H w_i u_H w_i^{-1} w_i, m') = K^{\mathrm{twist}}_{\chi\otimes\gamma}(w_0^H w_i u_H w_i^{-1}, m')\,.$$

This completes the proof. □

We let $I_{\mathbb{G}}, I_H, I_M$ and $I_{H'}$ be the Iwahori subgroups of $\mathbb{G}, H, M$ and H' respectively. We recall that the group K_X = the maximal compact subgroup of X is a disjoint union of $I_X w I_X$ where w ranges over the representatives of the Weyl group W_X. Moreover we know that $I_X w_0^X I_X = B_X^0 w_0^X U_X^0$ where w_0^X = the longest element in W_X and B_X^0 and U_X^0 represent $K_X\cap B_X$ and $K_X\cap U_X$.

The above comments then allows us to compare $T^{\mathbb{G}}_{\chi\otimes\gamma}(\varphi_1^{\mathbb{G}}\otimes\varphi_2)$ and $T_{\chi\otimes\gamma}(\varphi_1^H\otimes\varphi_2)$ when $\varphi_1^{\mathbb{G}}$ = characteristic function of $I_{\mathbb{G}} w_{\mathbb{G}}^0 I_{\mathbb{G}}$ (φ_1^H = characteristic function of $I_H w_H^0 I_H$).

Proposition 8.1. *Let $\varphi_1^{\mathbb{G}}$ and φ_1^H as above. Then let φ_2 be any function in $S(M)$ so that $\varphi_2(w_i x w_i^{-1}) = \varphi_2(x)$ for all $x\in M$. We have the identity:*

$$T^{\mathbb{G}}_{\chi\otimes\gamma}(\varphi_1^{\mathbb{G}}\otimes\varphi_2) = T^H_{\chi|_{T_H}}(\varphi_1^H\otimes\varphi_2)\,.$$

Proof. We first note that $K_{\mathbb{G}} \cap \mathbb{U}_i B_{GL_i} B_H = (K_{\mathbb{G}} \cap \mathbb{U}_i)(K_{\mathbb{G}} \cap B_{GL_i})(K_{\mathbb{G}} \cap B_H)$ and $K_{\mathbb{G}} \cap U_H N^i \mathbb{U}_i = (K_{\mathbb{G}} \cap U_H)(K_{\mathbb{G}} \cap N^i)(K_{\mathbb{G}} \cap \mathbb{U}_i)$. Then we use **Corollary 1** to **Lemma 8.1** applied to $\varphi_1^{\mathbb{G}} \otimes \varphi_2$. We obtain that

$$\begin{aligned} &T^{\mathbb{G}}_{\chi\otimes\gamma}(\varphi_1^{\mathbb{G}} \otimes \varphi_2) \\ &= c_H^{\text{twist}} \int\limits_{U_H^0 \times M'} \varphi_2(m') K^{\text{twist}}_{\chi|T_H \otimes \gamma}(w_0^H w_i u_H w_i^{-1}, m') du_H^0 dm' . \end{aligned}$$

Now we make change of variable on $U_H^0 = K_{\mathbb{G}} \cap U_H = K_H \cap U_H$, $u_H \to w_i u_H w_i^{-1}$. The set U_H^0 is stable under w_i, i.e., $w_i U_H^0 w_i^{-1} = w_i(K_G \cap U_H) w_i^{-1} = w_i K_G w_i^{-1} \cap w_i U_H w_i^{-1} = K_{\mathbb{G}} \cap U_H = U_H^0$. The module of the map is one. Thus we have that

$$\begin{aligned} &T^{\mathbb{G}}_{\chi\otimes\gamma}(\varphi_1^{\mathbb{G}} \otimes \varphi_2) \\ &= c_H^{\text{twist}} \int\limits_{U_H^0 \times M'} \varphi_2(m') K^{\text{twist}}_{\chi|T_H \otimes \gamma}(w_0^H u_H, m') du_H^0 \otimes dm' . \end{aligned}$$

On the other hand, we now observe that $T_{\chi\otimes\gamma}(\varphi_1 \otimes \varphi_2) = T^{\text{twist}}_{\chi\otimes\gamma}(\varphi_1 \otimes \varphi_2)$ where φ_2 satisfies the hypothesis of the Proposition and φ_1 is arbitrary in $S(H)$. Indeed,

$$\begin{aligned} T_{\chi\otimes\gamma}(\varphi_1 \otimes \varphi_2) = & \int\limits_{B_H \times B_M \times M} \varphi_1(b_H w_0^H \eta_H m) \varphi_2(b_M m) \\ & \chi^{-1} \delta_{B_H}^{\frac{1}{2}}(b_H) \gamma^{-1} \delta_{B_M}^{\frac{1}{2}}(b_M) db_H db_M dm \\ = & \int\limits_{B_H \times B_M \times M} \varphi_1(b_H w_0^H \eta_H m) \varphi_2(w_i^{-1} b_M w_i w_i^{-1} m w_i) \\ & \chi^{-1} \delta_B^{\frac{1}{2}}(b_H) \gamma^{-1} \delta_{B_M}^{\frac{1}{2}}(b_M) db_H db_M dm . \end{aligned}$$

By change of variables $m \to w_i m w_i^{-1} = m^{w_i}$, $b_M \to w_i^{-1} b_M w_i$ we have that the last integral equals

$$\int\limits_{B_H \times B_M \times M} \varphi(b_H w_0^H \eta_H m^{w_i}) \varphi_2(b_M m) \chi^{-1} \delta_{B_H}^{\frac{1}{2}}(b_H) \gamma^{-1} \delta_{B_M}^{\frac{1}{2}}(w_i b_M w_i^{-1}) db_H db_M dm .$$

But $\gamma^{-1} \delta_{B_M}^{\frac{1}{2}}(w_i b_M w_i^{-1}) = \gamma^{-1} \delta_{B_M}^{\frac{1}{2}}(b_M)$ and hence the last term equals

$$T^{\text{twist}}_{\chi\otimes\gamma}(\varphi_1 \otimes \varphi_2) .$$

However, we can write the above as

$$c_H^{\text{twist}} \int\limits_{B_H \times U_H \times M'} \varphi(b_H w_0^H u_H) \varphi_2(m') K^{\text{twist}}_{\chi\otimes\gamma}(b_H w_0^H u_H, m') db_H du_H dm' .$$

Now we choose $\varphi = \varphi_1^H$ and in the above integral we have

$$c_H^{\text{twist}} \int_{U_H^0 \times M'} \varphi_2(m') K_{\chi\otimes\gamma}^{\text{twist}}(w_0^H u_H, m') du_H^0 dm' .$$

Thus we have the equality stated in **Proposition 8.1.** $\square$

We now consider a simple root α of X. The corresponding root subgroups are $U_{\pm\alpha}$ of $U_X^{\pm}$. Moreover there is the factorization of $I_X = (I_X \cap U_X^-)B_X^0$ as groups. In particular the set $I_X w_\alpha w_0^X I_X$ admits a factorization as

$$B_X^0 \cdot (I_X \cap U_{-\alpha}) w_\alpha w_0^X U_X^0$$

We are going to compute the functional $T_{\chi\otimes\gamma}^{\mathbb{G}}$ applied to the specific function of the form $\varphi_\alpha^{\mathbb{G}} \otimes \varphi_2$ where $\varphi_\alpha^{\mathbb{G}} =$ the characteristic function of the set $I_X w_\alpha w_0^X I_X$.

Proposition 8.2. *Let $\varphi \in S(M)$ be any function which satisfies $\varphi(w_i x w_i^{-1}) = \varphi(x)$ for $x \in M$. (i) Let α be a simple positive root that belongs to $GL(R_i)$. Then*

$$T_{\chi\otimes\gamma}^{\mathbb{G}}(\varphi_\alpha^{\mathbb{G}} \otimes \varphi_2) = -q^{-1-\langle\chi,\check{\alpha}\rangle} T_{\chi\otimes\gamma}^{\mathbb{G}}(\varphi_1^{\mathbb{G}} \otimes \varphi_2)$$

(ii) Let α be a simple positive root that belongs to H. Then

$$T_{\chi\otimes\gamma}^{\mathbb{G}}(\varphi_\alpha^{\mathbb{G}} \otimes \varphi_2) = T_{(\chi|_{T_H})\otimes\gamma}^{H}(\varphi_\alpha^H \otimes \varphi_2)$$

Proof. The idea of the proof follows the methods given in [C-S]. Indeed we use the formula of $T_{\chi\otimes\gamma}^{\mathbb{G}}$ expressed as an integral over $\mathbb{G} \times M$ against the kernel function $K_{\chi\otimes\gamma}^{\xi}$. We in fact integrate $\varphi_\alpha^{\mathbb{G}} \otimes \varphi_2$ over such a space. This integral in the $\mathbb{G}$ variable has the domain $B_{\mathbb{G}}^0 \times (I_{\mathbb{G}} \cap U_{-\alpha}) \times U_{\mathbb{G}}^0$ relative to the product Haar measure $db_G^0 \otimes dt \otimes du$. That is, $T_{\chi\otimes\gamma}^{\mathbb{G}}(\varphi_\alpha^{\mathbb{G}} \otimes \varphi_2) =$

$$c_H^{\text{twist}} \int_{B_{\mathbb{G}}^0 \times (I_{\mathbb{G}} \cap U_{-\alpha}) \times U_G^0 \times M} \varphi_\alpha^{\mathbb{G}}(b u_{-\alpha}(t) w_\alpha w_0^{\mathbb{G}} u) \varphi_2(m')$$
$$K_{\chi\otimes\gamma}^{\xi}(b u_{-\alpha}(t) w_\alpha w_0^{\mathbb{G}} u, m) d_\ell b_{\mathbb{G}}^0 \otimes dt \otimes du) \otimes dm'$$

We note from [C-S] that the set $I_{\mathbb{G}} \cap U_{-\alpha}$ is given as follows: $U_{-\alpha}(t) \cong k$ or K (where K is the unique unramified extension of k) and $I_X \cap U_{-\alpha}(t) = \pi\mathcal{O}_k$ or $\pi\mathcal{O}_K$ ($\mathcal{O}_k$ and $\mathcal{O}_K$ the ring of integers in k and K respectively). Moreover we use the specific Bruhat decomposition of an element $u_{-\alpha}(t) \in I_{\mathbb{G}} \cap U_{-\alpha}$ as given in [C-S]:

$$u_{-\alpha}(t) = n_1(t) \cdot a_\alpha^{-m} \cdot m_0 w_\alpha^{-1} n_2(t) (|t| = q^{-m})$$

where $n_1(t) \in \{U_\alpha(t) \| t| \leq q^{+m}\}, n_2(t) \in \{U_\alpha(t) \| t| = q^{+m}\}, m_0 \in T_H \cap K_H$ and a_α^{-m} a certain prescribed element in T_H.

The set $(I_{\mathbb{G}} \cap U_{-\alpha})_m = \{u_\alpha(t) \mid |t| = q^m\}$ and in fact we know from [CS] that the map $u_\alpha(t) \to n_2(t)$ from $\{u_{-\alpha}(t) \mid |t| = q^{-m}\}$ to $I_{\mathbb{G}} \cap U_{-\alpha}$ is a bijection transforming normalized Haar measure to normalized Haar measure.

On the other hand

$$w_\alpha^{-1} n_2(t) w_\alpha w_0^{\mathbb{G}} = w_0^{\mathbb{G}} (w_\alpha w_0^{\mathbb{G}})^{-1} n_2(t) (w_\alpha w_0^{\mathbb{G}})$$

Moreover we note that

$$n_2^*(t) = (w_\alpha w_0^{\mathbb{G}})^{-1} n_2(t) (w_\alpha w_0^{\mathbb{G}}) \in U_{\overline{\alpha}}(t)$$

Here $\overline{\alpha}$ is a positive root in $\mathbb{G}$. In fact we know that $\overline{\alpha} = \alpha$ except in the case where $\mathbb{G} \cong SO(n,n)$ with n odd. In the case where n is odd this

$$\begin{cases} \overline{\alpha} = \alpha & \text{if } \alpha \text{ is a simple root of } G \text{ not of the form } \epsilon_{n-1} \pm \epsilon_n \\ \overline{\epsilon_{n-1} - \epsilon_n} = \epsilon_{n-1} + \epsilon_n & \\ \overline{\epsilon_{n-1} + \epsilon_n} = \epsilon_{n-1} - \epsilon_n & \end{cases}$$

Thus we have that $n_2^*(t) \in U_\alpha(t)$ except when $H = SO(m,m)$ with m odd and $\alpha = \epsilon_{m-1} - \epsilon_m$ or $\epsilon_{m-1} + \epsilon_m$. Moreover in the case that the root α lies in H we have that

$$w_i (w_\alpha w_0^{\mathbb{G}})^{-1} n_2(t) (w_\alpha w_0^{\mathbb{G}}) w_i^{-1} = (w_\alpha w_0^H)^{-1} n_2(t') (w_\alpha w_0^H) = (\tilde{n}_2)^*(t')$$

($t' = \lambda t$ with λ a unit).

Thus we deduce that

$$K^\xi_{\chi\otimes\gamma}(b u_{-\alpha}(t) w_\alpha w_0^{\mathbb{G}} u, m) = (\chi^{-1} \delta_{B_{\mathbb{G}}}^{\frac{1}{2}})(b a_\alpha^{-m}) K^\xi_{\chi\otimes\gamma}(w_0^{\mathbb{G}} n_2^*(t') u, m)$$

If the root α lies in $GL(R_i)$ then following the same construction as in [C-S] (where $u = u_{GL_i} \cdot u_H \cdot \tilde{u}$ with $u_{GL_i} \in N_i^0$, $u_H \in U_H^0$ and $u_H \in U_H^0$ and $\tilde{u} \in \mathbb{U}_i^0$)

$$\begin{aligned} K^\xi_{\chi\otimes\gamma}(w_0^{\mathbb{G}} n_2^*(t') u_{GL_i} u_H \tilde{u}, m) &= \Lambda_\xi(n_2^*(t') u_{GL_i} \cdot \tilde{u}) \\ K^\xi_{\chi\otimes\gamma}(w_0^{\mathbb{G}} u_H, m) &= \Lambda_\xi(n_2^*(t)) K^\xi_{\chi\otimes\gamma}(w_0^H w_i u_H w_i^{-1} w_i, m) \\ &= \Lambda_\xi(n_2^*(t)) K^{\text{twist}}_{\chi\otimes\gamma}(w_0^H w_i u_H w_i^{-1}, m)\,. \end{aligned}$$

Thus in the case α lies in $GL(R_i)$

$$\begin{aligned} &T^{\mathbb{G}}_{\chi\otimes\gamma}(\varphi^{\mathbb{G}}_\alpha \otimes \varphi_2) = \\ &c_H^{\text{twist}} \Bigg(\sum_{m\geq 1} \int_{B^0_{\mathbb{G}} \times (I_{\mathbb{G}} \cap U_{-\alpha})_m} (\chi^{-1} \delta_{B_{\mathbb{G}}}^{\frac{1}{2}})(b a_\alpha^{-m}) \Lambda_\xi(n_2^*(t')) \\ &\qquad db^0_{\mathbb{G}} \otimes dt \otimes dn_i^0 \otimes du_i^0 \Bigg) \Bigg(\int_{M \times U_H^0} \varphi_2(m) K^{\text{twist}}_{\chi\otimes\gamma}(w_0^H w_i u_H w_i^{-1}, m) dm \otimes du_H^0 \Bigg) \end{aligned}$$

Then from [CS] the first term in the above calculation equals $-q^{-1-\langle\chi,\check{\alpha}\rangle}$. We note here that by use of [CS], the above calculation does not depend on the choice of ψ (up to multiplication by a unit).

Next we note that the term (from the proof of **Corollary to Lemma 8.1** and **Proposition 8.1**).

$$c_H^{\text{twist}} \int\limits_{M\times U_H^0} \varphi_2(m) K_{\chi\otimes\gamma}^{\text{twist}}(w_0^H u_H, m) dm \otimes du_H^0 = T_{\chi\otimes\gamma}^{\mathbb{G}}(\varphi_1^{\mathbb{G}} \otimes \varphi_2)$$

Thus (i) of **Proposition 8.2** is valid.

Next we let α be a root belonging to H. The above considerations imply that

$$\begin{aligned} &K_{\chi\otimes\gamma}^{\xi}(bu_{-\alpha}(t)w_\alpha w_0^{\mathbb{G}} u, m) = \chi^{-1}\delta_{B_{\mathbb{G}}}^{\frac{1}{2}}(ba_\alpha^{-m}) \\ &K_{\chi\otimes\gamma}^{\xi}(w_0^{\mathbb{G}} n_2^*(t)u_H, m) = \chi^{-1}\delta_{B_{\mathbb{G}}}^{\frac{1}{2}}(ba_\alpha^{-m}) \\ &K_{\chi\otimes\gamma}^{\xi}(w_0^H w_i((\tilde{n}_2)^*(t')u_H)w_i^{-1}w_i, m) \end{aligned}$$

Thus in the case α lies in H

$$\begin{aligned} T_{\chi\otimes\gamma}^{\mathbb{G}}(\varphi_\alpha^{\mathbb{G}} \otimes \varphi_2) = c_H^{\text{twist}} \Bigg(\sum_{m\geq 1} \int\limits_{B_{\mathbb{G}}^0 \times (I_{\mathbb{G}}\cap U_{-\alpha})_m \times (N^i\times \mathbb{U}_i)^0 \times U_H^0 \times M} & (\chi^{-1}\delta_{B_{\mathbb{G}}}^{\frac{1}{2}})(a_\alpha^{-m})\varphi_2(m) \\ & \cdot K_{\chi\otimes\gamma}^{\xi}(w_0^H w_i(\tilde{n}_2^*(t)' u_H)w_i^{-1}w_i, m) \\ & \cdot db_{\mathbb{G}}^0 \otimes dt \otimes dn_i^0 \otimes du_i^0 \otimes du_H^0 \otimes dm \\ \equiv c_H^{\text{twist}} \sum_{m\geq 1} \int\limits_{(I_{\mathbb{G}}\cap U_{-\alpha})_m \times U_H^0 \times M} & \chi^{-1}\delta_{B_{\mathbb{G}}}^{\frac{1}{2}}(a_\alpha^{-m})\varphi_2(m')K_{\chi\otimes\gamma}^{\xi} \\ & \cdot (w_0^H(w_i\tilde{n}_2^*(t)'w_i^{-1})w_i u_H^0 w_i^{-1}w_i, m'\Bigg) dt du_H^0 dm' \end{aligned}$$

Now since $t' = \lambda t$ (λ a unit), we make the change of variables $t \to t' = \lambda t$ on each fiber $(I_{\mathbb{G}} \cap U_{-\alpha})_m$ where the module equals 1. We also make the changes of variables $u_H \to w_i u_H w_i^{-1}$ on U_H^0 again. In any case, we obtain

$$\begin{aligned} c_H^{\text{twist}} \sum_{m\geq 1} \int\limits_{(I_{\mathbb{G}}\cap U_{-\alpha})_m \times U_H^0 \times M} & (\chi^{-1}\delta_{B_G}^{\frac{1}{2}})(a_\alpha^{-m})\varphi_2(m') \\ & K_{\chi\otimes\gamma}^{\xi}(w_0^H \tilde{n}_2^*(t)u_H w_i, m) dt \otimes du_H^0 \otimes dm'. \end{aligned}$$

But

$$K_{\chi\otimes\gamma}^{\xi}(w_0^H \tilde{n}_2^*(t)u_H w_i, m) = K_{\chi\otimes\gamma}^{\text{twist}}(w_0^H \tilde{n}_2^*(t)u_H, m).$$

On the other hand from the proof of **Proposition 8.1** $T_{\chi\otimes\gamma}(\varphi^H_\alpha \otimes \varphi_2)$ equals

$$T^{\text{twist}}_{\chi\otimes\gamma}(\varphi^H_\alpha \otimes \varphi_2) = c^{\text{twist}}_H \sum_{m\geq 1} \int\limits_{B^0_H\times(I_H\cap U_{-\alpha})_m\times U^0_H\times M'} \chi^{-1}\delta^{\frac{1}{2}}_{B_H}(a_\alpha^{-m})\varphi_2(m')$$
$$K^{\text{twist}}_{\chi\otimes\gamma}(w^H_0\tilde{n}^*_2(t)u_H, m')db^0_H \otimes dt \otimes du^0_H \otimes dm\,.$$

But we know $\delta^{\frac{1}{2}}_{B_H} = \delta^{\frac{1}{2}}_{B_G}$ on B_H and that $I_{\mathbb{G}} \cap U_{-\alpha} = I_H \cap U_{-\alpha}$ (i.e., $I_H \subseteq I_G$). Thus we have established (ii) of **Proposition 8.2.** □

Now we must handle the delicate case where the simple root α lies in $\mathbb{U}_i$. In particular following the calculations in the proof of **Proposition 8.2** we deduce that

$$T^{\mathbb{G}}_{\chi\otimes\gamma}(\varphi_\alpha \otimes \varphi_2) =$$
$$c^{\text{twist}}_H \int\limits_{M\times U^0_H} \varphi_2(m)K^{\xi}_{\chi\otimes\gamma}(w^{\mathbb{G}}_0 u_H, m)$$
$$\left\{ \sum_{m\geq 1} \int\limits_{(I_{\mathbb{G}}\cap U_{-\alpha})_m} (\chi^{-1}\delta^{\frac{1}{2}}_{B_{\mathbb{G}}})(a_\alpha^{-m})\Lambda_\xi(u_H^{-1}n^*_2(t)u_H)dt \right\} du_H dm$$

Thus the first issue is to evaluate the inner integral. For this we require a more precise structural embedding of the group U_α in $\mathbb{U}_i$. Indeed we know

$$u_\alpha(t) = N(((tZ^*_\nu, 0, \cdots, 0), 0))$$

(for $t \in k$). Here $Z^*_\nu \in W_1(R_i)$ is the last element of the subspace $Sp\{Z^*_1, \cdots Z^*_\nu\}$ where $q_{W_1}(Z_i, Z^*_j) = \delta_{ij}$. Then using results from §3

$$z^{-1}u_\alpha(t)z = z^{-1}N((tZ^*_\nu, 0, \cdot, 0), 0))z = N((tz^{-1}(Z^*_\nu), 0, \cdots, 0), 0)$$

for any $z \in H$. Thus we have

$$\Lambda_\xi(u_H^{-1}u_\alpha(t)u_H) = \psi(tq_{W_1}(Z^*_\nu, u_H(\xi)))$$

Thus we have the following **Lemma**

Lemma 8.2. *Let α be a simple root of $\mathbb{G}$ which lies in $\mathbb{U}_i$ so that $\overline{\alpha} = \alpha$. Suppose that $|q_{W_1}(Z^*_\nu, u_H(\xi)| \leq 1$. Then the sum*

$$\sum_{m\geq 1}\left(\int\limits_{(I_{\mathbb{G}}\cap U_{-\alpha})_m} \chi^{-1}\delta^{\frac{1}{2}}_{B_{\mathbb{G}}}(a_\alpha^{-m})\Lambda_\xi(u_H^{-1}n^*_2(t)u_H)dt \right) =$$
$$\left(\frac{\zeta(\chi_i - \chi_{i+1})}{\zeta(\chi_i - \chi_{i+i} + 1)}\right)\left[1 - |q_{W_1}(Z^*_\nu, u_H(\xi))|^{\chi_i - \chi_{i+1}} q^{\chi_{i+1} - \chi_i}\right] - 1$$

Proof. Following the argument as in [C-S] we see that the sum of the above **Lemma** has the form

$$\int_{|x|<1} \psi(q_{W_1}(Z_\nu^*, u_H(\xi))x^{-1})|x|^{\chi_i-\chi_{i+1}-1}dx$$

Then since ψ has order zero the identity in the **Lemma** follows/**Q.E.D.**

We note here that since $u_H \in \mathbb{U}_H^0$, it follows that $|q_{W_1}(Z_\nu^*, u_H(\xi))| \leq 1$ for all $u_H \in U_H^0$.

Thus we deduce from the above **Lemma** that

$$\begin{aligned}
&T_{\chi\otimes\gamma}^{\mathbb{G}}(\varphi_\alpha \otimes \varphi_2) = \\
&\left[\frac{\zeta(\chi_i - \chi_{i+1})}{\zeta(\chi_i - \chi_{i+1} + 1)} - 1\right] T_{\chi\otimes\gamma}^{\mathbb{G}}(\varphi_1^{\mathbb{G}} \otimes \varphi_2) \\
&- \frac{\zeta(\chi_i - \chi_{i+1})}{\zeta(\chi_i - \chi_{i+1} + 1)} q^{\chi_{i+1}-\chi_i} c_H^{\mathrm{twist}} \int_{M\times U_H^0} \varphi_2(m) K_{\chi\otimes\gamma}^{\xi}(w_0^{\mathbb{G}} u_H, m) \\
&|q_{W_1}(Z_\nu^*, u_H(\xi))|^{\chi_i-\chi_{i+1}} du_H dm.
\end{aligned}$$

Then we have again

$$\begin{aligned}
K_{\chi\otimes\gamma}^{\xi}(w_0^{\mathbb{G}} u_H, m) &= K_{\chi\otimes\gamma}^{\xi}(w_0^H w_i u_H w_i^{-1} w_i, m) \\
&= K_{\chi\otimes\gamma}^{\mathrm{twist}}(w_0^H w_i u_H w_i^{-1}, m)\,.
\end{aligned}$$

Moreover, if we make the change of variables $u_H \to w_i u_H w_i^{-1}$ on U_H^0 and note that $w_i(Z_\nu^*) = Z_\nu^*$ $(\nu > 1)$ and $w_i(\xi) = \pm\xi$, then the inner integral above has the form

$$\int_{M\times U_H^0} \varphi_2(m') K_{\chi\otimes\gamma}^{\mathrm{twist}}(w_0^H u_H, m) |q_{W_1}(Z_\nu, w_0^H u_H(\xi))|^{\chi_2-\chi_{i+1}} du_H dm$$

(here we have used $w_0^H(Z_\nu^*) = Z_\nu$, see **Appendix 3 to §5** for this specific choice of w_0^H).

Then for $g_1 \in H$, $g_2 \in M$, we consider the function

$$K_{\chi\otimes\gamma}^{\mathrm{twist}}(g_1, g_2)|q_{W_1}(Z_\nu, g_1(\xi))|^{\chi_i-\chi_{i+1}}\,.$$

If $g_1 = b_H w_0^H \eta_H m^{w_i}$, $g_2 = b_M m$, then we evaluate

$$\begin{aligned}
&K_{\chi\otimes\gamma}^{\mathrm{twist}}(g_1, g_2)|q_W(Z_\nu, g_1(\xi))|^{\chi_i-\chi_{i+1}} \\
&\qquad = \chi^{-1}\delta_{B_H}^{\frac{1}{2}}(b_H)\rho_{\chi_i-\chi_{i+1}}(b_H)\gamma^{-1}\delta_{B_M}^{\frac{1}{2}}(b_M) \\
&\qquad\quad |q_{W_1}(Z_\nu^*, \eta_H(\xi))|^{\chi_i-\chi_{i+1}}\,.
\end{aligned}$$

Here $\rho_{\chi_i-\chi_{i+1}}(b_H) = |\rho(b_H)|^{\chi_i-\chi_{i+1}} = |s_1|^{\chi_{i+1}-\chi_i}$ (see above for this notation). In particular, $(\chi^{-1}\rho_{\chi_i-\chi_{i+1}})(b_H) = (\tilde{\chi})^{-1}(b_H)$ where $\tilde{\chi}\,|_{T_H} = \chi_i\varepsilon_{i+1} + \chi_{i+2}\varepsilon_{i+2} + \cdots + \chi_r\varepsilon_r$ (note χ_i is the coefficient of ε_{i+1} and not χ_{i+1}). In any case we have the identity:

$$K^{\text{twist}}_{\chi|_{T_H}\otimes\gamma}(g_1,g_2)|q_{W_1}(Z^*_\nu, g_1(\xi))|^{\chi_i-\chi_{i+1}} = K^{\text{twist}}_{\tilde{\chi}|_{T_H}\otimes\gamma}(g_1,g_2)|q_{W_1}(Z^*_\nu, \eta_H(\xi))|^{\chi_i-\chi_{i+1}}.$$

Thus we have that

$$\begin{aligned} c^{\text{twist}}_H &\left(\int_{M\times U^0_H} \varphi_2(m')K^{\text{twist}}_{\tilde{\chi}\otimes\gamma}(w^H_0 u_H, m')du^0_H dm\right) |q_{W_1}(Z^*_\nu, \eta_H(\xi))|^{\chi_i-\chi_{i+1}} \\ &= T^{\text{twist}}_{\tilde{\chi}|_{T_H}\otimes\gamma}(\varphi^H_1\otimes\varphi_2) \\ &= T_{\tilde{\chi}|_{T_H}\otimes\gamma}(\varphi^H_1\otimes\varphi_2). \end{aligned}$$

Thus we can finally compute $T^{\mathbb{G}}_{\chi\otimes\gamma}(\varphi_\alpha\otimes\varphi_2)$.

Proposition 8.3. *Let α be a simple root of $\mathbb{G}$ which lies in $\mathbb{U}_i$ and satisfies $\overline{\alpha} = \alpha$. Then with φ_2, satistying $\varphi_2(w_i x w_i^{-1}) = \varphi_2(x)$ for $x \in M$,*

$$\begin{aligned} &T^{\mathbb{G}}_{\chi\otimes\gamma}(\varphi_\alpha\otimes\varphi_2) = \\ &\left[\frac{\zeta(\chi_i-\chi_{i+1})}{\zeta(\chi_i-\chi_{i+1}+1)} - 1\right] T^{\mathbb{G}}_{\chi\otimes\gamma}(\varphi^{\mathbb{G}}_1\otimes\varphi_2) \\ &- \frac{\zeta(\chi_i-\chi_{i+1})}{\zeta(\chi_i-\chi_{i+1}+1)} q^{\chi_{i+1}-\chi_i} T^{\mathbb{G}}_{\tilde{\chi}\otimes\gamma}(\varphi^{\mathbb{G}}_1\otimes\varphi_2). \end{aligned}$$

Proof. From the comments above it suffices to show that $|q_{W_1}(Z^*_\nu, \eta_H(\xi))| = 1$. We note that this statement is verified in the **Appendix** to this section./**Q.E.D.**

Remark 2. It is straightforward to verify that $\tilde{\chi}\,|_{T_H} = w_\alpha(\chi)\,|_{T_H}$ (w_α = the reflection permuting $(i, i+1)$). Thus in the **Proposition** above

$$\begin{aligned} T^{\mathbb{G}}_{\chi\otimes\gamma}(\varphi_\alpha\otimes\varphi_2) = &\left(\frac{\zeta(\chi_i-\chi_{i+1})}{\zeta(\chi_i-\chi_{i+1}+1)} - 1\right) T^{\mathbb{G}}_{\chi\otimes\gamma}(\varphi^{\mathbb{G}}_1\otimes\varphi_2) \\ &- \frac{\zeta(\chi_i-\chi_{i+1})}{\zeta(\chi_i-\chi_{i+1}+1)} q^{-\langle w_\alpha(\chi),\check{\alpha}\rangle} T^{\mathbb{G}}_{w_\alpha(\chi)\otimes\gamma}(\varphi^{\mathbb{G}}_1\otimes\varphi_2) \end{aligned}$$

The strategy of computing $\lambda^{(\mathbb{G},M)}_{w,e}$ is to use the ideas of [C-S]. Indeed we note for the specific functions $\varphi^{\mathbb{G}}_1$ and φ^H_2 defined above (**Proposition 1**).

$$T^{\mathbb{G}}_{\chi\otimes\gamma}(\varphi^{\mathbb{G}}_1\otimes\varphi_2) = T_{\chi|_{T_H}\otimes\gamma}(\varphi^H_1\otimes\varphi_2)$$

(with φ_2 as given above).

Secondly we note that for w_α a simple reflection in $\mathbb{G}$ we have the following (**Proposition 2** and **Lemma 2**.)

(i) If α lies in $GL(R_i)$ then

$$T^{\mathbb{G}}_{\chi\otimes\gamma}((\varphi_1^{\mathbb{G}}+\varphi_\alpha)\otimes\varphi_2)=(1-q^{-1-\langle\chi,\check{\alpha}\rangle})T^{\mathbb{G}}_{\chi\otimes\gamma}(\varphi_1^{\mathbb{G}}\otimes\varphi_2)$$

(ii) If α lies in $\mathbb{U}_i$ then

$$T^{\mathbb{G}}_{\chi\otimes\gamma}((\varphi_1^{\mathbb{G}}+\varphi_\alpha)\otimes\varphi_2)=\left(\frac{\zeta(\chi_i-\chi_{i+1})}{\zeta(\chi_i-\chi_{i+1}+1)}\right)$$
$$\left[T^{\mathbb{G}}_{\chi\otimes\gamma}(\varphi_1^{\mathbb{G}}\otimes\varphi_2)-q^{-\langle w_\alpha(\chi),\check{\alpha}\rangle}T^{\mathbb{G}}_{w_\alpha(\chi)\otimes\gamma}(\varphi_1^{\mathbb{G}}\otimes\varphi_2)\right]$$

(iii) If α lies in H, then

$$T^{\mathbb{G}}_{\chi\otimes\gamma}((\varphi_1^{\mathbb{G}}+\varphi_\alpha)\otimes\varphi_2)=T^{H}_{\chi|_{T_H}\otimes\gamma}((\varphi_1^{H}+\varphi_\alpha^{H})\otimes\varphi_2).$$

On the other hand if w_α is a simple reflection then

$$I^{\mathbb{G}}_{w_\alpha}(F_1(\chi,\varphi_1^{\mathbb{G}}+\varphi_\alpha))=c_{w_\alpha}(\chi)$$
$$F_1(w_\alpha(\chi),\varphi_1^{\mathbb{G}}+\varphi_\alpha)$$

This is the eigenfunction property of $F_1(\chi,\varphi_1^{\mathbb{G}}+\varphi_\alpha)$ relative to the intertwining operator $I^{\mathbb{G}}_{w_\alpha}$. Here $c_{w_\alpha}(\chi)=\frac{\zeta(\langle\check{\alpha},\chi\rangle)}{\zeta(\langle\check{\alpha},\chi\rangle+1)}$.

Thus we deduce from (i), (ii), and (iii) above that

(i) If α lies in $GL(R_i)$

$$\gamma^{(\mathbb{G},M)}_{w_\alpha,e}(\chi,\gamma)=\frac{\zeta(\langle\check{\alpha},\chi\rangle+1)\quad\zeta(\langle\check{\alpha},\chi\rangle)}{\zeta(\langle\check{\alpha},w_\alpha(\chi)\rangle+1)\zeta(\langle\check{\alpha},\chi\rangle+1)}\left[\frac{T_{(w_\alpha\chi)|_{T_H}\otimes\gamma}(\varphi_1^H\otimes\varphi_2)}{T_{\chi|_{T_H}\otimes\gamma}(\varphi_1^H\otimes\varphi_2)}\right]$$

(ii) If α lies in $\mathbb{U}_i$

$$\gamma^{(\mathbb{G},M)}_{w_\alpha,e}(\chi,\gamma)=\left(\frac{\zeta(\langle\check{\alpha},\chi\rangle)}{\zeta(\langle\check{\alpha},\chi\rangle+1)}\right)\left[\frac{\zeta(\chi_{i+1}-\chi_i)\zeta(\chi_i-\chi_{i+1}+1)}{\zeta(\chi_i-\chi_{i+1})\zeta(\chi_{i+1}-\chi_i+1)}\right]$$
$$\left[\frac{T^{\mathbb{G}}_{w_\alpha(\chi)\otimes\gamma}(\varphi_1^G\otimes\varphi_2)-q^{\langle\chi,\check{\alpha}\rangle}T^{\mathbb{G}}_{\chi\otimes\gamma}(\varphi_1^{\mathbb{G}}\otimes\varphi_2)}{T^{\mathbb{G}}_{\chi\otimes\gamma}(\varphi_1^G\otimes\varphi_2)-q^{-\langle\chi,w_{\dot{\alpha}}(\check{\alpha})\rangle}T^{\mathbb{G}}_{w_\alpha(\chi)\otimes\gamma}(\varphi_1^G\otimes\varphi_2)}\right]$$

(iii) If α lies in H, then

$$\gamma^{(\mathbb{G},M)}_{w_\alpha,e}(\chi,\gamma)=\left(\frac{\zeta(\langle\check{\alpha},\chi\rangle)}{\zeta(\langle\check{\alpha},\chi\rangle+1)}\right)\left[\frac{T^H_{(w_\alpha\chi)|_{T_H}\otimes\gamma}((\varphi_1^H+\varphi_\alpha^H)\otimes\varphi_2)}{T^H_{\chi|_T\otimes\gamma}((\varphi_1^H+\varphi_\alpha^H)\otimes\varphi_2)}\right]$$

We note here that in (i) and (iii) we can choose φ_2 so that $T_{\chi|_{T_H}\otimes\gamma}(\varphi_1^H \otimes \varphi_2) \neq 0$. For instance, such a choice would be $\varphi_2 = \varphi_{K_M^{(n)}}$, $K_M^{(n)}$ = the n-th congruence subgroup of $K_M = \{\gamma \in K_M | \gamma \equiv I \mod \Pi^n\}$. It is clear that $\varphi_2(w_i x w_i^{-1}) = \varphi_2(x)$ since $w_i K_M^{(n)} w_i^{-1} = K_M^{(n)}$. In any case, the integral defining $T_{\chi\otimes\gamma}(\varphi_1^H \otimes \varphi_2)$ has the form (using **Corollary 3** to **Lemma 7.2**)

$$\int_{B_H \times B_M} \varphi_1^H(b_H w_0^H \eta_H b_M]\chi^{-1}\delta_{B_H}^{\frac{1}{2}}(b_H)\gamma\delta_{B_M}^{\frac{1}{2}}(b_M)d_\ell(b_H)d_\ell(b_M).$$

First the set $B_H w_0^H \eta_H B_M$ is open dense in $\mathbb{G}$ and hence the set $B_H^0 w_0^H U_H^0 \cap B_H w_0^H \eta_B B_M$ has positive measure. Since $\varphi_1^H > 0$ and if we choose χ^{-1} and $\gamma > 0$, we get that the above integral is nonvanishing! This in case (i), $T_{\chi T_H \otimes\gamma}(\varphi_1^H \otimes \varphi_2) \neq 0$. Moreover, in case (iii), we note that $\varphi_\alpha^H > 0$ and hence the integral defining $T^H_{\chi|_{T_H}\otimes\gamma}(\varphi_1^H + \varphi_\alpha^H) > 0$.

We compute $T^H_{\chi|_{T_H}\otimes\gamma}(\varphi_1^H \otimes \varphi_2)$ more exactly in **Appendix 2 to §8** to determine the nonvanishing of the term

$$T^H_{\chi\otimes\gamma}(\varphi_1^H \otimes \varphi_2) - q^{-\langle\chi, w_\alpha(\check{\alpha})\rangle} T^H_{w_\alpha(\chi)\otimes\gamma}(\varphi_1^H \otimes \varphi_2)$$

with φ_2 as above.

Then by direct calculation in cases (i) and (ii) above we have

(I) If α lies in $GL(R_i)$ or $\mathbb{U}_i$ then

$$\gamma_{w_\alpha,e}^{(\mathbb{G},M)}(\chi,\gamma) = \frac{\zeta(\langle\check{\alpha},\chi\rangle}{\zeta(\langle\check{\alpha}, w_\alpha(\chi)\rangle + 1)}$$

Remark 3. We note here that the $\gamma_{w_\alpha,e}^{(\mathbb{G},M)}(\chi,\gamma)$ in (i) above depends only on χ Moreover this factor equals the γ factor on the group $\mathbb{G}$ associated to the "Whittaker functional" on $\mathbb{G}$ given by the Jacquet type integral

$$F_1(\chi,\varphi) \rightsquigarrow \int_{U_\mathbb{G}} F_1(\chi,\varphi)[w_0^\mathbb{G} u_G]\psi_{U_\mathbb{G}}(u)du_\mathbb{G}$$

On the other and it is clear that in (iii) above $\gamma_{w_\alpha,e}^{(\mathbb{G},M)}(\chi,\gamma) = \gamma_{w_\alpha,e}^{(H,M)}(\chi|_{T_H}\gamma)$. That is, $\gamma_{w_\alpha,e}^{(H,M)}(\chi|_{T_H},\gamma)$ equals exactly the ratio (by definition) given on the right hand side of (iii) above. Thus

(iii) If α lies in H, then

$$\gamma_{w_\alpha,e}^{(\mathbb{G},M)}(\chi,\gamma) = \gamma_{w_\alpha,e}^{(H,M)}(\chi|_{T_H},\gamma)$$

Thus we have established the following **Theorem** (See **Theorem 1** of §6).

Theorem 8.1. (Determination of γ-factors)

(1) Let w_α be a reflection corresponding to a simple root in $GL(R_i)$ or $\mathbb{U}_i$. Then

$$\lambda_{w_\alpha,e}^{(\mathbb{G},M)}(\chi,\gamma) = \lambda_{w_\alpha}^{i+1}(\chi|_{T_{i+1}}).$$

Here $\lambda_{w_\alpha}^{i+1}(\chi|_{T_{i+1}})$ is the γ- factor

of the Jacquet type Whittaker function on the group $GL(R_{i+1})$ where $R_{i+1} = Sp\{v_1, \cdots, v_i, z_1\}$ with the associated representation $ind_{B_{i+1}}^{GL(R_{i=1})}(\chi|_{T_{i+1}})$. In concrete terms $\lambda_{w_\alpha}^{i+1}(\chi|_{T_{i+1}}) = \frac{\zeta(\langle\check{\alpha},\chi\rangle)}{\zeta(\langle\check{\alpha},w_\alpha(\chi)\rangle+1)}$

(2) Let w_α be a reflection corresponding to a simple root α on H. Then

$$\lambda_{w_\alpha,e}^{(\mathbb{G},M)}(\chi,\gamma) = \lambda_{w_\alpha,e}^{(H,M)}(\chi,\gamma).$$

Appendix 1. (Proof of a statement in **Proposition 3**)

We adopt here the notation used in **Appendix** 1 (Proof of **Lemma 1**) in §7.

We must have the choice of η_H above so that $q_{W_1}(Z_\nu^*, \eta_H(\xi))$ is a unit and η_H satisfies the requirement in the four (4) cases in **Appendix 3** to §5.

We choose the basis $\{Z_1, \ldots, Z_\nu\}$ to be compatible with the one given in the four (4) cases in **Appendix 3** to §5.

(a) Case (i) of §7 (Case (a), Case (b) of **Appendix 3** to §5). In case (1) we choose η_H so that $\eta_H(\xi) = \xi + \sum_{i\geq 2} \beta_i z_i$ and in case (2) we choose $\eta_H(\xi) = \xi + rX + \sum_{i\geq 2} \beta_i z_i$ (where all the β_i and r are units). In any case, we deduce easily that $q_{W_1}(Z_\nu^*, \eta_H(\xi))$ is a unit!

(b) Case (2) of §7 (Case (c) of **Appendix 3** of §5). Here we choose η_H so that $\eta_H(\xi_0) = \xi_0 + rX + \sum_{i\geq 2} \gamma_i z_i$ (again with r and all the γ_i units). Thus again we have that $q_{W_1}(Z_\nu^*, \eta_H(\xi))$ is a unit!

(c) Case (3) of §7 (Case (d) of **Appendix 3** of §5). Here we choose η_H so that $\eta_H(\xi_0) = \xi_0 + \sum_{i\geq 2} \alpha_i z_i$ (α_i units). Then $q_{W_1}(Z_\nu^*, \eta_H(\xi_0))$ is again a unit!

Appendix 2.

We note first that if we take $W_1 = R_i \oplus W_1(R_i) \oplus R_i^*$ and $\mathcal{L}$ an $\mathcal{O}$ lattice defined in W_1 (given in **Appendix 1 to §5**, see **Remark** $(*)'$) so that $\mathcal{L} = \mathcal{L} \cap R_i \oplus \mathcal{L} \cap W_1(R_i) \oplus \mathcal{L} \cap R_i^*$. That is, $\mathcal{L} \cap R_i = \mathcal{O} -$ Span of $\{v_1, \dots, v_i\}$, $\mathcal{L} \cap R_i^* = \mathcal{O} -$ *span* of $\{v_1^*, \dots, v_i^*\}$ and $\mathcal{L} \cap W_1(R_i) =$ the $\mathcal{O} -$ *span* given in **Remark** $(*)'$ of **Appendix 1** of **§5**.

Then $K_{\mathbb{G}} = \text{stabilizer}_{SO(W_1, q_{W_1})}(\mathcal{L})$.

First we characterize $K_{\mathbb{G}} \cap \mathbb{U}_i$.

Indeed we assert that

$$K_{\mathbb{G}} \cap \mathbb{U}_i = \{N_i((\xi_1, \dots, \xi_i), s)) \mid \xi_1, \dots, \xi_i \in \mathcal{L} \cap W_1(R_i) \text{ and } s_{ij} \in \mathcal{O}\}.$$

Then since $q(\mathcal{L}, \mathcal{L}) \in \mathcal{O}$, the identity above follows. Indeed, if $q(\mathcal{L}, \mathcal{L}) \in \mathcal{O}$, then N_i satisfying $\xi_i \in \mathcal{L} \cap W_1(R_i)$ and $s_{ij} \in \mathcal{O}$ clearly presserves $\mathcal{L}$ and N_i^{-1} has the same property!

On the other hand, let

$$N_i^{(\alpha)} = \{N_i((0, \dots, \xi, \dots, 0), 0) \mid \xi \in W_1(R_i)\}$$

(ξ appears in the α-th slot). Now $N_i^{(\alpha)}$ is a subgroup of N_i and we have the equality:

$$\mathbb{U}_i = \{N_i|(0, s)|s \in \text{Skew}\} N_i^{(1)} N_i^{(2)} \cdots N_i^{(i)}.$$

Also we note that a given element $n \in \mathbb{U}_i$ admits a unique decomposition relative to the above structural statement.

Moreover, it is easy to see that

$$K_{\mathbb{G}} \cap \mathbb{U}_i = (K_{\mathbb{G}} \cap \{N_i(0, s) \mid s \, \text{skew}\}) \prod_{\alpha} K_{\mathbb{G}} \cap N_i^{(\alpha)}.$$

We note the order of product can be reversed here. If $Z \in \mathcal{L} \cap W_1(R_i)$ with $q_{W_1}(Z) \neq 0$, then we also note that

$$\text{Stab}_{\mathbb{U}_i}(Z) = \{N_i(0, s) \mid s \, \text{skew}\} \cdot \prod \{N_i((0, \dots, \xi, \dots, 0, 0) \in N_i^{(\alpha)} \mid q_{W_1}(\xi, Z) = 0\}.$$

Now $\text{Stab}_{\mathbb{U}_i}(Z) \cap K_{\mathbb{G}} = (\{N_i(0, s) \mid s \, \text{skew}\} \cap K_{\mathbb{G}}) \prod \{N_i(0, \dots, \xi, 0, \dots) \in N_i^{(\alpha)} \mid \xi \in \mathcal{L}, q_{W_1}(\xi, Z) = 0\}$.

Now we are in a position to compute the integral

$$\int \varphi_1^H(b_H w_0^H \eta_H b_M) \chi^{-1} \delta_{B_H}^{\frac{1}{2}}(b_H) \gamma \delta_{B_M}^{\frac{1}{2}}(b_M) db_H \otimes db_M.$$

We must now determine the set

$$B_H^0 w_0^H U_H^0 \cap B_H w_0^H \eta_H B_M$$

more precisely.

Now we follow the notation from **Appendix 1 of §7**. We consider cases (1) and (2) first. Case (3) is handled separately below.

In cases (1) and (2) we let $i = r$. Moreover, $B_M = T_M U_M = T_M V_H Y_H^M = T_M V_H \mathrm{Stab}_{\mathbb{U}_r}(Z)$ (for the specific choice of Z given there). Thus $K_H \cap B_M = K_M \cap B_M = (K_M \cap T_M)(K_M \cap V_H)(K_M \cap \mathrm{Stab}_{\mathbb{U}_r}(Z))$.

Also we note here that we choose $\eta_H \in U_H$ to have the form

$$\eta_H = \prod N_i((0, m \ldots, \alpha_z Z, 0, \ldots, 0), 0)$$

with α_z a nonzero unit. In any case,

$$B_H w_0^H \eta_H t_M u_M = B_H w_0^H t_M^{-1} \eta_H t_M u_M\,.$$

Then

$$\begin{aligned} t_M^{-1} \eta_H t_M u_M &= \prod_\alpha N_i((0, \ldots, t_\alpha^{-1} \alpha_z Z, \ldots, 0), 0) u_M \\ &= \left[\prod_\alpha N_i(0, \ldots, t_\alpha^{-1} \alpha_z Z + \xi_\alpha, 0, \ldots, 0)\right] N_i((0, s')) v_H' \end{aligned}$$

with $v_H' \in V_H$, $\xi \in W_1(R_r)$ with $(\xi_\alpha, Z) = 0$.

Thus for the above element to belong to $U_H^0 = K_H \cap U_H$, we have that

$$t_\alpha^{-1} \in \mathcal{O}\,,\ \xi_\alpha \in \mathcal{L}\,,\ s' \in \mathrm{Skew}(\mathcal{O}) \text{ and } v_H' \in V_H \cap K_H\,.$$

(We use here that $\mathcal{L} \cap W_1(R_i) = \mathcal{O} \cdot Z \oplus \mathcal{O}((Z)^\perp \cap \mathcal{L})$).

In fact, the condition above is necessary and sufficient.

Thus we compute our given integral. In fact, we disintegrate the measure $db_M = dt_M \otimes du_M = dt_M \otimes dY_H^M \otimes dv_H$. The basic integral is

$$\begin{aligned} &\int \varphi_1^H(b_H w_0^H \eta_H t_M y_H^M v_H) \chi^{-1} \delta_{B_H}^{\frac12}(b_H) \gamma \delta_{B_M}^{\frac12}(t_M) d(b_H) \otimes dt_M \otimes dy_H^M \otimes dv_H \\ &= \int \varphi_1^H(b_H w_0^H t_M^{-1} \eta_H t_M y_H^M v_H)(\chi \delta_{B_H}^{\frac12})(w_0^H t_M (w_0^H)^{-1}) \chi^{-1} \delta_{B_H}^{\frac12}(b_H) \gamma \delta_{B_M}^{\frac12}(t_M) \\ &\qquad db_H \otimes dt_M \otimes dy_H^M \otimes dv_H \\ &= \int \varphi_1^H(w_0^H(\prod_\alpha N_i(0, \ldots, t_\alpha^{-1} Z, \ldots, \xi_\alpha)) N_1((0, s)) v_H) \\ &\qquad \chi \delta_{B_M}^{\frac12}(w_0^H t_M (w_0^H)^{-1}) \gamma \delta_{B_M}^{\frac12}(t_M) d^x(t_M) ds dv_H\,. \end{aligned}$$

We have normalized ds, $d\xi_\alpha$ so that an $\mathrm{Skew}_r(\mathcal{O})$ and $d^X(t_M)$ to have mass 1 on the units in each case. Thus we find that the above integal equals a fixed integral of the type (Δ nonzero, independent of χ and γ)

$$\Delta \int\limits_{|t_1|\leq 1, |t_2|\leq 1, \ldots, |t_r|\leq 1} |t_1|^{(\chi_{i+1}-\gamma_1+A_1)}|t_2|^{(\chi_{i+2}-\gamma_2+A_2} \ldots |t_r|^{\pm\chi_{i+r}-\gamma_r+A_r}$$
$$d^x(t_1)\ldots d^x(t_r)$$
$$=\Delta\left(\prod_{t\neq r}\zeta_v(\chi_{i+t}-\gamma_t+A_t)\right)(\zeta(\pm\chi_{i+r}-\gamma_r+A_r))\,.$$

Thus we note that

$$T^H_{\chi\otimes\gamma}(\varphi_1^H\otimes\varphi_2)=\Delta(\prod_{j\neq r}\zeta_v(\chi_{i+j}-\gamma_j+A_j))\zeta_v(\pm\chi_{i+r}-\gamma_r+A_r))$$

and thus

$$T^H_{\chi\otimes\gamma}(\varphi_1^H\otimes\varphi_2)-q^{\chi_i-\chi_{i+1}}T^H_{w_\alpha(\chi)\otimes\gamma}(\varphi_1^H\otimes\varphi_2)$$
$$=\Delta(\prod_{j\neq 1,\neq r}\zeta_v(\chi_{i+j}-\gamma_j+A_j)\zeta_v(\pm\chi_{i+r}-\gamma_r+A_r))$$
$$[\zeta_v(\chi_{i+1}-\gamma_1+A_1))-q^{\chi_i-\chi_{i+1}}\zeta_v(\chi_i-\gamma_1+A_1)]\,.$$

However, we note that

$$\chi\to q^\chi\zeta_v(\chi-\gamma_1+A_1)$$

is not a constant function!

For case (3) of the **Appendix 1 to §7** we note that similar arguments work as above. The difference here is that $B_Hw_0^H\eta_HB_M$ is not contained in just the open cell $B_Hw_0^HB_H$ but a bigger cell $B_Hw_0^HP_H$ where P_H is the rank one parabolic which contains $T_MxSO(2,1)$ as a Levi component.

However, similar considerations as above (cases (1) and (2)) apply nevertheless. In fact, we get exactly the same calculation for our basic integral and also the nonvanishing.

BIBLIOGRAPHY

[B-F-G] Bump, D., Friedberg, S., Ginzburg, D., *Whittaker Orthogonal Models, functoriality and the Rankin Selberg Method*, Inventiones Math. **109** (1992), 55-96.

[B-Z] Bernstein, J., Zelevinski, A., *Induced Representations of reductive p-adic groups I*, Ann. Scient. Ec. Norm. Sup. **10** (1977), 441-472.

[C-PS] Cogdell, J., Piatetski-Shapiro, I., *The Converse Theorem for GL_n*, (to appear in IHES Journal).

[C-S] Casselman, W., Shalika, J., *The unramified principal series of p-adic groups II, the Whittaker function*, Composition Math. **41** (1980), 207-231.

[F] Furosawa, M., *On Fourier coefficients of Eisenstein series on $SO(5,2)$*, American Journal of Math. **115** (1993), 823-860.

[G] Ginzburg, D., *L functions for $SO_n \times GL_k$*, J. Reine Angew.Math **405** (1990).

[G-J] Godemont, R., Jacquet, H., *Zeta Functions of Simple Algebras*, Springer Lecture Notes in Math **260** (1972).

[G-PS-R] Gelbart, S., Piatetski-Shapiro, I., Rallis, S., *Explicit Construction of Automorphic L functions*, Springer Lecture Notes in Math **1254** (1986).

[G-P] Gross, B., Prasad, D., *On the Decomposition of a Representation of SO_n when restricted to SO_{n-1}*, Can. J. Math. Soc. **44** (1992), 974-1002.

[J-PS-S] Jacquet, H., Piatetski-Shapiro, I., Shalika, J., *Rankin Selberg Convolutions*, American Journal of Math. **105** (1983), 367-483.

[K-R] Kudla, S., Rallis, S., *Poles of Eisenstein Series and L functions*, in honor of Piatetski-Shapiro, Israel Math Conference Proceedings (1990), 81-111.

[M-S] Murase, A., Sugano, T., *Shintani functions and its Applications to automorphic L functions*, Math. Ann. **299** (1994), 17-56.

[M-W] Moeglin, C., Waldspurger, J. L., *Le spectrum résidual de $GL(n)$*, Ann. Scient. Ec. Norm. Sup. (IV) **22** (1989), 605-674.

[N] Novodvorsky M., *Functions $\mathcal{J}$ pour des groupes orthgonaux*, C. R. Acad. Sci. Par. A. **280** (1975), 1421-1422.

[PS-R] Piatetski-Shapiro, I., Rallis, S., *ε factor of Representations of Classical Groups*, Proc. of Nat. Acad. of Science **83** (1986), 4589-4593.

[R] Rallis, S., *Gelfand Pairs Associated to L functions*, (in preparation).

[S] Soudry, D., *Rankin Selberg Convolutions for $SO_{2\ell+1} \times GL_n$: Local Theory*, Memoirs of American Math. Soc. **500** (1993).

[W] Waldspurger, J. L., talk at ICM (Zurich) (1994).

DEPARTMENT OF MATHEMATICS, THE OHIO STATE UNIVERSITY, COLUMBUS, OHIO 43210, USA

Editorial Information

To be published in the *Memoirs*, a paper must be correct, new, nontrivial, and significant. Further, it must be well written and of interest to a substantial number of mathematicians. Piecemeal results, such as an inconclusive step toward an unproved major theorem or a minor variation on a known result, are in general not acceptable for publication. *Transactions* Editors shall solicit and encourage publication of worthy papers. Papers appearing in *Memoirs* are generally longer than those appearing in *Transactions* with which it shares an editorial committee.

As of March 31, 1997, the backlog for this journal was approximately 8 volumes. This estimate is the result of dividing the number of manuscripts for this journal in the Providence office that have not yet gone to the printer on the above date by the average number of monographs per volume over the previous twelve months, reduced by the number of issues published in four months (the time necessary for preparing an issue for the printer). (There are 6 volumes per year, each containing at least 4 numbers.)

A Copyright Transfer Agreement is required before a paper will be published in this journal. By submitting a paper to this journal, authors certify that the manuscript has not been submitted to nor is it under consideration for publication by another journal, conference proceedings, or similar publication.

Information for Authors and Editors

Memoirs are printed by photo-offset from camera copy fully prepared by the author. This means that the finished book will look exactly like the copy submitted.

The paper must contain a *descriptive title* and an *abstract* that summarizes the article in language suitable for workers in the general field (algebra, analysis, etc.). The *descriptive title* should be short, but informative; useless or vague phrases such as "some remarks about" or "concerning" should be avoided. The *abstract* should be at least one complete sentence, and at most 300 words. Included with the footnotes to the paper, there should be the 1991 *Mathematics Subject Classification* representing the primary and secondary subjects of the article. This may be followed by a list of *key words and phrases* describing the subject matter of the article and taken from it. A list of the numbers may be found in the annual index of *Mathematical Reviews*, published with the December issue starting in 1990, as well as from the electronic service e-MATH [**telnet e-MATH.ams.org** (or **telnet 130.44.1.100**). Login and password are **e-math**]. For journal abbreviations used in bibliographies, see the list of serials in the latest *Mathematical Reviews* annual index. When the manuscript is submitted, authors should supply the editor with electronic addresses if available. These will be printed after the postal address at the end of each article.

Electronically prepared papers. The AMS encourages submission of electronically prepared papers in $\mathcal{A}\mathcal{M}\mathcal{S}$-TeX or $\mathcal{A}\mathcal{M}\mathcal{S}$-LaTeX. The Society has prepared author packages for each AMS publication. Author packages include instructions for preparing electronic papers, the *AMS Author Handbook*, samples, and a style file that generates the particular design specifications of that publication series for both $\mathcal{A}\mathcal{M}\mathcal{S}$-TeX and $\mathcal{A}\mathcal{M}\mathcal{S}$-LaTeX.

Authors with FTP access may retrieve an author package from the Society's Internet node `e-MATH.ams.org` (130.44.1.100). For those without FTP access, the author package can be obtained free of charge by sending e-mail to `pub@math.ams.org` (Internet) or from the Publication Division, American Mathematical Society, P.O. Box 6248, Providence, RI 02940-6248. When requesting an author package, please specify $\mathcal{A}\mathcal{M}\mathcal{S}$-TEX or $\mathcal{A}\mathcal{M}\mathcal{S}$-LATEX, Macintosh or IBM (3.5) format, and the publication in which your paper will appear. Please be sure to include your complete mailing address.

Submission of electronic files. At the time of submission, the source file(s) should be sent to the Providence office (this includes any TEX source file, any graphics files, and the DVI or PostScript file).

Before sending the source file, be sure you have proofread your paper carefully. The files you send must be the EXACT files used to generate the proof copy that was accepted for publication. For all publications, authors are required to send a printed copy of their paper, which exactly matches the copy approved for publication, along with any graphics that will appear in the paper.

TEX files may be submitted by email, FTP, or on diskette. The DVI file(s) and PostScript files should be submitted only by FTP or on diskette unless they are encoded properly to submit through e-mail. (DVI files are binary and PostScript files tend to be very large.)

Files sent by electronic mail should be addressed to the Internet address `pub-submit@math.ams.org`. The subject line of the message should include the publication code to identify it as a Memoir. TEX source files, DVI files, and PostScript files can be transferred over the Internet by FTP to the Internet node `e-math.ams.org` (130.44.1.100).

Electronic graphics. Figures may be submitted to the AMS in an electronic format. The AMS recommends that graphics created electronically be saved in Encapsulated PostScript (EPS) format. This includes graphics originated via a graphics application as well as scanned photographs or other computer-generated images.

If the graphics package used does not support EPS output, the graphics file should be saved in one of the standard graphics formats—such as TIFF, PICT, GIF, etc.—rather than in an application-dependent format. Graphics files submitted in an application-dependent format are not likely to be used. No matter what method was used to produce the graphic, it is necessary to provide a paper copy to the AMS.

Authors using graphics packages for the creation of electronic art should also avoid the use of any lines thinner than 0.5 points in width. Many graphics packages allow the user to specify a "hairline" for a very thin line. Hairlines often look acceptable when proofed on a typical laser printer. However, when produced on a high-resolution laser imagesetter, hairlines become nearly invisible and will be lost entirely in the final printing process.

Screens should be set to values between 15% and 85%. Screens which fall outside of this range are too light or too dark to print correctly.

Any inquiries concerning a paper that has been accepted for publication should be sent directly to the Editorial Department, American Mathematical Society, P. O. Box 6248, Providence, RI 02940-6248.

Editors

Selected Titles in This Series

(*Continued from the front of this publication*)

(See the AMS catalog for earlier titles)